Studies in Logic
Volume 42

Meta-argumentation
An Approach to Logic and Argumentation Theory

Volume 31
Nonmonotonic Reasoning. Essays Celebrating its 30th Anniversary
Gerhard Brewka, Victor W. Marek and Miroslaw Truszczynski, eds.

Volume 32
Foundations of the Formal Sciences VII. Bringing together Philosophy and Sociology of Science
Karen François, Benedikt Löwe, Thomas Müller and Bart van Kerkhove, eds.

Volume 33
Conductive Argument. An Overlooked Type of Defeasible Reasoning
J. Anthony Blair and Ralph H. Johnson, eds.

Volume 34
Set Theory
Kenneth Kunen

Volume 35
Logic is not Mathematical
Hartley Slater

Volume 36
Understanding Vagueness. Logical, Philosophical and Linguistic Perspectives
Petr Cintula, Christian G. Fermüller, Lluís Godo and Petr Hájek, eds.

Volume 37
Handbook of Mathematical Fuzzy Logic. Volume 1
Petr Cintula, Petr Hájek and Carles Noguera, eds.

Volume 38
Handbook of Mathematical Fuzzy Logic. Volume 2
Petr Cintula, Petr Hájek and Carles Noguera, eds.

Volume 39
Non-contradiction
Lawrence H. Powers, with a Foreword by Hans V. Hansen

Volume 40
The Lambda Calculus. Its Syntax and Semantics
Henk P. Barendregt

Volume 41
Symbolic Logic from Leibniz to Husserl
Abel Lassalle Casanave, ed.

Volume 42
Meta-argumentation. An Approach to Logic and Argumentation Theory
Maurice A. Finocchiaro

Studies in Logic Series Editor
Dov Gabbay   dov.gabbay@kcl.ac.uk

Meta-argumentation
An Approach to Logic and Argumentation Theory

Maurice A. Finocchiaro

© Individual author and College Publications 2013.
All rights reserved.

ISBN 978-1-84890-097-4

College Publications
Scientific Director: Dov Gabbay
Managing Director: Jane Spurr
Department of Informatics
King's College London, Strand, London WC2R 2LS, UK

http://www.collegepublications.co.uk

Original cover design by Orchid Creative   www.orchidcreative.co.uk
Printed by Lightning Source, Milton Keynes, UK

All rights reserved. No part of this publication may be reproduced, stored in a retrieval system or transmitted in any form, or by any means, electronic, mechanical, photocopying, recording or otherwise without prior permission, in writing, from the publisher.

Contents

Acknowledgments viii

Introduction: A Special Class of Arguments 1

PART I: THE META-ARGUMENTATION APPROACH

Chapter 1: Methodological Considerations: Toulmin's Applied Logic 4
1.1 *Argumentative Reality* 4
1.2 *Theory vs. Practice* 5
1.3 *Theory vs. Method* 7
1.4 *Theory vs. Observation* 11
1.5 *Theory vs. Argumentation* 15
1.6 *Summary* 17

Chapter 2: Elementary Principles of Interpretation and Evaluation 18
2.1 *Reasoning and Argumentation* 18
2.2 *Interpretation and Structure* 20
2.3 *Evaluation and Criticism* 22
2.4 *Example 1: The Observational Argument for Heavenly Unchangeability* 26
2.5 *Example 2: The Anti-Copernican Argument from Vertical Fall* 30
2.6 *Summary* 32

Chapter 3: Basic Types of Meta-argumentation 34
3.1 *Ground-level Argumentation* 34
3.2 *Argument Analysis* 34
3.3 *Self-Reflective Argumentation* 35
3.4 *Example: The Simplicity Argument for Terrestrial Rotation* 36
3.5 *Argumentation Theory* 39
3.6 *Summary* 41

PART II: THEORETICAL META-ARGUMENTS

Chapter 4: Dialectical Definitions of Argument 42
4.1 *Concepts of Dialectic* 42
4.2 *Concepts of Argument* 44
4.3 *Goldman's Argument* 48
4.4 *Johnson's Argument: Its Illative Tier* 50
4.5 *Johnson's Argument: Its Criticism of Alternative Positions* 53
4.6 *Johnson's Argument: Replies to Objections* 57
4.7 *Conclusion: A Moderately Dialectical Conception* 60
4.8 *Summary* 64

Chapter 5: The Hyper Dialectical Definition of Argument 65
5.1   Eemeren's Hyper Dialectical Conception   65
5.2   The Pragma-dialectical Argument   67
5.3   The Symmetry of the Dialectical and Illative Tiers   69
5.4   Another Argument   70
5.5   Mill's Hyper Dialectical Argumentative Practice?   71
5.6   Summary   73

Chapter 6: Common Methods of Argument Criticism 75
6.1   Krabbe's Formal-Fallacy Criticism   75
6.2   Govier's Refutation by Logical Analogy   78
6.3   Woods's Parity of Reasoning   81
6.4   Conclusion: Common Types of Critical Meta-arguments   83
6.5   Summary   84

Chapter 7: Deep Disagreements, Fierce Standoffs, Etc. 85
7.1   Fogelin on Deep Disagreements   85
7.2   Critiques of Fogelin's Argument   86
7.3   Woods on Standoffs of Force Five   97
7.4   Criticism of Woods's Argument   104
7.5   Johnstone on Philosophical Controversies   110
7.6   Conclusion: The Role of Meta-argumentation   116
7.7   Empirical Testing   119
7.8   Summary   122

Chapter 8: Conductive Arguments, Pro-and-Con Reasoning, Etc. 123
8.1   Introduction   123
8.2   Wellman's Invention of Conduction   125
8.3   Hitchcock on Conductive Adequacy   128
8.4   Govier's Synthesis   132
8.5   Critiques of Govier's Argument   136
8.6   David Brooks on the Health-care Bill   144
8.7   Logical Structure and Visual Representation   148
8.8   Other Views: Conduction under Various Aliases   149
8.9   A Classic Example: Galileo's Dialogue   152
8.10   Conclusions: Progress, Problems, Prospects   157
8.11   Summary   159

Chapter 9: Self-referential Arguments 162
9.1   Blair on 'Argument' and 'Logic' in Textbooks: Argumentum Ad Hominem?   162
9.2   Eemeren on Fallacies as Strategic Derailments: Tu Quoque Ad Hoc?   167
9.3   Johnson on Anticipating Objections: Caveat Emptor?   171
9.4   Conclusions: Varieties of Self-referential and Ad Hominem Arguments   176

PART III: FAMOUS META-ARGUMENTS

Chapter 10: Mill on Liberty of Argument   178
10.1 Introduction   178
10.2 Discovering Truth and Appreciating Fallibility   179
10.3 Considering Objections and Appreciating Reasons   181
10.4 Avoiding One-sidedness and Appreciating Partial Truths 184
10.5 Conclusions: Structure, Contents, and Epistemology 186
10.6 Summary   191

Chapter 11: Mill on Women's Liberation   192
11.1 Introduction   192
11.2 Argument Unnecessary and Counterproductive?   193
11.3 Causal Undermining   194
11.4 Predictive Extrapolation   196
11.5 Conclusion: Meta-argumentative vs. Dialectical vs. Illative Tiers 197
11.6 Summary   199

Chapter 12: Hume on Intelligent Design   201
12.1 Introduction: Barker's Meta-argumentative Analysis 201
12.2 Multiplicity of Barker's Claims and of Hume's Meta-arguments 203
12.3 Cleanthes's vs. Philo's Ground-level Argument   204
12.4 Philo's Constructive and Critical Meta-arguments 206
12.5 Conclusion: The Manageable Complexity of Meta-argumentation 217
12.6 Summary   218

Chapter 13: Galileo on the Motion of the Earth   219
13.1 Introduction: A Galilean Case Study and Test Case 219
13.2 Observational Description of the Galilean Argumentation 221
13.3 Merits and Defects of Ground-arguments and Meta-arguments 223
13.4 Structure of Circular and Question-begging Arguments 227
13.5 Conclusions: Logic vs. Rhetoric, and Theorizing vs. Meta-argumentation 234
13.6 Summary   239

Conclusion: Argumentation Theory as Meta-argumentation   242

Bibliography   248

Index   268

Acknowledgments

In researching, writing, and publishing this book, I have benefited from the support and encouragement of many persons and institutions, and they deserve acknowledgment here. For comments, feedback, or various kinds of assistance on individual chapters of this book, I thank Tony Blair, Daniel Cohen, Albert DiCanzio, Darin Dockstader, Frans van Eemeren, Tom Fischer, Alec Fisher, James Freeman, David Godden, Trudy Govier, Hans Hansen, David Hitchcock, Peter Houtlosser, Ralph Johnson, Erik Krabbe, Jan Albert van Laar, Peter Loptson, Robert Pinto, Agnès van Rees, Henrique Ribeiro, Peter Tan, Wanda Teays, Harald Wohlrapp, John Woods, and Frank Zenker. I owe a special debt of gratitude to the University of Nevada, Las Vegas, the Philosophy Department there, its chairman David Beisecker, my colleagues Ian Dove, Todd Jones, Bill Ramsey, and James Woodbridge, and the other departmental colleagues; they have continued to provide institutional and moral support, even after I decided to retire from formal teaching to work full time on research, scholarship, and writing. Similarly, at the same institution, Nick Panissidi of the Campus Computing Services was immensely helpful with technical issues at the very last stages of my manuscript production.

For the opportunity to present, disseminate, and discuss earlier versions of the key ideas of this book, and for assistance with the necessary travel expenses and practical arrangements, I am grateful to the following organizations: Association for Informal Logic and Critical Thinking (AILACT); Centre for Research on Reasoning, Argumentation and Rhetoric (CRRAR), at the University of Windsor; International Society for the Study of Argumentation (ISSA), and University of Amsterdam; Ontario Society for the Study of Argumentation (OSSA); and Faculty of Letters, University of Coimbra, Portugal. Acknowledgments are also due to the following publishers or copyright holders for various parts of this book that were published in earlier versions reflecting various occasions and contexts: Cambridge Scholars Publishing for chapters 1 and 13; University of California Press for ch. 2 and 3; the journal *Informal Logic* and Cambridge University Press for ch. 4; Lawrence Erlbaum Associates for ch. 5; Springer Science and Business Media and the journal *Argumentation* for ch. 6; College Publications for ch. 8; and Vale Press for ch. 10.

Introduction
A Special Class of Arguments

A meta-argument is an argument about one or more arguments, or about argumentation in general. It may be contrasted to a ground-level argument, which is typically about such things as natural phenomena, historical events, human actions, etc. Sometimes, a ground-level argument is an argument which, in a particular discussion, is the subject matter of some meta-argument; in this case, the argument is ground-level vis-à-vis the meta-argument. And in general, an argument may be defined (cf. chapter 4, below) as a set of statements that attempts to justify a claim by supporting it with reasons, or by defending it from objections, or both.

Meta-arguments are ubiquitous in ways that may not be prima facie apparent. The most obvious point in this regard is perhaps the fact that meta-arguments occur commonly even in the context of everyday argumentation when one is evaluating ground-level arguments, at least insofar as even in such a context the process of evaluation consists not only of advancing an evaluative claim about the ground-level argument, but also of justifying that claim. Furthermore, in some fields, such as the history of philosophy and the history of science (and intellectual history in general), meta-argumentation is prevalent insofar as one of their principal aims is the interpretation of arguments, and meta-argumentation is obviously required to justify such interpretive claims. Last but not least, it is important to stress that meta-arguments occur crucially in the context of logic and argumentation theory, since such theorizing is by definition about arguments and consists in large part of the argumentative justification of theoretical claims.

Such ubiquity is, of course, relative to the domain of argumentative discourse, and is not meant to apply to the domain of human discourse in general. That is, to claim that meta-arguments are ubiquitous does not mean that they are omnipresent in ordinary language or lay discourse, as they are in technical, scholarly, and scientific speech and discussions. Rather the claim means that meta-arguments are as common in argumentative discourse as ground-level arguments are in lay discourse. For, on the one hand, I would hold that ground-level argumentation is not as explicit in ordinary discourse as some logicians and argumentation theorists would like to believe. On the other hand, even when not explicitly present, argumentation is often implicit in the sense that the logical theorist can reconstruct the ordinary discourse in argumentative terms; and it is possible for such a reconstruction to be carried out sufficiently carefully so as to avoid inaccuracy or distortion. Similarly, within the domain of argumentative discourse, meta-argumentation is not always explicitly present, but can often be reconstructed by the theorist or analyst from the material that is explicit, and this reconstruction can be done accurately and fairly. In this book, the emphasis will be on explicit (as distinct from implicit) argumentation, both at the ground-level and the meta-level, in the belief that logic and argumentation theory must follow a robustly empirical approach. This will be made clear at the outset, in chapter 1, which begins with an ostensive definition of argumentative reality, in the sense of the domain of explicit argumentation.

Given the ubiquity of meta-argumentation, it is not surprising that much scholarly work on argumentation is of some relevance to meta-arguments. Indeed, any scholarly discussion of interpretive or evaluative principles for arguments is likely to itself exemplify meta-argumentation in various ways, as well as to be indirectly relevant to meta-argumentation insofar as the interpretation and evaluation of arguments are themselves examples of meta-argumentation.

Nor should it be surprising that some scholars have begun to study meta-argumentation explicitly. Here, what I have found most inspiring has been the work of Erik Krabbe (2003) on meta-dialogues. Additionally, I have found encouraging the work of some scholars in computer science and artificial intelligence on the metacognitive aspects of argumentation and reasoning. For example, it is difficult to over-estimate the suggestiveness and potential fertility of the idea "that arguments and dialogues are inherently meta-logical processes. By this we mean that the arguments made by protagonists in a debate must refer to each other" (Wooldridge, McBurney, and Parsons 2005, section 1). Similarly, it is uncanny to find an analysis of argumentation in terms of meta-argumentation, attempting to "show how to use meta-argumentation as a general methodology for modeling argumentation" (Boella, Gabbay, van der Torre, and Villata 2009, 50).[1] Such encouragement remains despite the significant methodological differences between their approach (which tends to be formal and mathematical) and mine (which is informal and philosophical).

However, although meta-arguments are common in all areas of human cognitive *practice*, and although *implicit studies* of them can be found in many works, and although a *few explicit* scholarly studies do exist, meta-argumentation has never (to my knowledge) been studied explicitly, directly, and systematically in a book-length treatment. This lacuna is especially unfortunate because a focus on meta-argumentation can offer not only an understanding of a special class of arguments, but also a promising way of doing logic and argumentation theory in general. The crucial reason for this promise is that (as already suggested) to theorize about arguments is essentially and ultimately to argue about arguments, and hence to do logic and argumentation theory as meta-argumentation is merely to practice in one's own field what one preaches about others. In short, meta-arguments are special in two senses: they constitute a particular class of arguments, worth studying to learn about their nature and their similarities and differences with other particular types of arguments; and they are especially important because logic and argumentation theory are themselves particular instances of meta-argumentation.

This book is a first monograph-length attempt at a systematic study of meta-arguments. It consists of three parts. The first focuses on methodology and fundamentals. It elaborates an approach to logic and argumentation theory partly inspired by and adopted from Stephen Toulmin's applied logic, which is however distinguished from his substantive model of argument. It defines and illustrates some basic concepts and principles for the interpretation and evaluation of argumentation in general. And it distinguishes and inter-relates a number of basic types of meta-argumentation.

The second part focuses on theoretical meta-arguments, in the sense that it examines from the viewpoint of meta-argumentation several topics and problems that are not only intrinsically interesting and important, but also have received wide discussion in recent logic and argumentation theory. They include: how the concept of argument is to be defined, focusing on the meta-arguments of Ralph Johnson, Alvin Goldman, and Frans van Eemeren (chapters 4-5); how to demonstrate the formal incorrectness of arguments, centering on the meta-arguments of Erik Krabbe, Trudy Govier, and John Woods (ch. 6); whether deep disagreements and intractable quarrels can be resolved by argumentation, stressing the meta-arguments of Robert Fogelin and his many critics, as well as the related arguments of Woods and Henry Johnstone (ch. 7); whether pro-and-con arguments are best conceived as a distinctive class of so-called conductive arguments, what their nature is, and how they are to be evaluated, concentrating on the meta-arguments of Carl Wellman, David Hitchcock, and Govier, and their many critics (ch. 8); and when and how some arguments

---

[1] See also Perlis 1988; Jakobovits and Vermeir 1999; Costantini 2002; Modgil and Bench-Capon 2011.

can or should be interpreted or evaluated in terms of their own criteria, highlighting meta-arguments by Tony Blair, van Eemeren, and Johnson (ch. 9).

The third part focuses on famous meta-arguments, in the sense that it reconstructs several meta-arguments that have achieved classic status and have retained perennial interest and relevance, mostly by various great thinkers in the history of thought. These include: John Stuart Mill's eloquent plea for freedom of discussion, thought, and argument, found in his essay *On Liberty* (ch. 10); Mill's preliminary argument against the subjection of women and in favor of women's liberation and equality, advanced in his book on *The Subjection of Women* (ch. 11); David Hume's critique of the design argument for the existence of God as the intelligent designer of the universe, elaborated in his *Dialogues Concerning Natural Religion* (ch. 12); and Galileo Galilei's critique of the geostatic argument from vertical fall, contained in a brief but crucial passage of his *Dialogue on the Two Chief World Systems, Ptolemaic and Copernican* (ch. 13).

In relation to other books, although (as mentioned) there seems to be no existing book-length treatment of meta-argumentation, some basic intuitions in this direction are contained in my own *Arguments about Arguments: Systematic, Critical, and Historical Essays in Logical Theory* (2005) and *Galileo and the Art of Reasoning: Rhetorical Foundations of Logic and Scientific Method* (1980). However, those books are concerned primarily with ground-level argumentation and do not provide a systematic treatment of meta-argumentation. Moreover, the present book applies the systematized approach to new material.

Furthermore, although there are apparently no other books on meta-argumentation, the present book does not claim to cover completely unchartered territory. Rather it belongs to a cluster of fields that, besides including a few classics (such as the *Port-Royal Logic* in the seventeenth century), has recently witnessed a growing body of literature. What I have in mind is such books as these: Stephen Toulmin, *The Uses of Argument* (1958); Ch. Perelman and L. Olbrechts-Tyteca, *La Nouvelle Rhetorique: Traité de l'Argumentation* (1958); Henry Johnstone, Jr., *Philosophy and Argument* (1959); Charles Hamblin, *Fallacies* (1970); Michael Scriven, *Reasoning* (1976); Robert Fogelin, *Understanding Arguments* (1978); Alec Fisher, *The Logic of Real Arguments* (1988); Trudy Govier, *The Philosophy of Argument* (1999); Ralph Johnson, *Manifest Rationality* (2000); Larry Wright, *Critical Thinking* (2001); Frans van Eemeren and Rob Grootendorst, *A Systematic Theory of Argumentation* (2004); and John Woods, *Errors of Reasoning: Naturalizing the Logic of Inference* (2013). My book shares with these works many of the aims, approaches, concepts, and terms of that cluster of fields known by such labels as logical theory, argumentation theory, informal logic, practical logic, applied logic, naturalized logic, argument analysis, reasoning, critical thinking, and applied epistemology. In fact, my book is addressed to scholars and students in this cluster of fields. And it should be noted that to these fields there correspond various branches of such disciplines as philosophy, communication studies, rhetoric, cognitive psychology, cognitive science, linguistics, and artificial intelligence.

PART I
THE META-ARGUMENTATION APPROACH

Chapter 1
Methodological Considerations: Toulmin's Applied Logic

Works in logic and argumentation theory usually discuss, as the subject matter of their theorizing, arguments which are so simple that one often misses the crux of the matter. On the other hand, such works also typically advance, to justify their own theoretical claims, arguments which are so complex that one often loses the crucial argumentative thread. Both of these excesses will hopefully be avoided in the present work. The remedy to the second problem will be a constant and ongoing challenge throughout the course of this investigation. However, to remedy the former problem is relatively easier, and to that end I proceed immediately to mention some realistic material, indeed some real and significant historical cases.

1.1 Argumentative Reality

Let us begin with an ostensive definition of argumentation. Here are some paradigm examples.

In his book *On the Heavens*, Aristotle argued that the earth stands still at the center of the universe. A crucial piece of argumentation is contained in the following passage:

The natural movement of the earth, part and whole alike, is to the center of the whole—whence the fact that it is now actually situated at the center—but it might be questioned, since both centers are the same, which center it is that portions of the earth and other heavy things move to. Is this their goal because it is the center of the earth or because it is the center of the whole? The goal, surely, must be the center of the whole. For fire and other light things move to the extremity of the area which contains the center. It happens, however, that the center of the earth and of the whole is the same. Thus, they do move to the center of the earth, but accidentally, in virtue of the fact that the earth's center lies at the center of the whole. That the center of the earth is the goal of their movement is indicated by the fact that heavy bodies moving towards the earth do not move parallel but so as to make equal angles, and thus to a single center, that of the earth. It is clear, then, that the earth must be at the center and immovable, not only for the reasons already given, but also because heavy bodies forcibly thrown quite straight upward return to the point from which they started, even if they are thrown to an infinite distance. From these considerations then it is clear that the earth does not move and does not lie elsewhere than at the center.[1]

Of course, many questions arise about this passage, but for now its function is merely to help us ostensively define or identify what may be called argumentative reality. In a similar vein, but for historical and topical variety, let us also focus on the following argument for the existence of God found in Thomas Aquinas's *Summa theologica*:

The fifth way is taken from the governance of things. We see that things which lack knowledge, such as natural bodies, act for an end, and this is evident from their acting always, or nearly always, in the same way, so as to obtain the best result. Hence it is plain that they achieve their end not by chance,

---

[1] Aristotle, *On the Heavens*, book 2, chapter 14, 296b8-27 (1941, 434).

but by design. Now whatever lacks knowledge cannot move towards an end, unless it be directed by some being endowed with knowledge and intelligence, as the arrow is directed by the archer. Therefore some intelligent being exists by whom all natural things are ordered to their end; and this being we call God.[2]

Again, now is not the time to interpret, analyze, and evaluate this argument. Suffice it to have added what may be called a metaphysical argument by a thirteenth century thinker to the earlier one dealing with a physical, astronomical, or cosmological topic and advanced by an ancient Greek of the fourth century BC. Finally, let us complete our stock with an argument dealing with social, political, and moral issues and formulated by a nineteenth-century Englishman. The passage comes from John Stuart Mill's book on *The Subjection of Women*:

The object of the Essay is to explain as clearly as I am able, the grounds of an opinion which I have held from the very earliest period when I formed any opinions at all on social or political matters ...: That the principle which regulates the existing social relations between the two sexes—the legal subordination of one sex to the other—is wrong in itself, and now one of the chief hindrances to human improvement; and that it ought to be replaced by a principle of perfect equality, admitting no power or privilege on the one side, nor disability on the other.[3] ... It will not do for instance[4] ... to say that the *nature* of the two sexes adapts them to their present functions and position, and renders these appropriate to them. Standing on the ground of common sense and the constitution of the human mind, I deny that any one knows, or can know, the nature of the two sexes, as long as they have only been seen in their present relation to one another. If men had ever been found in society without women, or women without men, or if there had been a society of men and women in which the women were not under the control of men, something might have been positively known about the mental and moral differences which may be inherent in the nature of each. What is now called the nature of women is an eminently artificial thing—the result of forced repression in some directions, unnatural stimulation in others.[5]

I wanted to begin by quoting these arguments in order to focus our attention on that part or aspect of reality which I want to examine in the present work. In fact, when reading works in logical theory and argumentation theory, it is not always clear that everyone is talking about the same things, and so some of the differences and disagreements may be incommensurable. But this is not to say that all are, and thus we should make an effort to sort them out. Let us undertake some of these sortings.

1.2 Theory vs. Practice

I must confess that one of the reasons why I began with an ostensive definition of the topic at hand is that I wanted to model myself on the same technique used by a great twentieth century logician and analytical philosopher at the beginning of one of his books. I am referring to Willard Quine and his *Philosophy of Logic*. However, I would be less than candid if I did not add that this particular Quinean technique offered me an opportunity for some critical comments. In any case, besides the imitation and the criticism, there is also a higher motive here, namely to clarify the distinction between theory and practice, and to underscore the fact that the relevant sense of logic here is that of theory, discipline, or science of argumentative reality.

---

[2] Thomas Aquinas, *Summa theologica*, First Part, question 2, article 3 (1952, 1: 13).
[3] John S. Mill, *On the Subjection of Women*, chapter 1, paragraph 1 (1988, 1).
[4] Mill, *Women*, ch. 1, paragraph 18 (1988, 21).
[5] Mill, *Women*, ch. 1, par. 18 (1988, 22).

Quine began his essay on the *Philosophy of Logic* with a quotation from Lewis Carroll's *Through the Looking-Glass and What Alice Found There*. It comes from chapter 4, entitled "Tweedledum and Tweedledee," and it reads as follows: " 'I know what you are thinking about,' said Tweedledum; 'but it isn't so, nohow.' [¶] 'Contrariwise,' continued Tweedledee, 'if it was so, it might be; and if it were so, it would be, but as it isn't, it ain't. That's logic.' "[6] Quine uses Tweedledee's utterance as the epigraph to the book's preface, and he then begins his own exposition with the following inimitable words:

We shall be occupied in this book with the philosophy of logic in substantially Tweedledee's sense of the word 'logic'. This is not the invariable sense of the word. Precedent could be cited for applying the word collectively to two dissimilar studies: deductive and inductive logic. The philosophy of inductive logic, however, would be in no way distinguishable from philosophy's main stem, the theory of knowledge. What arrogates a distinctive bit of philosophy to itself is deductive logic, the discipline that Tweedledee had in mind. [¶] If pressed to supplement Tweedledee's ostensive definition of logic with a discursive definition of the same subject, I would say that logic is the systematic study of the logical truths. Pressed further, I would say that a sentence is logically true if all sentences with its grammatical structure are true. Pressed further, I would say to read this book. [Quine 1970, xi; 1986, vii]

Here Quine is making a useful distinction between deductive and inductive logic; he is plausibly equating the philosophy of inductive logic with epistemology in general; he is giving his own original definition of logic; and he is succinctly and eloquently summarizing in one sentence his own philosophy of logic. None of this is, or should be, objectionable in the present context.

However, what I want to take issue with is Quine's interpretation of Tweedledee's utterance. With all due respect, I think Quine is equivocating. For Tweedledee is indeed giving an ostensive definition, but of logic in the sense of logical practice, rather than of logic in the sense of logical theory. That is, Tweedledee is pointing to an example of (allegedly) logical reasoning, or deductive reasoning, or logical argumentation, or ratiocinative practice; he is not pointing to a piece of logical theorizing, or the discipline that studies deductive reasoning. Of course, I do not deny that in ordinary language one sense of the word 'logic' is to mean actual reasoning or real argumentation or ratiocinative practice; so, as an ostensive definition of logical reasoning, Tweedledee's last clause is unobjectionable. What I am saying is that another ordinary meaning of the word 'logic' is that of the theory of reasoning or argumentation, and that this is the sense which Quine has in mind. Moreover, I would claim that logic in the sense of logical theory is the relevant meaning of the word in the present context. On the other hand, logic in the sense of ratiocinative practice or argumentative reality is the subject matter studied by logical theory.

There is another criticism I would make of Quine's appropriation of Tweedledee's ostensive definition, namely that it is misleading. That is, even after we have clarified that Tweedledee's main utterance is a piece of deductive reasoning studied by logical theory, it encourages one to think that the subject matter of logical theory is trivial arguments like Tweedledee's argument why "contrariwise, ... it ain't." This difficulty might be called the problem of the trivialization of logic.

Unfortunately, this criticism is to some extent also applicable to works in logic and argumentation theory that try to provide an alternative to the paradigm of formal deductive logic. For example, in Toulmin's book *The Uses of Argument* and in the works of many commentators, there is considerable discussion of an argument that reads as follows: Harry

---

[6] Carroll 2000, 181; cf. Quine (1970, xi; 1986, vii).

was born in Bermuda, so he is a British subject.[7] And in many works by and about John Pollock, the following example seems to be ubiquitous: this object is red because it looks red; the point here is to understand that this argument advances a good prima facie reason, which is nevertheless defeasible because it can be defeated by the rebuttal that the object is being observed under red light.[8]

On the other hand, Quine knows very well that logic is *not* a trivial discipline, and some of his own work contributes to demonstrating this nontriviality. For me, the key reason for the nontriviality of deductive logic is that it provides a viable and indeed unsurpassed theory of mathematical reasoning, a point to be elaborated below. And the consequence I would draw from this is that the examples used for introductory elucidations or ostensive definitions should be proofs of mathematical truths. For instance, there are sufficiently brief and accessible proofs that the square root of two is an irrational number (Salmon 1984, 32-33); that there are only five regular convex solids;[9] and that the Pythagorean Theorem holds (Malone and Sherry 1998, 9-10). Similarly, Toulmin knows very well that his alternative approach must deal with important and substantial arguments, and he does precisely this in several other works.[10] And Pollock too realizes that the adequacy of his theory of defeasible reasoning needs to be demonstrated more tangibly; his own self-imposed requirement is that his theory should be implementable by the construction of a machine with artificial intelligence, and he goes a considerable way toward fulfilling it (Pollock 1995, 4, 49-50, 299-361).

Thus, although my point should not be misunderstood or exaggerated, nevertheless we should be clear that logic is some kind of theorizing, and that it is or ought to be first and foremost about significant cases of argumentation and reasoning. We are not dealing primarily, let alone exclusively, with the Tweedledee type of trivialities.

1.3 Theory vs. Method

Another distinction which is generally significant, but also specifically relevant in this context, is that between substantive theory and methodological approach. This is especially important because it needs to be appreciated in order to us to understand properly the work and legacy of Toulmin in this field.

Toulmin ended his epoch-making book on *The Uses of Argument*, with a chapter entitled "Conclusion," in which he summarized the main points of the approach he was advocating. The label which he chose for this approach was "applied logic" (Toulmin 1958, 255; 2003, 235). However, he did not insist much on this label, the reason being, I conjecture, that such a label might be misleading by conveying the wrong impression that it involves the application to real argumentation of principles somehow obtained or available independently of, or prior to, the application. Nevertheless, here I shall adopt this phrase. Labels aside, let us examine how Toulmin described his approach.

The first feature is that "a radical re-ordering of logical theory is needed in order to bring it more nearly into line with critical practice" (Toulmin 1958, 253; 2003, 234). We

---

[7] For example, Toulmin (1958, 99-107; 2003, 92-100); Verheij 2006, 185-201. As far as I can tell, Toulmin's (2003) updated edition of *The Uses of Argument* reprints the text of the original (1958) edition, and merely adds a new two-page preface.
[8] For example, Pollock (1974, 40-43; 1995, 39-41); Prakken 2006, 236.
[9] Cf.. Toulmin, Rieke, and Janik 1979, 87-89; Aberdein 2006, 330-32.
[10] Toulmin (1953, 1972, 2001); Toulmin and Goodfield 1961; Jonsen and Toulmin 1988. Thus, in reflecting on Toulmin, Woods (2006a) wisely and shrewdly considers several other works besides Toulmin's (1958, 2003).

might say that here Toulmin is advocating a pragmatic approach, although of course the word 'pragmatic' is a much used term, in many senses that do not necessarily coincide with Toulmin's. For example, among scholars who describe their own approach as at least partly pragmatic are Robert Brandom (1994, 3-64; 2004, 4), Frans van Eemeren and Rob Grootendorst (2004, 31-37), Ralph Johnson (2000a, 102), and Douglas Walton (1996, 37-41). I believe their meanings overlap in part with Toulmin's. His own meaning seems to be that argumentative practice is more important vis-à-vis logical theory than ordinarily or traditionally supposed. This does not mean that practice is primary, but rather it is merely a *denial* of the primary of theory, and an affirmation that theory and practice are equally important and in mutual interaction.

Second, for Toulmin, there is a "need for a *rapprochement* between logic and epistemology ... The question, 'How does our cognitive equipment (our understanding) function?', must be treated for philosophical purposes as equivalent to the question, 'What sorts of arguments could be produced for the things we claim to know?'—so leaving aside the associated psychological and physiological questions" (Toulmin 1958, 254; 2003, 234). That is, Toulmin is proposing an epistemic or epistemological approach. However, although this word is another loaded term, his own particular meaning here is clear: epistemology is essentially a branch of logical theory insofar as it is the part that studies a special class of arguments, namely arguments that justify knowledge claims. Thus, it should be noted that Toulmin's epistemological approach amounts to a reversal of the program apparently advocated by those scholars who speak of an epistemic approach in the study of argumentation.[11]

The third element of Toulmin's approach is relatively uncontroversial and widely shared. That is, logical theory should be descriptive as well as prescriptive, or factual as well as evaluative, or analytical as well as normative. In his own words, "the proper business of both [logic and epistemology] is to study the structures of our arguments in different fields, and to see clearly the nature of the merits and defects characteristic of each type of argument" (Toulmin 1958, 255; 2003, 235). The key words here are: structures, merits, and defects; and in what follows I shall exploit these notions. An interesting aspect of Toulmin's conception of the normative is that it pertains to both merits and defects.

The fourth component of the Toulminian approach has already been indirectly and implicitly mentioned, but he gives it a special emphasis and explicit label: the "comparative method." That is, Toulmin stresses "the importance in logic of the comparative method—treating arguments in all fields as of equal interest and propriety, and so comparing and contrasting their structures without any suggestion that arguments in one fields are 'superior' to those of another" (Toulmin 1958, 254; 2003, 234). In my opinion, this is best interpreted as a recommendation that logical theorists should strive to reach a judicious balance between the similarities and differences among the various fields, or in Toulmin's own words: "broad similarities there may be between arguments in different fields, both in the major phases of the arguments ... and in their micro-structure ... it is our business, however, not to insist on finding such resemblances at all costs, but to keep an eye open quite as much for possible differences" (Toulmin 1958, 256; 2003, 236). On the contrary, Toulmin's comparative method should *not* be interpreted as one-sidedly stressing what seems to have become a dogma among some Toulminian followers and scholars, namely: "that validity is an intra-field, not an inter-field notion. Arguments within any field can be judged by standards appropriate within that field, and some will fall short; but it must be expected that the standards will be field-dependent, and that the merits to be demanded of

---

[11] See, for example, Biro and Siegel (1992, 2006a, 2006b), Freeman 2005, Lumer (2005a, 2005b), Pinto 2001, Siegel 1994.

an argument in one field will be found to be absent ... from entirely meritorious arguments in another" (Toulmin 1958, 255; 2003, 235). This widely appropriated and somewhat relativistic thesis should be viewed in the context of a balanced and judiciously comparative orientation.

Fifthly, for Toulmin, "logic ... may have to become less of an *a priori* subject than it has recently been ... Accepting the need to begin by collecting for study the actual forms of argument current in any field, our starting point will be confessedly empirical" (Toulmin 1958, 257; 2003, 236-38). Again, these terms (empirical, *a priori*) are as loaded as any of the other descriptions of Toulmin's approach. Some scholars, for example Else Barth,[12] have also spoken of an "empirical logic," to mean something very different. Others, e.g. Pollock, do not accept the usual contrast (endorsed by Toulmin) between the *a priori* and the empirical, but rather allow that logical truths are *a priori* in one sense, but also empirical in another.[13] Still others, such as John Woods, speak of "abstract sciences," as the proper contrast to empirical sciences "for which the criterion of empirical adequacy is a legitimate standard" (Woods 2003, 2). Nevertheless, Toulmin's notion of the empirical is contextually clear. For one thing, it should be noted that this remark speaks of an empirical "starting point," and there is no reason to equate this with the end point or with everything else in between. Moreover, it is clear that this notion subsumes or overlaps with both the pragmatic and the comparative methods already mentioned, as well as the next two that follow.

In fact, there is a sixth component to Toulmin's approach: "not only will logic have to become more empirical; it will inevitably tend to be more historical ... In the natural sciences, for instance, men such as Kepler, Newton, Lavoisier, Darwin and Freud have transformed not only our beliefs, but also our ways of arguing and our standards of relevance and proof ... Grotius and Bentham, Euclid and Gauss, have performed the same double feat for us in other fields" (Toulmin 1958, 257; 2003, 237). That is, here Toulmin is saying that the history of thought in general, and the history of science in particular, constitute a uniquely important and importantly unique database for logical theory. He is also saying that this historical database contains two main lines of materials that were originally novel and are now established: substantive claims about some subject matter, be it the physical world, numbers and geometrical figures, life, society, and the human mind; and methods and principles of argument. In short, he is suggesting that the history of thought should be seen as the history of argumentation, and so it should be accordingly mined by the logical theorist.

Finally, Toulmin describes a seventh component to his approach, a method appropriated from the enterprise or discipline of natural history: "scrutinise the logical history, structure and *modus operandi* of the sciences using the eye of a naturalist, without preconceptions or prejudices imported from outside. This will mean seeing and describing the arguments in each field as they are, recognising how they work; not setting oneself up to explain why, or to demonstrate [how] they necessarily must work. What is required, in a phrase, is not epistemological *theory* but epistemological *analysis*" (Toulmin 1958, 258; 2003, 238). Here Toulmin's contrast between theory and analysis together with a plea for analysis is very telling and extremely important. However, by itself the term 'analysis' or

---

[12] Barth 1985a; cf. Finocchiaro 2005a, 46-64.

[13] Pollock's point is that "*a priori* truths are truths that can be established on the basis of our logical intuitions" (1974, 332), but "logical intuitions do not provide us with conclusive reasons for *a priori* judgments, any more than our sight provides us with conclusive reasons for judging the colors of things" (1974, 320); rather, "our logical intuitions ... provide us with prima facie reasons for 'self-evident' *a priori* judgments" (1974, 327).

'analytical' does not usually convey the contrast to theory, and so it would be misleading to speak of an analytical approach. For this reason, and because of his reference to natural history, I shall speak of it as the naturalist component of Toulmin's approach. What he seems to be proposing is that there should be more careful description of argumentative reality, as long as such description is not contrasted to prescription, but rather is taken to include a description of merits and defects, as well as structures, to use Toulmin's earlier language; such naturalist description is prior to the theoretical explanation, interpretation, and evaluation of that reality.

In summary, Toulmin's so-called "applied logic" is a methodological approach to logical theorizing that has the following features: it is pragmatic, in the sense of taking practice as essential (i.e., co-equal with theory); epistemological, in the sense of attaching great importance to the arguments for knowledge claims; simultaneously normative and descriptive; comparative, in the sense of sensitive to both the differences and similarities of different fields; empirical, in the sense of the opposite of *a priori*; historical, in the sense of oriented toward the history of thought; and naturalist, in the sense of emphasizing description à la natural history, or analysis as contrasted to theory.

It is important to note that this summary of his main conclusions which Toulmin gives in the final chapter of his book does not include his model of the layout of arguments which he discusses in an earlier chapter (Toulmin 1958, 94-145; 2003, 87-134). That model formulates several principles for the interpretation of argumentation that distinguish and interrelate various elements labeled as follows: claim, data, warrant, backing, qualifier, and rebuttal. I think Toulmin is right not to include his model of argument layout in his book's final "Conclusion." For his main aim is methodological, namely to advocate a new approach to the study of argumentation along the lines of the applied logic which we have just described, whereas his model of the layout of arguments is a particular substantive theory toward which his commitment is more tentative and less robust.

In fact, when the model is presented, elaborated, and explained in other parts of the book, Toulmin make no serious attempt to show that he has arrived at his model by following the applied-logic approach he advocates. This remains true despite the fact that he claims to have derived the model from the field of legal and jurisprudential argumentation, for this derivation amounts to mere subjective inspiration. Moreover, the introduction and discussion of the model in the book strikes the critical reader as amounting to little more than a terminological or linguistic variant of the standard framework. That is, by and large, and as a first approximation, Toulmin's 'claim' corresponds to the notion of conclusion; his 'data' to 'minor premise'; 'warrant' to 'major premise'; 'backing' to support for the major premise; and rebuttal to objection, con reason, or counter-evidence. This initial impression and intuition of substantive equivalence under different jargon has lately been demonstrated in great detail and with technical skill in an important paper by Chris Reed and Glenn Rowe.[14]

However, here I do not wish to stress or elaborate these reservations about the Toulmin model of argument layout and diagramming. For I certainly find it sufficiently intriguing and potentially fruitful to place it in the category of ideas that would be worth pursuing if (1) they were not already being pursued at great length by others, and if (2) their pursuit were more worthwhile than the pursuit of other ideas. However, with regard to first proviso, it is undeniable that the Toulmin model has attracted many followers and commentators, who typically focus primarily or exclusively on the model; and even when they do not focus on the model, they focus on other particular, substantive theses such as

---

[14] Reed and Rowe 2006; see also the same judgment expressed by Verheij (2006, 194).

the relativistic thesis about validity being an intra-field rather than an inter-field norm.[15] And with regard to the second proviso, I believe his applied-logic approach is more worthy of pursuit.

In fact, my main point here and now is that there is an important distinction to be made between Toulmin's applied-logic approach and his model of argument layout: the former represents the method which he is proposing in the field of logical theory and the study of argumentation; the latter is a substantive theory about how arguments can be analyzed, interpreted, laid out, and diagrammed.[16] Furthermore, there is little correspondence between these two elements of Toulmin's work, in the sense that his model is not really an example, let alone a good example, of the utilization of his applied-logic approach. Finally, it is Toulmin's applied-logic approach that has been relatively neglected and that deserves to be pursued more explicitly, more seriously, and more systematically.

1.4 Theory vs. Observation

It is obvious that several of the components of Toulmin's "applied-logic" approach share a family resemblance that makes them cluster together; I am referring to the so-called pragmatic, comparative, empirical, historical, and naturalist components. Subsuming these under the notion of observation, we can say that Toulmin is making a plea for more and better observation of argumentative reality. And exploiting the common contrast between theory and observation, we have that he is advocating that logic and argumentation theory should be more firmly and robustly based on observation.

In pursuing and advocating such an observational approach, there is no reason why one should fail to appreciate the Kantian insight that whereas concepts without percepts are empty, percepts without concepts are blind. Nor is there any reason why, using more modern terminology, one should disregard the principle that observation is normally theory-laden in some sense.[17] The point is that logic and argumentation theory should be conceived not as empty theorizing, but rather as observationally grounded; the observation of argumentative reality should have a greater and more important role.

Let us now give some historical or empirical (observational!) content to this abstract distinction between theory and observation. The first thing I want to do is to retrieve the

---

[15] For substantiation of these generalizations, see the 25 papers in Hitchcock and Verheij 2006, as well as Freeman 2011 and Allen 2011. But of course, there are exceptions among them, and I would cite Woods 2006a and Verheij (2006, 194-202) as examples of exceptions. Occurring in a different context, another exception is Benthem 2009. Another important exception is Johnson 1981b, which also has the merit of displaying an attitude toward Toulmin that is both appreciative and critical.

[16] The distinction between method and theory is, of course, valid and valuable in general, and so it is applicable to many other cases; and the application may yield a result that is the reverse of my attitude toward Toulmin, in the sense that whereas I am endorsing Toulmin's method but not his model, in other cases one may endorse the theory but not the method. An especially instructive and relevant example for the present investigation is the model of argumentation and meta-argumentation elaborated by Boella, Gabbay, van der Torre, and Villata (2009); they "introduce the meta-argumentation viewpoint on argumentation, which conceptualizes argumentation together with arguing about argumentation ... In meta-argumentation, arguments ... are interpreted as meta-arguments which are mapped to 'argument *a* is accepted' for some argument a" (p. 351). This is a thesis whose substance, if I understand it correctly, I find congenial and acceptable and will elaborate in my own way later (chapter 11.5), when I discuss the possibility that all arguments have implicitly a meta-argumentative element. However, I would characterize their approach as formal, mathematical, and aprioristic, and hence the opposite of the method I advocate.

[17] See Brown (1979, 1987) for a general discussion of this principle.

observational basis of standard formal deductive logic (hereafter abbreviated FDL[18]).

Many authors have criticized formal deductive logic (FDL) for its apriorist approach in the study of argumentation. I believe such criticism is justified to the extent that FDL is regarded as a *general* theory of argumentation. However, it is equally important to note that if FDL is taken to be a theory of *mathematical* argument and reasoning, then it remains a viable and unsurpassed account, at least as a first approximation. That is, mathematical proof provides the observational basis for FDL. But we must be careful to interpret such observational grounding properly. Such grounding should not be viewed as the oversimplified enumerative induction caricatured in many critiques of empiricism and inductivism. I believe such grounding may be seen as hypothetico-deductive confirmation. In short, if the argumentative reality under investigation is mathematical proof, then FDL may be viewed as elaborating an hypothetico-deductive confirmation of its claims about mathematical proof.

Excellent support for this claim can be found in Alfred Tarski's work, *Introduction to Logic and to the Methodology of Deductive Sciences*. The book begins with an exposition and elaboration of some basic elements of logic: constants vs. variables, sentential calculus, theory of identity, monadic predicate calculus, and theory of relations. This part makes up about one third of the book. Then there is another part dealing with the fundamentals of deduction and the deductive method, discussing such topics as: primitive vs. defined terms, axioms vs. theorems, models, proofs, consistency, completeness, and decidability. Finally, the third part consists of an application of the ideas previously elaborated abstractly ("theoretically," so to speak) to the case of elementary arithmetic. Tarski examines several axiomatizations of arithmetic, beginning with simpler versions, and leading up to Peano's axiomatization of the arithmetic of natural numbers, and then also to Hilbert's axiomatization of the arithmetic of real numbers. I am interpreting this third section of Tarski's book to be partly an observational description of mathematical practice, and partly a theoretical interpretation and evaluation of it in terms of the fundamental principles of the logic and the theory of deduction elaborated in the earlier sections of the book.[19] Those principles are thereby confirmed, and mathematical practice within arithmetic is thereby explained in the sense of rendered intelligible and comprehensible.

Earlier, in introducing this account, I remarked that it is viable and unsurpassed *as a first approximation*. It is now time for a brief elaboration of that cryptic qualification. This account need not deny, and does not deny, that there are aspects of mathematical practice whose understanding and evaluation require the elaboration of principles other than standard FDL, and more in line with other less formal and less deductive aspects of argumentation theory. I am thinking of such processes as the discovery of new mathematical proofs, the use of visual diagrams in proofs, epistemological controversies surrounding the discussion of some problems in the foundations of mathematics, and the application of mathematical truths to physical reality.[20] However, to exaggerate the importance or prevalence of such non-formal or non-deductive processes in mathematical practice would be as injudicious as to deny their existence altogether. In my view, they represent simply refinements, or finer approximations, to the basic FDL account of mathematical argumentation.

---

[18] I adapt this abbreviation from Johnson (2000a, 31-32) and Woods, Johnson, Gabbay, and Ohlbach (2002, 4).

[19] These three parts or sections of Tarski's book (1965) correspond, respectively, to pp. 3-116, 117-54, and 155-226.

[20] Cf. Aberdein (2006, 2007), Dove (2007, 2009), Finocchiaro (2010b, 97-120), Franklin 1987, Krabbe 1997, Lakatos 1976, Mancosu (2005, 2008).

One reason why I wanted to elaborate the robust observational or pragmatic basis of standard FDL is to contrast it to what I feel is the relative poverty of observational description on the part of many argumentation theories that seek to provide an alternative and are sounder in other ways. But let us first repeat that once we leave the field of mathematical argumentation, and we enter the fields of most other academic disciplines or of everyday argumentation, FDL becomes excessively apriorist, unrealistically prescriptivist, too impracticable, and observationally inadequate, as many alternative theories have fairly objected. In these other fields, FDL becomes the wrong first approximation, and at best it can serve to deal with these fields' secondary aspects that represent refinements to a first approximation to be provided by some alternative account. Now, although there are notable exceptions,[21] the observational basis of many of these alternative theories is rather weak, and indeed it pales in comparison to FDL's empirical basis from the field of mathematical practice.

Here, the example I shall take as a target of my criticism is Perelman's "new rhetoric." Let me recall, however, that I have already expressed a similar criticism of Toulmin's model of the layout of arguments, which strikes me as infinitely less well founded on everyday argumentative practice or even legal argumentation than FDL is on mathematical practice. But, let us focus on Perelman.

There are many sound and acceptable elements in Perelman's substantive theory of argumentation. For example, he is right that the exclusive reliance on the formal logic of demonstration has the irrationalistic tendency of leaving most of the human sciences and human affairs in the realm of the arbitrary. And he is right in wanting to take the study of persuasion phenomena and rhetorical problems out of the hands of propaganda ministers, Madison Avenue firms, public relation experts, and preachers from the pulpit. More specifically, he makes a valuable contrast between argumentation and demonstration, defining the former as the domain of "discursive techniques allowing us to induce or to increase the mind's adherence to the theses presented for its assent" (Perelman and Olbrechts-Tyteca 1969, 4). And more specifically still, he is correct in holding the key thesis that the crucial distinction between demonstration and argumentation can best be appreciated by appreciating the notion of begging the question, or *petitio principii*: this "is not an error of logic, but of rhetoric" (Perelman and Olbrechts-Tyteca 1969, 112); it is "an error in argumentation" (Perelman and Olbrechts-Tyteca 1969, 114); indeed it is "the most serious mistake in argumentation" (Perelman 1979, 69). That is, once we define "begging the question, that is, to base one's argument on premises that the audience rejects ... it is only within the framework of a theory of argumentation that one can take an accusation of *petitio principii* into consideration and examine whether or not the implied criticism is legitimate" (Perelman and Olbrechts-Tyteca 1969, 114).

However, if we look for the justification of such theoretical claims, arguments are very difficult to find. Or to be more charitable and precise, we find few if any observational arguments, based on the description of some robust body of evidence or meticulous observation of some aspect of argumentative reality. Instead, more frequently we find what might be called deductive or analytic arguments. For example, the claim that begging the question is an error in argumentation is essentially a deductive consequence of the definitions of argumentation and of begging the question, given in the passages just quoted. And the claim that begging the question is not an error of logic is a consequence of those definitions, plus some additional assumptions filling in more details about the process of begging the question. Even if such deductive arguments were otherwise unobjectionable, one would want to know how the various conclusions relate to argumentative reality. When

---

[21] Such as Fisher (1988, 2004), Fogelin 1978, Govier 1999, Woods 2013.

a connection is made, it turns out to be weak, as we can see from the following case, involving the admittedly crucial notion of *petitio principii*.

In fact, the only example of begging the question which we find in Perelman's *New Rhetoric* is the following passage from Antiphon's speech on the murder of Herodes: "I would have you know that I am much more deserving of your pity than of punishment. Punishment is indeed the due of the guilty, while pity is the due of those who are the object of an unjust accusation" (Perelman and Olbrechts-Tyteca 1969, 113). Perelman interprets this as an argument with the first sentence as conclusion and the second sentence as major premise; and there is a missing minor premise, namely the claim that "I am the object of an unjust accusation." But this claim cannot be accepted by the judges before deciding the case; thus, the conclusion is being based on a premise which is not acceptable before accepting the conclusion; and so the argument begs the question.

However, I would object that the passage could also be interpreted as consisting of two different ways of stating the same claim: the first sentence would be stating it more specifically and explicitly, by referring to the speaking defendant; the second sentence would be formulating the same claim more generally and implicitly; and the lack of an explicit reasoning indicator would favor this interpretation. In this case, the passage as such would not be an argument, and so could not be a question-begging argument.

Moreover, even if we accept Perelman's interpretation, it is obvious that the missing premise is not identical with the conclusion; thus, if the defendant proceeds to give a subargument supporting that missing premise, and this subargument is independent of the conclusion in Perelman's reconstruction, then there would be no begging of the question. That is, the passage quoted by Perelman could be the last step of a long and complex argument that does not assume the proposition "I am much more deserving of your pity than of punishment," which is the final conclusion. Of course, in order to ascertain this possibility, we would have to look at the rest of the speech; hence, I am not concluding that Perelman's interpretation is wrong, but rather than it is not justified by his discussion. This in turn means that the passage is not a good example of what he is trying to illustrate.

Unfortunately, this is a general difficulty affecting Perelman's *New Rhetoric*. That is, although the examples are more numerous, substantive, and realistic than the Quine-Tweedledee type of example, nevertheless the passages are usually taken out of context and the propriety and strength of their illustrative and justifying power depend on assumptions that are not made explicit and that may or may not be accurate. What we would need is something which, *mutatis mutandis*, would have the realism and robustness which Peano's axiomatization of arithmetic has for a theorist of mathematical argumentation, such as Tarski.

To summarize this section, many elements of the applied-logic approach stemming from Toulmin can be viewed as requiring a greater and more determined emphasis on observation. In turn, the observational approach can and should be interpreted as advocating that logic and argumentation theory be grounded more firmly and more robustly on the observational description of argumentative reality. Accordingly, formal deductive logic can be seen as a theory of mathematical argumentation, possessing a robust observational basis, by means of an hypothetico-deductive confirmation of its basic principles. By contrast, alternative argumentation theories such as Perelman's new rhetoric seem to lack a comparably adequate observational basis in some aspect of non-mathematical argumentative reality. In this regard, what is needed is theorizing which (like Perelman's) takes seriously the subject matter of argumentation as distinct from demonstration, but which (like Tarski's) also takes seriously the observational description of its chosen subject matter.

1.5 Theory vs. Argumentation

We have seen that one component of the method advocated by Toulmin proposes that epistemology focus primarily on studying arguments justifying knowledge claims. In any case, for the logical theorist who takes seriously the study of argumentation, such a move is only a natural one. In a similar vein, the same focus on argumentation suggests a natural reformulation or reconceptualization of the observational emphasis of the Toulminian method, which subsumes the pragmatic, comparative, empirical, historical, and naturalist components, as we have just seen. Let us say, then, that in advancing and justifying its own claims, logical theory should employ *observational arguments* to a much greater extent than is common.

Here what I am calling observational arguments correspond partly to what are commonly called inductive arguments. However, since the notion of inductive argument is notoriously vague and imprecise (or at least controversial), and since it is often taken in a narrow sense (namely as induction by enumeration), it is important to note that observational arguments in the present context are also meant to correspond to arguments that are variously called abductive arguments, conductive arguments, inferences to the best explanation, hypothetico-deductive confirmations, arguments from analogy, statistical syllogisms, and the like. Another cluster of arguments I am subsuming under this notion of observational arguments are ampliative, probable, plausible, and defeasible arguments. Observational arguments are meant to be contrasted to arguments that are commonly called deductive, analytic, deductively valid, formally valid, truth-preserving, conclusive, and entailment-achieving.

This argumentational reformulation of Toulmin's method in logical theory carries with it another important feature. Because the claims advanced by logicians and argumentation theorists are first and foremost, ultimately and essentially, about arguments and argumentation, it follows that their own observational arguments are meta-arguments, in the simple sense of being arguments about arguments. Here, meta-arguments are meant to be contrasted to ground-level arguments, which are arguments about topics other than arguments (such as numbers, natural phenomena, historical events, and human actions), or arguments which in a given context are the subject matter of meta-arguments. The methodological claim is that logical theory and argumentation theory are or ought to be instances of meta-argumentation.

If we take the point of view of meta-argumentation, then the emphasis on observational arguments means in part an emphasis on arguments some of whose premises are observational claims about real arguments. Such observational meta-claims could then be used in meta-arguments of the form of enumerative induction to arrive at broader generalizations about arguments; or they could be used in meta-arguments from analogy to reach conclusions about similar arguments; or they could be used in hypothetico-deductive confirmations as facts to be explained by deducing them from hypothetical principles that are thereby justified; or they could be used in statistical syllogisms to infer particular empirical claims from more general empirical claims.

However, some theoretical meta-argumentation could and would be deductive, in the sense of exploring the entailment relations among observations, or among concepts, or among theoretical principles. In fact, the meta-argumentative step is a distinct proposal from the observational or empirical emphasis. The meta-argumentative step is in a sense the result of wanting to stress in one's own activity of theorizing the practice which one is studying or reflecting upon. Here, once again the model of Tarski's procedure is perhaps more productive than Perelman's. For, in his theorizing about mathematics, Tarski was interpreting mathematics in terms of the deductive method and in the process he was himself using primarily the mathematical method; the result was properly conceived as meta-mathematics. On the other hand, with regard to Perelman, in his own theorizing about a subject matter generally characterized as argumentation distinct from demonstration, I don't see that he was paying sufficient attention to the observation of argumentative reality, and so his theorizing exhibits more attempted demonstration than actual argumentation.

So far in this section, I have been saying that Toulmin's applied-logic approach can and should be also reconceptualized as advocating, first, a greater use of observational arguments, and, second, a more self-reflective awareness of meta-arguments; that is, logic and argumentation theory should be practiced as observationally-grounded meta-argumentation. With this reconceptualization, formal deductive logic can be seen as a theory of mathematical argumentation, itself consisting largely of meta-argumentation based on observational premises that describe various aspects of mathematics, and employing primarily (although not exclusively) deductive arguments that mirror mathematical reasoning. By contrast, an alternative argumentation theory such as Perelman's new rhetoric, while consisting inevitably of meta-argumentation, seems to fail to model itself sufficiently on the nondemonstrative argumentative reality which is its subject matter. What is needed is theorizing consisting of meta-argumentation which, like Perelman's, takes seriously the subject matter of argumentation as distinct from demonstration, but which, like Tarski's, models its own argumentation on that of its subject matter.

More generally, so far in this chapter I have elaborated an approach to logical theory and the study of argumentation that may be characterized as Toulminian. However, I have made it clear that it is Toulminian only in the sense that it corresponds to the approach which he explicitly or reflectively advocated, but not in the sense that it builds upon Toulmin's own substantive model of the layout of arguments. Moreover, I have hinted at the fact, and now want to make clear, that my approach also overlaps with those of other authors and derives from other sources. I am referring to various approaches that are known under such labels as: informal logic, practical logic, naturalized logic, logic of real arguments, argument analysis, philosophy of argument, argumentation theory, and immanent dialectical approach.[22] It is also important to add that elsewhere I have advocated such an approach under the label of "historical-textual" approach (Finocchiaro 2005a, 14, 34-45), and that I have already put it into practice by studying an extremely rich and varied collection of arguments: that is, the ground-level arguments for and against the motion of the earth discussed in a book by Galileo Galilei published in 1632, his *Dialogue on the Two Chief World Systems, Ptolemaic and Copernican*.[23] What I want to do in this book is to continue with that type of investigation, but do so with a new twist that has emerged as being especially fruitful.

This twist amounts to a focus on meta-arguments, as distinct from just arguments in the sense of ground-level arguments. In fact, in my methodological considerations so far, meta-argumentation has already emerged as a key and emblematic component of the applied-logic approach à la Toulmin; that is, theorizing about argumentation is best conceived and practiced as arguing about argumentation. Such meta-argumentation should also strive to be pragmatic, comparative, empirical, historical, and naturalist, as Toulmin recommends in general. I plan to satisfy these requirements while studying a special class of arguments, namely arguments about arguments, i.e., meta-arguments. Now, there happen

---

[22] Cf., *respectively*, Johnson and Blair 2002, Walton 1989; Woods, Johnson, Gabbay, and Ohlbach 2002; Woods 2013; Fisher 1988; Scriven 1976; Govier 1999; Eemeren, Grootendorst, and Snoeck Henkemans 1996; Krabbe 1999, Houtlosser and van Laar 2007.
[23] Finocchiaro 1980a; 1997, 309-72; 2005a, 65-91, 128-47, 386-408.

to be two domains or contexts that provide excellent material that will be examined in the course of this work. One is the arguments advanced by logicians and argumentation theorists to justify their theories and theoretical claims; these might be called theoretical meta-arguments. The other is arguments from the history of thought which for various reasons have acquired classic status; these might be called famous meta-arguments.

Some of the latter are worth identifying immediately here also because they happen to relate to the paradigm examples of argumentative reality which I used earlier to provide my ostensive definition. Parts of Mill's *Subjection of Women* consist of critical meta-argumentation against the traditional arguments for subjection; in fact, the passage quoted earlier is such a meta-argument, rather than a simple ground-level argument. The design argument for the existence of God, quoted above from Aquinas's formulation, generated many meta-argumentative critiques, the best and most celebrated of which is Hume's *Dialogues concerning Natural Religion*. And the arguments for geocentrism and geostaticism, such as those quoted above from Aristotle's *On the Heavens*, were famously refuted by Galileo in his *Dialogue on the Two Chief World Systems*.

1.6 Summary

This chapter has undertaken a methodological elaboration of an approach to the study of argumentation which Toulmin called "applied logic," and which I have previously labeled the "historical-textual" approach. This elaboration was meant to motivate and justify a novel proposal to use that approach to study arguments about arguments, called meta-arguments. The elaboration consisted of distinctions between logical theory on one side and in turn: argumentative reality, ratiocinative practice, methodological approach, observation, argumentation, and meta-argumentation. The discussion also contains a three-fold critical appreciation: it endorses Toulmin's applied-logic approach, but criticizes his substantive model of argument layout; it defends the adequacy of formal deductive logic as observationally grounded in mathematical practice, but questions it as a general theory of argumentation; and it supports the aim of general theories of argumentation, such as Perelman's "new rhetoric," but faults its observational grounding.

Chapter 2
Elementary Principles of Interpretation and Evaluation

The last chapter elaborated an approach to the study of argumentation that is pragmatic, empirical, historical, comparative, naturalist, and both descriptive and normative. It corresponds to Toulmin's idea of an applied logic, as well as to the historical-textual approach which I have advocated and practiced on other occasions. From the point of view of such a methodological approach, the conceptual and substantive details of a corresponding theory are relatively open. The purpose of this chapter is to introduce the basic details of such a theory.

2.1 Reasoning and Argumentation

Reasoning is the activity of the human mind that consists of giving reasons for conclusions, reaching conclusions on the basis of reasons, or drawing consequences from premises. In other words, it is the interrelating of thoughts in such a way as to make some thoughts dependent on others, and this interdependence can take the form of some thoughts being based on others or some thoughts following from others. Reasoning is thus a special kind of thinking; all reasoning is thinking, but not all thinking is reasoning.

The occurrence of reasoning is normally indicated, and can always be explicitly indicated, by the use of *reasoning indicators*. These are words like the following (or phrases synonymous with such words): 'therefore', 'thus', 'so', 'hence', 'consequently', 'because', 'since', 'for'. Reasoning indicators, however, are only hints, since it is possible to express simple reasoning without them and for them to have other meanings that do not indicate reasoning. Nevertheless, reasoning indicators enable us to formulate an operational definition: reasoning is the type of thinking that occurs whenever there is a high incidence of words such as 'therefore', 'because', and 'consequently'.

Reasoning is linguistically expressed in *arguments*. An argument is a basic unit of reasoning in the sense that it is a piece of reasoning sufficiently self-contained to constitute by itself a more or less autonomous instance of reasoning.

Reasoning indicators serve to interconnect the *propositions* of an argument. A proposition is any part of an argument that is capable of being accepted or rejected by itself. It is also capable of being stated as a complete sentence, so that it can stand by itself. Propositions may also be called assertions, statements, claims, or theses; for our purposes, we will not distinguish among these five terms. An argument may thus be conceived as a series of propositions some of which are based on others, where the interconnections are expressed by means of reasoning indicators.

The simplest possible argument contains two propositions and can always be expressed in either one of two standard forms that are logically equivalent: (1) A, therefore B; or (2) B because A. In both (1) and (2), B is the *conclusion* and A is the *reason* or *premise*. In other words, although both words 'therefore' and 'because' are reasoning indicators, they indicate different ways to express reasoning; the proposition preceding 'therefore' is the reason or premise, the one following it is the conclusion; whereas the proposition preceding 'because' is the conclusion, and the one following it is the reason. The conclusion of an argument is thus the proposition that is based on the others, whereas the reasons or premises are the propositions on which the conclusion is based. For our purposes, the words 'reason' and 'premise' will be used interchangeably.

Reasons and conclusions are mutually interdependent concepts: a proposition can be a

Elementary Principles

conclusion only in a context where it is being based on some reason, and a proposition can be a reason only in a context where a conclusion is being based on it. The relationship between reasons and conclusion claimed in a given argument can be expressed by a number of terms that will be regarded as synonymous for our purposes. The conclusion may be said to be based or grounded on, to be justified or supported by, and to be inferred or derived from the reasons.

There are several standard ways of referring to arguments. Sometimes we speak of *the argument that S* or of the fact that someone *argued that S*, where S is a sentence; this refers to an argument whose conclusion is the proposition expressed by S, and whose premises are being left unspecified, perhaps because their identity is obvious in the context. Sometimes we speak of an *argument for N*, where N is a noun or noun phrase; this means an argument whose conclusion is some proposition obtained from N in a contextually obvious manner, and whose premises again are being left unspecified. Similarly, to speak of an *argument against N* is to speak of the argument whose conclusion is the denial or negation of such a proposition easily constructed from N. Sometimes it is contextually obvious what the conclusion of an argument is, and then one may want to identify a particular argument by a brief description of the most important premise; one would then speak of the *argument from N*, where N is again a noun phrase out of which one can easily form a proposition that serves as a premise from which that conclusion is drawn.

Objections are negative counterparts of arguments in the sense that they are arguments whose conclusion is the denial or negation of some controversial proposition. An *objection to or against N* is an argument whose conclusion is the negation or denial of some proposition constructed from N in a contextually obvious manner; that is, an objection to or against N is the same as an argument against N. A special and more complicated case arises when N refers to an argument rather than a proposition, for then the objection is an argument about another argument; that is, an *objection to an argument* is an argument whose conclusion is a proposition stating that the original argument has some flaw; this leads to the topic of the evaluation of arguments (which will be discussed later in this chapter), as well as the topic of meta-argumentation (which will be discussed in the next chapter). An *objection from N* means an argument whose main premise is a proposition easily formed from N, and whose conclusion is both negative and easily identified; an objection from N is essentially the same as an argument from N, except that the 'objection' designation is more likely used by someone who rejects the conclusion of the argument, whereas the 'argument' designation is more likely used by someone who accepts it. The *objection that S*, where S is a sentence, means an argument whose main premise is a proposition expressed by S and whose conclusion is a proposition that is negative and easily identified in the context.

Counterarguments are special kinds of objections, namely, objections to the conclusions of the original arguments. That is, suppose we begin by considering an argument for N; then a counterargument to such an argument would be an objection to N or an argument against N. In other words, a counterargument to a given argument is an argument whose conclusion is the denial or negation of the conclusion of the given argument.

The differences among arguments, objections, and counterarguments are differences in perspective; that is, the differences relate to whether one is affirming or denying a given controversial proposition. However, normally there is nothing intrinsically positive or negative about a proposition; the same thought can usually be expressed either positively or negatively. The point is that in a normal controversial situation there are arguments for both sides of the dispute. Each side is affirmative from its own viewpoint and negative from the opposite viewpoint. Let us call two propositions, P and Q, contrary or inconsistent when they cannot both be true (though they could perhaps both be false); in such a case, the arguments for P are also arguments against Q, and the arguments for Q are arguments against P. In other words, the arguments for P are objections to Q, and the arguments for Q are objections to P. Or again, the arguments for P have counterarguments consisting of the arguments for Q, and the arguments for Q have counterarguments consisting of the arguments for P.

2.2 Interpretation and Structure

An *argument with serial structure* (for short a *serial argument*) is made up of at least two subarguments combined so that the conclusion of one is simultaneously a reason of the other.[1] The simplest serial argument has the form: A because B, and B because C (or equivalently: C, therefore B; therefore A). Here B is the reason of the subargument "A because B" (or "B, therefore A") and also the conclusion of the subargument "B because C" (or, "C, therefore B").

Every proposition in a serial argument falls into one and only one of the following categories: intermediate proposition, final reason, or final conclusion. An *intermediate proposition* in a given serial argument is a proposition that serves as the conclusion of one subargument and as a reason of another subargument. A *final reason* in a given serial argument is a proposition that is a reason of some subargument but not the conclusion of any subargument. The *final conclusion* in a given serial argument is a proposition that is the conclusion of some subargument but not the reason of any subargument. In the example here, A is the final conclusion, B is the one and only intermediate proposition, and C is the one and only final reason.

The *propositional structure* of an argument or piece of reasoning refers to the interrelationships among its various elements, namely, among its subarguments and among its propositions. Such structure may be pictured in a *structure diagram* constructed in accordance with the following rules:[2]

(1) Label each proposition with some number, letter, or symbol.
(2) When one proposition is a reason supporting another, write the reason under the conclusion and indicate the fact by a solid line leading up from the first to the second.
(3) Place at the top the proposition that is supported by one or more other propositions but that does not itself support any others; this proposition at the top is the final conclusion of the argument.
(4) Place at the lowest level those propositions that support other propositions but are not themselves supported by anything else; such propositions are the final reasons of the argument.
(5) Structure diagrams typically have propositions that both support and are supported by other propositions; such propositions are the intermediate propositions of the argument. That is, they are propositions which are logically placed between the final reasons and the final conclusion of the argument, and which are reasons from the viewpoint of what they immediately support and conclusions from the viewpoint of what they are immediately supported by. Intermediate propositions have some support

---

[1] Here I adopt the terminology of Freeman (1991, 93-95) and Thomas (1986, 57-58); the concept, however, is the same as that elaborated in Finocchiaro (1980a, 313-14) under the label of "complex argument."
[2] Similar rules were adopted in Finocchiaro 1980a from Angell 1964; they are now common in the literature on informal logic and argumentation theory.

Elementary Principles

lines leading up to them from below and some leading up from them to other propositions above them.

Sometimes it is useful to label propositions with a standard numbering system. This *standard labeling* is done in accordance with the following rules:

(6) Label the final conclusion by some small number, for example '*1*'.
(7) Label reasons that directly support the final conclusion *c* by adding numerals to the number denoting it. For example, if the final conclusion is denoted by *c*, its directly supporting reasons are *c1*, *c2*, *c3*, and so on.
(8) Reasons that directly support the same intermediate proposition *p* are assigned numbers that begin with the same numeral *p*, and continue by adding digits in accordance with the previous rule (7). For example, reasons supporting *12* are labeled *121*, *122*, *123*, and so on.

Two reasons that immediately support the same proposition (and thus have the same number of digits in a diagram with standard labels) are *linked* to each other (or *interdependent* with each other) when each depends on the other to support that proposition and each alone is insufficient or incomplete to provide that support. When reasons are linked, the rules for standard labeling apply without change. When two reasons are not linked, they are *independent*. So two reasons are independent of each other when each does not depend on the other to support the conclusion based on them (whether or not each alone is sufficient to provide that support, that is, whether or not each is linked with *some other* reason). In other words, two reasons may be independent of each other and yet be linked with other reasons.

When an argument contains independent reasons, it may be useful to represent this fact in labeling the propositions and in drawing the structure diagram. The standard labeling of independent reasons is done as follows:

(9) To distinguish one *set* of reasons from another independent set, the lowercase letters *a*, *b*, *c*, and so on, are placed after the label of the proposition they support; these letters are carried for all lower propositions supporting these independent reasons when these lower propositions are themselves labeled by the previous rules.

And the distinction between linked and independent reasons is represented in a diagram as follows:

(10) The support lines from linked reasons always converge to some point below the conclusion they support; when there are no independent reasons, such lines converge directly to the symbol of this conclusion; when there are independent reasons, an horizontal bar is drawn between the supported conclusion and its immediately supporting reasons, and the various independent sets of linked reasons converge to distinct points on that horizontal line.

These rules will be illustrated and applied presently. One use of this idea of the independence of reasons is that it allows us to integrate separate arguments into a unified whole, if they have the same final conclusion.

The *latent* (or *implicit*) propositions of an argument are those propositions that are not explicitly stated in the argument but are implicitly assumed or taken for granted by the proponent of the argument. Latent propositions may also be called assumptions, presuppositions, missing premises, or implicit statements; for our purposes, such terms will be used interchangeably. The interrelationships among the latent propositions are called the latent propositional structure, or more simply latent structure; the latent structure may be contrasted with the explicit structure, namely the interrelationships of the explicitly stated propositions. There are two main types of latent propositions. In one case, a proposition is latent when for a particular step of the argument it is needed, in addition to the other explicit propositions involved in that step, to fully justify or to better justify that step; in this case, the latent proposition is always linked with and never independent of the reason(s) explicitly present in that step of the argument. In another case, a proposition is latent when it is one of the reasons being implicitly used to justify one of the final reasons in the explicit argument. Because of their position in structure diagrams, in the first case the propositions are called *horizontally latent*, in the second case *vertically latent*. In a structure diagram all latent structure is drawn in dotted lines.

Finally, many arguments are such that they need to be reconstructed. A *reconstruction* of an argument is a restatement of it such that no logically extraneous propositions are included and all logical interconnections among the stated propositions are explicitly and clearly indicated by means of reasoning indicators. A reconstructed argument normally makes explicit some of the propositions implicit or latent in the original formulation; however, the reconstruction cannot include all of them because their number is indefinitely large and the desirability of explicitness is subject to contextual constraints.

2.3 Evaluation and Criticism

So far I have discussed the basics of how reasoning and arguments are expressed and stated, how they are described and interpreted, what their structure is, and how they are reconstructed. Next, we must go on to the evaluation of reasoning and arguments. Evaluation is only one of a cluster of terms that are regarded as synonymous here; thus we may speak interchangeably of assessment, appraisal, judgment, and criticism.

Some caution is, however, needed in regard to the last term, for it can have a broad and a narrow meaning. Broadly construed, criticism is equivalent to evaluation; this is the relevant meaning when we speak, for example, of "critical reasoning." But criticism has also a negative connotation, for it can also mean negative or unfavorable evaluation; it provides, in fact, a handy term for the latter. Thus, I will often speak of criticism in its narrow meaning. However, I will use the term in both senses and, when needed, will give appropriate clues to avoid misunderstanding.

This point leads immediately to a fundamental fact about the concept of evaluation, namely that it can be of two opposite types: favorable, positive, approving, or appreciative; and unfavorable, negative, disapproving, or destructive. That is, the basic aim of evaluation is to determine whether an argument is good or bad, right or wrong, valid or invalid, correct or incorrect, sound or unsound, plausible or implausible, convincing or unconvincing, strong or weak, logical or illogical, and cogent or fallacious.

Another general point is that normally evaluation is a matter of degree, rather than an all-or-none affair. In a sense, evaluation is the judgment of value or worth, and value or worth is usually a nondiscrete or gradual notion, a spectrum embodying various shades of merit or demerit. One may, however, define particular evaluative categories referring to the extreme points along a particular dimension, and then one has a discrete evaluative concept.

Thirdly and perhaps most importantly, the evaluation of arguments is conceived here as involving two main things: an evaluative claim about some argument, and the articulation of the rationale for the evaluative claim. That is, evaluating an argument will be conceived as a special case of reasoning, namely reasoning about the argument; in short, the evaluation of an argument is a higher-order or metalevel argument about the original argument.

Combining these three points, we may say that an evaluation of an argument consists of a new argument whose conclusion attributes some merit or flaw to the original argument, whose reasons are meant to justify such an attribution, and where it is understood that normally such a justification can only be more or less strong rather than completely right or completely wrong.

A fourth important point is that negative evaluation (criticism in the narrow sense) is much more common than favorable evaluation. The causes of this fact are unclear; I believe that part of the explanation is that criticism is instructive in a way in which positive evaluation is not.[3] At any rate, a fruitful way of proceeding is to look for flaws in an argument, which is then regarded as successful if and insofar as no flaws are found or demonstrated. Thus, it is useful to catalogue some basic types of criticism.

One type of criticism is to criticize the conclusion as an individual claim, namely, to try to refute it. This is usually done by means of an objection to the original conclusion, which, as explained above, is equivalent to advancing a counterargument. In short, suppose the original argument claims "C because R"; this type of criticism objects that "not-C because R',", where R' is a proposition different from R. This is a common but not very perspicacious type of criticism because often the objecting reason has nothing to do with the original reason; so the criticism does not undermine the reason advanced in the original argument; and so it is of questionable relevance. As suggested earlier, the existence of an argument for a given proposition and of an objection against the same proposition provides the usual background for a controversy that generates argumentation; however, what is required for further progress in the discussion is to move to the level of evaluation by raising other questions, rather than by merely repeating the arguments that define the original controversy.

It is therefore not surprising that another common type of criticism of an argument is to criticize the reason adduced for the conclusion. One questions the individual truth of this reason by means of an objection against it, namely, an argument designed to show that the reason is not true. If successful, such an objection would undermine the conclusion of the original argument, for this conclusion was grounded on a reason, and this grounding disappears if this reason is not true. In short, to the original argument "C because R," this criticism responds by claiming "not-R because R'."

A third type of criticism tries to undermine an argument by undermining its reason, in the literal sense that this reason is alleged to be groundless or without foundation. The difference between this criticism and the previous one is the difference between claiming that a proposition is false and claiming that it is unsupported or not properly supported; this is the difference between saying that there is an argument concluding that the proposition is false, and saying that there is no argument concluding that the proposition is true. Thus, the proper answer to this type of criticism is to construct an argument supporting the truth of this proposition, or to strengthen or defend some existing argument supporting the same; whereas the proper answer to the former type of criticism is to point out what is wrong with the critical argument that tries to refute that proposition. In the simplest case, this type of criticism is a request that the argument "C because R" be strengthened by another argument to the effect that "R because R'." This type of criticism is related to the serial structure of an argument.

These three types of criticism share a common feature; none questions the relationship between the premise and the conclusion of the argument. In a sense, these criticisms

---

[3] For further details, see Finocchiaro 1980a, 338-41; 1988, 245-58; 2005a, 148-58.

involve, the relationship between the conclusion or the premise and the world; or, if you will, the relationship between the conclusion or the premise and other propositions that were not part of the original argument. Other types of criticism focus on the relationship between the given premise and conclusion.

The most obvious way of criticizing the relationship between a reason and its conclusion is to question what might be called the relevance[4] of the reason. That is, this criticism questions how and why the conclusion can be inferred from the reason; it may take two different forms. In the weaker form, the criticism claims that it is unclear how the conclusion can be inferred from the reason; for example, given the original argument "C because R," the connection between the two propositions may not be obvious. In this case the critical claim is weak enough as to require little or no justification; the criticism is really a request for an elaboration of the original argument. In turn, the answer would be to provide other reasons which, together with the original one, make it clear, or clearer, how the conclusion would be inferred. In other words, the structure of the original argument needs to be complicated by adding other *linked* reasons; the new argument would read "C because R, R′, R″, and so on."

In the stronger form, the criticism claims that the conclusion cannot be inferred from the reason. In this case a supporting critical argument must be given. That is, this criticism takes the form that the conclusion does not follow from the premise(s) because of some specified reason. The type of such a supporting critical reason then generates various subtypes of this criticism. If we call *disconnection*[5] the general flaw in reasoning corresponding to this type of criticism, the various special cases may be regarded as types of disconnection.

Let us begin with the most familiar subtype. It may happen that the conclusion of an argument does not follow from the premises because it is possible for the premises to be true while the conclusion is false. This possibility is commonly shown by constructing an argument with the same form as the original one, but having obviously true premises and obviously false conclusion;[6] such a constructed argument is called a counterexample. So in this first type of disconnection, the conclusion does not follow because an appropriate counterexample exists. Other well-known ways of describing the flaw are to say that the conclusion does not follow for one of several reasons: because it does not follow *necessarily*; because it does not follow in virtue of the *form* of the argument; because it does not follow ("analytically") in virtue of the meaning of the terms involved; or because it does not follow in virtue of the rules of *deductive* inference. Correspondingly, this first type of disconnection could be labeled formal, analytic, or deductive invalidity.

A second reason why one may be entitled to say that the conclusion does not follow from the premises is that it does not follow with any greater probability than some other specifiable proposition. Here, the critic produces another argument which has the same premises as the original argument but a different conclusion, and which appears of equal strength as the original. This occurs primarily with explanatory arguments whose conclusion is an explanation of what is stated in the premises, and the criticism amounts to providing an alternative explanation. Occasionally it may happen that the explanandum occurs because of both factors mentioned in the two allegedly alternative conclusions, but the point is that a given explanation has no force if there is no reason to prefer it to an

---

[4] Cf. Freeman (1988; 1991), Govier 1985b, and Johnson and Blair 1977.
[5] I adapt *disconnection* from Perkins 1989 and from Perkins, Allen, and Hafner 1983.
[6] It can also be shown by imagining a situation in which the premises of the original argument are clearly true and its conclusion is clearly false. For more details and clarifications, see chapter 6 below.

alternative. This may be called explanatory disconnection.[7]

To appreciate the third type of disconnection we need to stress that we are dealing with natural language argumentation as it occurs in ordinary life, and that such arguments are always incompletely stated and have latent propositions or tacit presuppositions. Then it is easy to see that a reason why the conclusion does not follow from the premises may be that one of the presuppositions is false. What does this falsity mean in this context? It really amounts to the existence of some ("sound") argument constructible in the context, whose conclusion is the denial of the presupposition. Even the groundlessness of such a presupposition would create trouble for the original argument, at least as long as such groundlessness is not merely asserted but demonstrated, that is, as long as one gives contextually sound arguments to show there is no good reason to assert the presupposition. In short, to the original argument "C because R," this criticism objects that in the context this argument is equivalent to "C because R and R'," but it so happens that "not-R' because R''." This third subtype may be said to involve a presuppositional disconnection and correspondingly presuppositional criticism.

Here, a pattern is beginning to emerge. In the first type of disconnection, the critical claim is grounded on the construction of an appropriate counterexample; in the second, on the production of an alternative explanation; in the third, on the construction of a presuppositional refutation. These three entities are arguments different from but suitably related to the original. This pattern allows us to define a fourth type of disconnection, where the conclusion does not follow because what does follow is some specifiable proposition inconsistent with it. Such a proposition may be called a counterconclusion, and the new argument supporting it a counterargument; but this is a special type of counterargument, whose premises contain some premises of the original argument and other propositions that are independently justifiable or contextually acceptable. In short, to the original argument "C because R", this criticism objects that "not-C because R, R', R'', and so on." This type of flaw may be called internal disconnection, and the corresponding criticism internal criticism, because the crucial element of the criticism is an expansion of the original argument that accepts some of the original reasons.[8]

The fifth type of disconnection may be called semantical and involves the problem of equivocation. It applies only to arguments having at least two linked reasons. Semantical disconnection occurs when the conclusion does not follow because the premises contain a term which has two meanings such that, if it is used in one sense, one of the premises is false (though they would imply the conclusion), whereas if the term is used in the other sense, the premises clearly do not imply the conclusion (though admittedly the previously problematic premise becomes true); in short, in the context the conclusion cannot follow from true premises. This disconnection is intimately related to presuppositional disconnection since the semantical ambiguity in question is normally not a self-subsisting property of a term, but rather something that must be argued in the context on the basis of (often latent) inferential relationships affecting the term.

Finally, the sixth subtype also involves presuppositional disconnection, but refers to the flaws of begging the question and circularity. I call it persuasive disconnection. Here, the conclusion does not *follow* from the premises because it *is* one of the premises; this is not meant literally in the sense that the conclusion is identical to an explicit premise, but in

---

[7] This is reminiscent of inductive incorrectness, and so it may also be called inductive disconnection; still, since the connection between explanation and induction is problematic, it may be advisable to avoid the latter label.

[8] This is also essentially equivalent to *ad hominem* argument in Johnstone's sense; see chapter 7 below.

the sense that it is identical to one of the latent propositions; moreover, this latency is frequently vertical and not always horizontal. That is, we seldom find arguments like "C because C" or "C because R, R', C, and R''," where the circularity is too small and obvious; rather we find "C because R," with C distinct from R but included in the latent structure.

I have defined six types of disconnection: formal, explanatory, presuppositional, internal, semantical, and persuasive. All involve arguments whose conclusion may be said not to follow from the premises. Thus the corresponding evaluations involve criticism of the relationship between premises and conclusion. Each disconnection corresponds to a type of criticism where the critic argues that in the original argument the reasons given do not properly connect with the conclusion, and in each case the reason for the disconnection involves the construction on the part of the critic of some other argument, or some additional part of the original argument. These six types of criticism should be added to the first four, which were examined earlier and to which short labels may now be given: an argument against the conclusion of the original argument may be called conclusion-refuting criticism; an argument against a premise of the original argument may be termed premise-refuting criticism; a claim or argument that a reason of the original argument is groundless may be labeled reason-undermining criticism; and a claim or argument that the relationship between the reason and the conclusion of the original argument is unclear may be named reason-relevance criticism. The last one should be distinguished from the criticism that the reason-conclusion relationship is missing, namely, that there is no connection or that the reason is irrelevant; to deny the existence of the connection is a stronger criticism which in turn generates the six subtypes just defined.

2.4 Example 1: The Observational Argument for Heavenly Unchangeability

It is now time to illustrate these general concepts by concrete examples. These illustrations will be taken from argumentation that was common during the Copernican Revolution and is now recorded in Galileo Galilei's *Dialogue on the Two Chief World Systems, Ptolemaic and Copernican*. For this work is not only a significant document of that epoch-making episode, but also a rich collection of arguments of perennial interest and relevance (cf. Finocchiaro 1980a; 1997; 2013b).

One of the simplest and most instructive arguments was the so-called observational argument for heavenly unchangeability. The original argument reads quite simply: no heavenly changes have ever been observed; therefore, the heavenly region is unchangeable (Galilei 1897, 71-72; 1997, 91-92). Its structure is so simple that its initial interpretation is easily completed. We have just two propositions, the final conclusion (1) and the final reason (11). However, a few additional comments are in order.

First, there is the question of the latent structure, which in this case happens to be extremely interesting. This question may be motivated by asking how and why the reason is relevant to the conclusion, which is a way of proposing the fourth type of criticism mentioned above, the reason-relevance criticism. That is, the original argument obviously presupposes that observation corresponds to reality, which is an extremely important epistemological principle. If such a missing premise is added, it constitutes a second reason linked to the first and should be assigned the label (12).

Note that this interpretation implicitly rejects the construal that the original argument is an appeal to ignorance. That is, the argument might be taken to have the form "no A's have been observed; therefore, no A's exist"; and this might be regarded as tantamount to arguing that because we are ignorant of something, namely, because we do not know that P is true, therefore we can conclude that P is false. Now, this is certainly a fallacious manner of reasoning, which is easily recognized as such when made explicit, but which nevertheless is common. Thus, to interpret the original argument as an appeal to ignorance would immediately generate this criticism, depriving it of any worth. Such criticism in turn might be criticized as attacking a straw man, namely, as being based on an untenable interpretation. We might also add that such criticism would seem to violate the principle of charity,[9] according to which, when evaluating an argument, one should interpret it in such a way that the argument avoids some of the most obvious errors. At any rate, the uncharitable interpretation would also involve taking the passage out of context, for the context makes it clear that the form of the argument is rather the following: "A's have been observed to have the property P; therefore, A's have property P"; that is, "the heavens have been observed to be devoid of (qualitative) change; therefore the heavens are devoid of (qualitative) change." Finally, it should be mentioned that Galileo does not take the argument as an appeal to ignorance; I mention the possibility merely for the sake of illustration.

Before proceeding to see how Galileo does criticize it, some more interpretation will be instructive. As he points out, the argument for heavenly unchangeability is a step in a longer argument that seeks to support a further conclusion, namely, the earth-heaven dichotomy; that is, the proposition that the earthly and the heavenly regions of the universe are radically different. In other words, the proposition that the heavens are unchangeable would be linked to the proposition that the earth is constantly changing to support the dichotomy; the new argument (for the earth-heaven dichotomy) would then have two linked reasons supporting its final conclusion, and it would also have serial structure because heavenly unchangeability would no longer be the final conclusion (as it is in the original argument with which we started), but rather an intermediate proposition. The point is that the simple argument in the original statement can be grafted onto other arguments to generate bigger and more complex arguments, and in the process the original would become just a subargument of the new argument.

One other interpretive point is worth making. The observational argument for heavenly unchangeability was only one of at least two independent arguments supporting the same conclusion; there was also an *a priori* or theoretical argument. That is, to support heavenly unchangeability, the Aristotelians advanced another independent reason, namely, that there is no contrariety in the heavenly region (Galilei 1897, 62-71; 1967, 38-47). Now, this reason immediately generates questions also illustrating many of the general concepts discussed earlier. Moreover, to keep proper track of the propositional structure, let us label this new reason (1b1) and change the label of the previous reason from (11) to (1a1), in accordance with the rules for the standard labeling of structure diagrams.

The first question about the alleged lack of heavenly contrariety (proposition 1b1) is why should we accept it. The traditional Aristotelian answer was: because (1b11) the heavenly region is the domain of natural circular motion, and (1b12) natural circular motion has no contrary (the way, for example, that straight upward motion is contrary to straight downward). The other main question is why the lack of heavenly contrariety is relevant to heavenly unchangeability; the answer is in terms of the theory of change as deriving from contrariety, namely because (1b2) change can exist if and only if there is contrariety.

Finally, let us assign labels to the other propositions mentioned previously, namely, the numeral (2) to the proposition that the earth is constantly changing, and the numeral (3) to the earth-heaven dichotomy; and let us assign the standard label of (1a2) to the latent proposition that observation corresponds to reality, in accordance with the rules that distinguish linked from independent reasons. Then we are in a position to picture all the arguments, subarguments, and propositions we have mentioned so far, and their

---

[9] See Scriven 1976, 71-73, for a general discussion of this principle.

interrelationships. The structure diagram would be as shown in Figure 2.1.

Figure 2.1

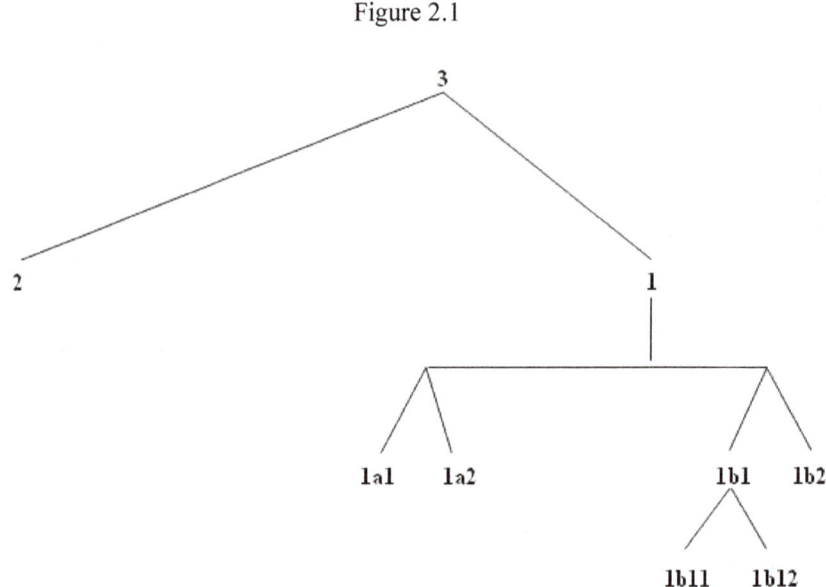

On the other hand, if proposition (3) were relabeled (1) and the rest relabeled accordingly, the structure would of course be unchanged, but the labels would be: 1 / 11, 12 / 12a1, 12a2, 12b1, 12b2 / 12b11, 12b12. Thus, although the explicit structure of the observational argument for heavenly unchangeability is very simple, if we wish to consider its latent structure or its relationship to other ideas and other arguments, then it should come as no surprise that the structure becomes more complicated.

Galileo criticizes the observational argument for heavenly unchangeability in five ways. One is to argue that in his own time the premise is no longer true, in the light of, for example, the telescopic observations of sunspots and of lunar mountains and the naked-eye observation of novas. Although this criticism targets directly only the individual truth value of the premise, it is the one that takes him the longest to articulate, since he has to argue every inch of the way through all sorts of controversial issues in order to refute the premise (Galilei 1897, 75-83; 1997, 96-107).

It is also important to point out that besides refuting the premise, Galileo criticizes the argument by refuting its conclusion. For in this case, the refutation of the premise is not a purely negative affair; to deny that no changes have ever been observed in the heavens is to affirm that some changes have been observed. But the observation of heavenly changes can be combined with the same plausible version of the principle that observation corresponds to reality to yield the conclusion that the heavenly region is changeable.

Moreover, this principle is not only intrinsically plausible, but also was accepted by the Aristotelians; indeed, we saw above that it is a missing premise of the argument being scrutinized. In short, here we have an example of what I called internal criticism: the original argument states "C because R"; to this we must add a latent and linked reason, say R'; so the argument is really "C because R and R'"; this may be reformulated as "C because R' and other reasons"; Galileo's criticism amounts to the counterargument "not-C because R' and R''," where R'' refers to observations of heavenly changes that had become incontrovertible. What makes this criticism internal is that it is crucially based on a proposition which the proponents of the original argument accept, although this proposition

Elementary Principles

is then combined with something else they do not accept or know anything about; but since this additional critical premise is demonstrably true, the counterargument is successful.

Other criticisms are based on the contextual distinction of two meanings for the phrase "heavenly changes": a heavenly change can mean the generation or decay of a heavenly body as a whole, and it can mean a partial change within a heavenly body. Equivalently, the ambiguity might be taken to involve the term body, which could mean either a whole heavenly body or a part of it (Galilei 1897, 74-75; 1997, 95-96).

When interpreted holistically, the original argument amounts to the following: no one has ever observed any generation or decay of heavenly bodies in the heavenly region; therefore, the heavenly region is unchangeable. It is then subject to the criticism that this way of reasoning would lead one to the following absurd argument: no one has ever observed any generation or decay of terrestrial globes in the terrestrial region; therefore, the terrestrial region is unchangeable.

When the argument is interpreted the other way, Galileo objects that it is still wrong because no terrestrial changes would be noticeable to an observer on the moon before some particular very large terrestrial change had occurred, and yet terrestrial bodies are obviously changeable and would have been so even before that occurrence.

What are we to make of the last two criticisms? They try to establish the invalidity or formal disconnection of these two versions of the observational argument. The technique used is that of the construction of counterexamples, i.e., arguments of the same form but with true premises and false conclusion. Notice that, despite the ambiguity mentioned by Galileo, he is not charging equivocation. So what we have are two examples of criticism by counterexample, charging invalidity or formal disconnection, not semantical disconnection.

At this point someone might interject that the very fact that this particular argument was called *a posteriori* shows that it was meant to be a probable or inductive rather than deductive argument, and hence the criticisms just made might be correct but of dubious force or relevance. Perhaps to anticipate this objection, Galileo has another criticism that can be taken to address precisely this issue (Galilei 1897, 72-74; 1997, 92-95).

His fifth criticism is directed to the more plausible particularistic (second) version involving parts of globes. It amounts to the following argument: if there were changes within the heavenly bodies, then most of them could not be observed from the earth, since the distances from the heavenly bodies to the earth are very great, and on earth changes can be observed only when they are relatively close to the observer; moreover, even if there were changes in the heavenly bodies large enough to observe from the earth, then they might not have been observed, since even large changes cannot be observed unless careful, systematic, exact, and continual observations are made, and no such observations have been made, at least not by the argument's proponents.

This criticism interprets the original argument as an explanatory argument; that is, it presents the conclusion about heavenly unchangeability as the explanation of the observational absence mentioned in the premise. Two other ways of explaining the fact are suggested: it may be due to the great distance between the earth and the heavenly bodies, and/or to the lack of sufficiently careful observations of the heavenly bodies. These alternative explanations do not *refute* the Aristotelian *explanation* but rather the Aristotelian *argument*; that is, this criticism does not prove the conclusion of the original argument false, but rather *weakens* the inferential link between premise and conclusion, for there is no reason to prefer the Aristotelian to the Galilean explanation. Thus, the point being made is a logical criticism, affecting primarily the premise-conclusion relationship in the original argument. The flaw being charged is explanatory disconnection. The type of criticism involved is criticism by construction of an alternative explanation.

To summarize, Galileo criticizes the observational argument for heavenly unchangeability by arguing in part that it has a false premise. He also shows that it has a false conclusion, which is to say he can demonstrate that the heavens are changeable. Moreover, he charges the argument with internal incoherence to some extent insofar as it presupposes that observation corresponds to reality, but in the light of new evidence this principle justifies a contrary conclusion. He argues further that the argument is formally invalid because the failure to observe heavenly changes shows merely that the heavenly bodies have undergone no changes so far, not that they are unchangeable by nature. Finally, he argues that even the probable version of the argument is problematic because it is questionable whether the failure to observe heavenly changes is due to their nonexistence, or instead to such factors as the great distance of the heavenly bodies, or the fact that past observations have been insufficiently precise and systematic. We may say that he criticizes the argument both factually and logically, as long as we understand that even factual criticism normally requires argument since part of what is in question is what the facts are; moreover, logical criticism should not be construed narrowly as referring merely to formal invalidity, but as referring to the evaluation of the relationship between premises and conclusion, and this evaluation again hinges crucially on argumentation.

2.5 Example 2: The Anti-Copernican Argument from Vertical Fall

Another instructive example of critical interpretation is Galileo's discussion of the anti-Copernican argument from vertical fall.[10] The original argument may be *reconstructed* as follows: (2) the earth does not rotate because (21) bodies fall vertically, and (22) this could not happen if the earth rotated; for (221) if the earth rotated and bodies fell vertically, then falling bodies would have a mixture of two natural motions, toward and around the center; but (222) this is impossible since (222a1) every body can have only one natural motion, and since (222b1) on a ship moving forward rocks dropped from the top of the mast fall behind and land away from the foot of the mast (towards the stern).

This is a reconstruction because it is meant to be not a mere synopsis of the relevant text but also an interpretation illustrating the concepts discussed earlier; for example, I used standard labels to reveal the propositional structure unambiguously. However, my interpretation is meant to be textually accurate. Given the labeling used, it is a mechanical task to represent the argument in a structure diagram, as in Figure 2.2.

Figure 2.2

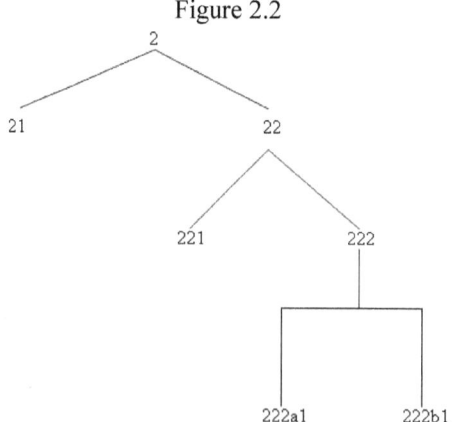

The first Galilean criticism (Galilei 1897, 164-66; 1997, 155-57) is that if the phrase

---

[10] Galilei 1897, 164-175; 1997, 155-70. For detailed discussions and analyses, see chapter 13 below and Finocchiaro (1980a, 115-17, 192-200, 208-13, 277-91, 329-30, and 387-91; 2010b, 124-29).

"vertical fall" is taken literally, then the argument begs the question because it presupposes that the earth is motionless, which is the conclusion it is trying to prove. The literal meaning of vertical fall is downward motion in a straight line along an extended terrestrial radius, namely perpendicular to the earth's surface. This may be called *actual* vertical fall and should be distinguished from *apparent* vertical fall, which means fall which *is seen* to be vertical to an observer on the earth's surface, as when a rock is dropped from the top of a tower and lands at its foot directly below with no visible deviation. The two would coincide on a motionless earth; but an important point easily agreed upon by both sides of the dispute is that *if* the earth were in axial rotation then the two would *not* coincide, since the appearance of vertical fall would imply an actually slanted path and thus an actual nonvertical fall.

This first criticism begins by noting that premise 21 of the original argument seems to refer to *actual* vertical fall; then there is no problem with the first inferential step from premises 21 and 22 to conclusion 2, since it is an instance of denying the consequent. However, the critic asks how the argument's proponents know that bodies do indeed fall vertically, namely, that premise 21 is true. The answer would be because (211) bodies are *seen* to fall vertically, namely, because of apparent vertical fall. Now, the truth of this observation is undeniable, but the critic next asks how (212) actual vertical fall follows from apparent vertical fall. This is a legitimate question since, as just explained, on a rotating earth apparent vertical fall would not imply actual vertical fall. It seems that the Aristotelians must assert this implication, and that the only way to justify it is to assume that (2121) the earth is motionless, for indeed (2122) if the earth is motionless then apparent vertical fall does imply actual vertical fall. Unfortunately, this assumption (proposition 2121) is precisely what the original argument is trying to prove. In other words, proposition 2121 happens to be identical to proposition 2, and so the bigger argument (namely, the argument that includes the latent structure unearthed by the above interpretation) is circular; thus the original argument, which overlaps with this bigger argument by sharing the subargument from 21 and 22 to 2 begs the question precisely at this point. This criticism is largely a presuppositional criticism, which exposes both presuppositional and persuasive disconnections.

However, perhaps the original argument was speaking of apparent rather than actual vertical fall. In that case the issue would reduce to the tenability of the impossibility of mixed motion, namely, proposition 222. This impossibility amounted to a denial of what later came to be known as the principle of the superposition or composition of motions; this principle was then one crucial issue in the controversy. In the reconstructed argument the impossibility of mixed motion is explicitly supported with two independent reasons, and so its correctness largely depends on whether the corresponding subarguments involve any difficulty. Galileo criticizes both of these subarguments, thereby providing a good example of reason-undermining criticism.

Consider first the subargument that (222) it is impossible for a falling body to have two natural motions, toward and around the center (i.e., vertical and horizontal) because (222a1) every body can have only one natural motion. This may be called the theoretical argument for the impossibility of mixed motion since its premise was a basic principle of Aristotelian physics. As one might expect, Galileo explicitly questions the empirical correctness of this premise, but he does not do that in this passage, and so it need not concern us here. Let us note simply that, once again, the empirical issue requires argumentation, and that we would have an example of reason-refuting criticism (Galilei 1897, 281-89, 423-42; 1967, 256-64, 397-416).

Galileo also implicitly objects that the two "natural" motions of falling bodies on a rotating earth would be "natural" in different senses: the downward fall would be natural in the sense of spontaneous (not in the sense of potentially everlasting), whereas the rotational motion would be natural in the sense of potentially everlasting (not in the sense of spontaneous). Now, the premise here (proposition 222a1) may be true of each of these two kinds of natural motion, but it would imply only that a body cannot have simultaneously two kinds of spontaneous motion or two kinds of everlasting motion; it clearly does not imply that it cannot have one of one kind, and one of the other kind.

This criticism involves a charge of equivocation and is an example of semantical criticism in the general classification elaborated earlier. But since it is, once again, not explicit in the passage being considered, the fuller demonstration of this point is left as an exercise for the reader.[11]

Galileo's explicit criticism is directed at the subargument that (222) a mixture of free fall and perpetual horizontal motion from west to east in accordance with diurnal rotation would be impossible because (222b1) on a forward moving ship rocks dropped from the top of the mast fall behind. This is clearly an observational argument insofar as the premise is an experimental report; it is also an explanatory argument insofar as the impossibility of mixed motion mentioned in the conclusion is advanced as the explanation of the experiment mentioned in the premise.

Galileo objects (Galilei 1897, 166-69; 1997, 158-62) partly that the rock's failure to move simultaneously in two directions is not the only possible explanation of the alleged fact: it might happen because the horizontal motion imparted by the ship to the rock is violent motion, which would be dissipated after the rock is left to itself; or it might happen because of air resistance, which would oppose the horizontal motion acquired by the rock. This is a charge of explanatory disconnection (as defined earlier) and uses the critical technique of constructing an alternative explanation.

He also criticizes this subargument (222b1 to 222) by objecting to the claim about the results of the ship experiment (proposition 222b1); he argues that this alleged experiment does not in fact happen this way (Galilei 1897, 169-74; 1997, 162-69). That is, he criticizes the truth of the premise. However, a simple appeal to the facts of observation was impossible in the context, which required instead an argument concluding with the denial of the proposition in question. This is a case of premise refuting criticism.

In summary, Galileo criticizes the argument from vertical fall in several ways. He begins by distinguishing between actual and apparent vertical fall. Insofar as the argument refers to actual vertical fall, it is subjected to presuppositional criticism and shown to beg the question. Insofar as it refers to apparent vertical fall, it depends on a premise asserting the impossibility of mixed motion; he undermines this premise as groundless by criticizing a theoretical and an observational subargument supporting it. The criticism of the theoretical subargument involves premise-refuting criticism and semantical criticism. The criticism of the observational subargument involves criticism by the construction of an alternative explanation and also premise-refuting criticism. Again, we may say that he engages in a mixture of empirical and logical criticism, as long as we understand that the empirical is essentially dependent on argumentation and not just brute facts, and the logical involves not abstract irrelevancies but questions about the relationship between premises and conclusions.

2.6 Summary

This chapter began with a general, abstract, and systematic elaboration of a number of

---

[11] This criticism can be derived from an earlier passage (Galilei 1897, 38-57; 1967, 14-32); cf. Finocchiaro 1980a, 349-53, 387-89.

elementary concepts and principles needed for the interpretation and evaluation of reasoning and argumentation. The large number of particular definitions can be grouped under some main headings. That is, I first explained what reasoning and argumentation are. Then I articulated what is meant by the structure of reasoning and argumentation, the description of such structure being the key task in the understanding and interpretation of reasoning. This was followed by a clarification of the concepts of evaluation and criticism, and in the process I focused on their fundamental features and developed a taxonomy of ten common types of argument criticism. After this general discussion I illustrated most of these general ideas by two extended Galilean examples, the critical interpretation of the observational argument for heavenly unchangeability, and the critical interpretation of the anti-Copernican argument from vertical fall.

All these general explanations and particular examples were meant to provide an introduction to the basic tools of interpretation and evaluation.

Four main concepts have been elaborated: reasoning, argument, interpretation, and evaluation. These are distinct but interrelated. Reasoning is the activity of basing one thought on another; argument is a piece of reasoning; interpretation is the description of details to provide understanding; and evaluation is the assessment of worth. One relationship is that interpretation and evaluation normally have to be based on argument and so are themselves reasoning. Another is that normally evaluation has to be also based on interpretation, or that normally interpretation leads to evaluation.

Chapter 3
Basic Types of Meta-argumentation

The last chapter described and illustrated the essential details of a theory intended to provide interpretation and evaluation of reasoning and argumentation. The theory was described in terms of a number of basic general principles involving such concepts as reasoning, argument, interpretation, and evaluation. The general description was then illustrated by some extended but concrete examples of reasoned interpretations and evaluations of arguments. That discussion now needs to be extended and refined in several directions.

3.1 Ground-level Argumentation

The purpose of this chapter is to discuss several forms which the reflection upon reasoning may take. That is, it aims to define, distinguish, and interrelate several kinds of reasoning about reasoning. Reasoning about reasoning may be labeled *metareasoning*. Similarly, an argument about one or more arguments may be labeled a *meta-argument*.

In other words, the most basic distinction in this context is the distinction between *ground-level* reasoning and argumentation and *metalevel* reasoning and argumentation. Ground-level reasoning is reasoning about such objects as mathematical entities, natural phenomena, historical events, and practical problems, whereas meta-level reasoning is reasoning about reasoning. Additionally, ground-level reasoning is reasoning that in a particular discussion is at a lower level as compared to other higher-level metareasoning. Finally, the term *object-level* will be regarded as synonymous to *ground-level*.

As an example of ground-level reasoning, I would like to quote an argument that was widely discussed by physicists, astronomers, and philosophers in the 16th and 17th centuries during the Copernican Revolution. This was a series of episodes that began in 1543 with the publication of Copernicus's book *On the Revolution of the Heavenly Spheres* and climaxed in 1687 with the publication of Newton's *Mathematical Principles of Natural Philosophy*. A key issue was whether the earth stands still at the center of the universe, or instead is a planet spinning daily on its axis and orbiting the sun once a year. The former was, of course, the geostatic, geocentric, Aristotelian, and Ptolemaic world view, which for thousands of years had been almost universally accepted on the basis of what I would regard as very plausible and convincing arguments. One of these involved the alleged experiment of dropping a rock from the top of the mast of a ship on two different occasions, when the ship is docked and anchored motionless in a harbor and when it is advancing forward traveling over the water. This is what the argument claimed: "because when the ship stands still the rock falls at the foot of the mast, and when the ship is in motion it falls away from the foot, therefore, inverting, from the rock falling at the foot one infers the ship to be standing still, and from its falling away one argues to the ship being in motion; but what happens to the ship must likewise happen to the terrestrial globe; hence, from the rock falling at the foot of the tower one necessarily infers the immobility of the terrestrial globe" (Galilei 1997, 163; cf. 1897, 169-70).

3.2 Argument Analysis

With these ideas in the background, we may say that one type of metareasoning is the reasoned interpretation or evaluation of an argument. Note, however, that this really

Basic Types of Meta-argumentation 35

involves two special cases, the reasoned interpretation of arguments and the reasoned evaluation of arguments; for interpretation and evaluation should be distinguished (as we saw in the last chapter). On the other hand, it is useful to have a single handy label to refer to one or the other or both, that is, *argument analysis*. Similarly, note that I speak of the reasoned interpretation or evaluation and not of mere interpretation or evaluation; that is, in these two special cases of metareasoning we must have not only an analytical claim about an argument, but also a justification of the claim by means of reasons.

For example, let us consider some interpretive claims about the argument just quoted. One might claim that the final conclusion of the argument is the proposition that the terrestrial globe is immobile; that the argument's last step is some kind of conditional argument using the premise that the rock dropped from the tower lands at its base (and not some distance to the west); that another step or subargument is an argument from analogy, beginning with a premise stating differential results for the ship experiment, drawing an analogy between a mast on an advancing ship and a tower on a rotating earth, and implicitly concluding that the rock dropped from the tower would land at its base if and only if the earth is motionless.

Such interpretive claims obviously have to do with the understanding of the argument, and their justification or criticism would involve argumentation about the argument. But it should be equally obvious that such claims are not evaluative and do not attempt to assess the strength or validity of various aspects of the argument. Examples of evaluative claims might be the following (cf. Galilei 1897, 167-75; 1997, 158-70). One might question the strength of the analogy by arguing that there are significant dissimilarities between the case of an advancing ship and that of a rotating earth: (1) on a rotating earth the rock would have an inherent natural tendency to move along with the tower, but on an advancing ship the rock would not have an inherent natural tendency to move along with the mast and ship; and (2) on a rotating earth the air surrounding the tower and the rock would be moving eastward carried by the earth's rotation, whereas on an advancing ship the surrounding air does not share its motion. Another evaluative claim is that the argument's key experimental premise is false: on an advancing ship the rock dropped from the top of the mast lands at the foot of the mast, in the same place where it lands when the ship is motionless. During the Copernican controversy, this experiment was a controversial issue whose resolution required not only observation and well designed experimentation, but also argumentation interpreting such empirical data.[1]

3.3 Self-Reflective Argumentation

So far, the first case of metareasoning (argument analysis) involves the critical interpretation of already constructed arguments advanced by others. However, sometimes it is proper and advisable to engage in some interpretation and/or evaluation of an argument that one is constructing and advancing oneself. If we subsume self-interpretation and self-evaluation under the notion of self-reflection, then we may say that at such times one is engaged in the self-reflective formulation or construction of one's own argument.

In other words, argument analysis occurs when the subject matter of our argument is itself an argument. However, the subject matter of most arguments is not arguments, but rather numbers, atoms, human affairs, social institutions, historical events, and so on. In such cases, one may engage in argument in a more or less self-reflective manner.

What is required for reasoning to be self-reflective, and how do we distinguish

---

[1] For more details on the ship analogy argument, see Finocchiaro (1980a, 36, 116-17, 389-91; 1997, 16-70; 2010a).

reasoning which is self-reflective from reasoning which is not? We should not expect that, when we are self-reflectively constructing our own argument, we will analyze and evaluate it with the same degree of explicitness and formality as when we are critically analyzing the arguments of others. But the degree of self-interpretation or self-evaluation cannot be too low, otherwise we would have an instance of mere ground-level reasoning and not of metareasoning.

Sensitivity to self-interpretation is normally shown by careful attention to such questions as what are our conclusions and what are our reasons, whether we are advancing just one reason or more than one, and how our reasons are meant to connect with our conclusions. Sensitivity to self-evaluation is normally shown by careful attention to such questions as whether we are advancing a conclusive or very strong or moderately strong or weak argument; this involves the degree of support the reasons lend to the conclusion. It is also shown by paying attention to possible criticism, objections, and counterarguments against our own argument, and to ways of rebutting these.

3.4 Example: The Simplicity Argument for Terrestrial Rotation

An extended example will give more definite meaning to these imprecise stipulations and concrete content to the general concept of self-reflective reasoning. It consists of one of Galileo's arguments for the earth's diurnal axial rotation; it is based on considerations of simplicity and is advanced as merely probable (Galilei 1897, 139-50; 1997, 128-42).

Since the number of propositions is large, and the structure is relatively complex, I shall use the technical apparatus of tree-root diagrams, introduced in the last chapter, to picture the propositional structure of the argument. The labels for the propositions are the ones which I shall be progressively assigning to them as they are introduced into the discussion. It will be easier to follow the discussion by referring to the diagram in Figure 3.1.

Figure 3.1

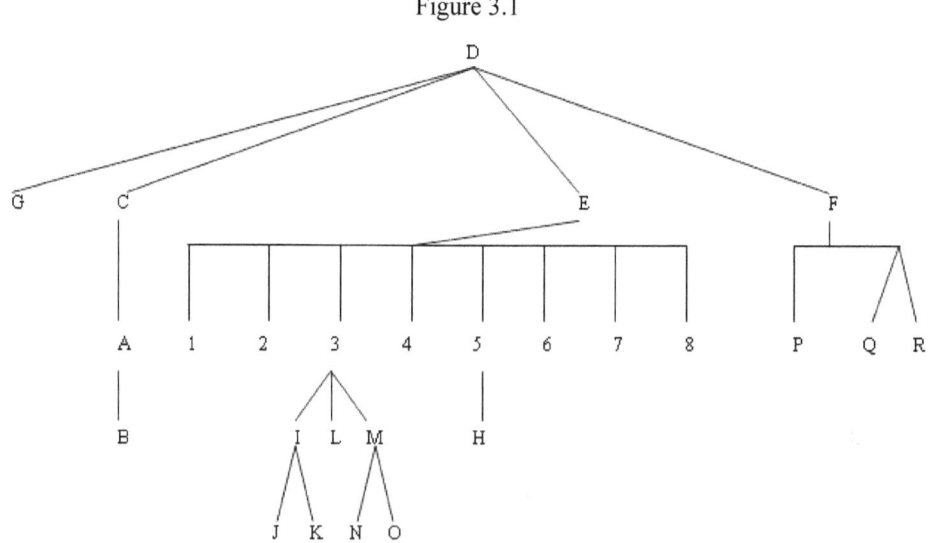

Galileo begins the discussion with a statement of the principle of the relativity of motion, namely the idea that motion exists only in relation to things lacking it, whereas shared motion has no effect on the relationship of things sharing it; let us abbreviate this

Basic Types of Meta-argumentation 37

proposition by A (and then we will progressively label other key propositions in alphabetical order, unless explicitly labeled otherwise in the text under consideration). In the course of the discussion he gives a partial justification supporting this principle in terms of familiar examples of shared and unshared motion, though he does not stress this justification and regards the principle as relatively uncontroversial; let us label this support B. His primary interest is to apply this principle to the problem of explaining apparent diurnal motion; that is, the relativity of motion is taken to imply that the apparent westward diurnal motion can be explained by saying either that all heavenly bodies revolve westward daily around a motionless earth or that the earth alone rotates eastward on its own axis every day; let C refer to this consequence of the principle.

Given this premise (that diurnal motion can be explained either way), Galileo goes on to argue that therefore (D) terrestrial axial rotation is more likely than universal revolution around a motionless earth because (E) the geokinetic explanation of apparent diurnal motion is simpler than the geostatic explanation, and (F) nature usually operates by the simplest possible means. The argument so far is: A because B; because A, therefore C; and because C, E, and F, therefore D.

Galileo makes it clear that he is asserting the final conclusion only with probability, and that his conclusion embodies an implicit comparison between the two contradictory views which generate the controversy. He also explicitly states that his conclusion needs to be qualified in another crucial way; that is, it depends on the assumption that all other relevant phenomena can also be explained either way, namely the assumption that all the many anti-Copernican objections can be refuted and all the phenomena on which they are based could occur on a moving earth. These Galilean remarks are explicit indications of the self-evaluating element needed for critical reasoning.

Let us label the assumption just mentioned G. The best place for this assumption in the overall structure of this argument is to have it as an additional premise directly supporting the final conclusion, alongside C, E, and F. We might ask at this point the more technical question whether this assumption is a reason linked with or independent of these other three. I would opt for a link on the grounds that it makes a point which is similar to, though of course more general than, the specific claim that diurnal motion can be explained either way (C); moreover, the simplicity considerations made in the other two premises (E and F) apply as much to the general as to the specific point. Note also that this assumption is a generalization about anti-geokinetic arguments, and that the rest of the book supports it by critically examining each in turn and attempting to refute them all. However, in this passage this assumption is unsupported and so constitutes a final reason.

The rest of this passage argues in support of the greater simplicity of terrestrial rotation (E) and in support of the principle of simplicity (F), which so far have just been stated. In its central part, Galileo lists seven numbered reasons why terrestrial axial rotation is simpler than universal geocentric revolution; then at the end he adds an eighth reason, without labeling it as such. That is, the geokinetic system involves (1) fewer moving parts, smaller bodies moving, and lower speeds; (2) only one direction of motion (eastward) rather than two opposite ones (eastward *and* westward); and (3) periods of revolution which follow a uniform pattern, namely that of increasing with the size of the orbit. Moreover, the geostatic system involves (4) a complex pattern for the size and location of the orbits of the fixed stars; (5) complex changes in the individual orbits of fixed stars (due to [H] the precession of the equinoxes); (6) an incredible degree of solidity and strength in the substance of the stellar sphere which holds fixed stars in their fixed relative positions; (7) a mysterious failure for motion to be transmitted to the earth after it has been transmitted all the way down from the outer reaches of the universe to the moon; and (8) the postulation of an *ad hoc primum mobile*.

It is probably best if we leave these numerals alone and let them stand respectively for the propositions supporting the comparative simplicity claim (E). Though all of these eight premises are propositions which implicitly compare and contrast the two alternatives in regard to a particular property, it should be noted that only the first two seem to involve matters of degree, whereas the other six seem to involve discrete properties which one of the two world systems possesses but the other one lacks. An interesting question here is whether these eight reasons are linked or independent. I would say, first, that each strengthens the others, and so the total amount of support they give to their conclusion is a function of all of them taken collectively; this suggests that they may be viewed as linked. However, each reason provides some degree of support independently of the others; moreover, each proposition is a perfectly natural answer to the question whether there is another reason why one should accept the comparative simplicity claim; thus, I am inclined to regard them as basically independent. Note also that the first proposition has three parts, each of which would need separate support; that the fifth one is provided a brief justification; and that the third is especially important and is supported by a relatively lengthy, novel, and strong argument.

As phrased above, proposition 3 states that the periods of revolution have a uniformity in the Copernican system which they lack in the geostatic one. In the supporting subargument Galileo begins with a claim which I call the law of revolution: (I) it is probably a general law of nature that, whenever several bodies are revolving around a common center, the periods of revolution become longer as the orbits become larger. He then supports this by the well-known fact that (J) the planets revolve in accordance with this pattern, and by his own discovery that (K) Jupiter's satellites also follow the pattern. The important point is that, though this feature of planetary revolutions was known to the Ptolemaics and incorporated into their system, before the discovery of Jupiter's satellites it would have been rash to generalize a single case into a general law; however, the completely different and unexpected case of Jupiter's satellites suggested that this was not an accidental coincidence but had general systemic significance; thus, while this inferential step ("J, K, so I") is not conclusive, it has considerable strength.

Given the law of revolution (I), Galileo goes on to combine it with the claim that, whereas (L) the earth's diurnal motion in the Copernican system is consistent with the law of revolution, (M) the diurnal motion of the universe in the Ptolemaic system is not. The point would be that this difference gives the Copernican system a uniformity or regularity lacking in the Ptolemaic system, which is what is asserted by Galileo's third reason (proposition 3) why the former has greater simplicity than the latter (proposition E). He considers the consistency between Copernican diurnal motion and the law of revolution to be sufficiently obvious to need no support. The argument would be that in the Copernican system the diurnal motion is the axial rotation of the earth, and this axial rotation is not an orbital revolution and does not involve a member of a series of increasingly large orbits; thus, the law is not even meant to apply to terrestrial rotation. To justify the inconsistency between the law of revolution and Ptolemaic diurnal motion, he explains that (N) in the geostatic system the diurnal motion corresponds to the revolution of the outermost sphere (whether stellar sphere or *primum mobile*) around the central earth, but (O) this outermost sphere involves both the largest orbit and the shortest period. This completes the subargument supporting the greater Copernican uniformity in regard to periods and orbits of revolution (proposition 3), namely the argument *from* the law of revolution, for short.

At the end of the passage, there is a discussion of the principle of simplicity (F), which was a crucial premise directly supporting the final conclusion about the greater likelihood of terrestrial rotation (D). Although there might be some question how this principle should be formulated, let us focus on the interpretation adopted above, namely that nature usually operates by the simplest possible means. When so stated, Galileo in part suggests what might be called a teleological justification for the principle of simplicity, namely the teleological principle that (P) it useless to do with more means what can be done with fewer. However, he also recognizes that (Q) the principle of simplicity is subject to the following theological objection: given a God who is all-knowing and all-powerful, God can create and operate a more complicated system as easily as a simpler system, and so a more complex world system is as likely to exist as a simpler one; in other words, it is false that nature usually operates by the simplest possible means because nature was created by an infinitely powerful God and such a God would be as likely to use more power to operate a more complex universe as to use less power to operate a simpler world. Galileo tries to address this objection directly, though I do not find his answer too relevant or convincing; however, his presentation of this objection indicates the kind of self-evaluation which is part of self-reflective reasoning. Finally, let us simply label his answer R, and let us note that we have labeled by the simple propositional label Q the whole admission of the existence of the theological objection, including a statement of its details as part of proposition Q; in terms of these labels, the last subargument has the structure "F because P, and because Q and R."[2]

This leads to a final comment about this instance of self-reflective argumentation. That is, though we have imposed some structure onto Galileo's argument for terrestrial axial rotation, his presentation of the argument is sufficiently self-reflective to make its overall structure relatively clear.

3.5 Argumentation Theory

The first two kinds of meta-argumentation examined so far (argument analysis and self-reflective argumentation) need now to be distinguished from and interrelated with a third type. This third kind of meta-argumentation is the *theory of argument*, which may also be called *argumentation theory*, or *logical theory* (in Toulmin's sense). For the latter too aims at the interpretation and evaluation of arguments. What then is the difference? I believe the difference is one of generality, systematicity, and conceptual explicitness, which are key elements of the mental activity called theorizing.

It should be noted, however, that such theorizing is not completely value-free or purely interpretive, but also normative. Thus, when needed, to avoid misunderstanding, we may also speak of the *normative theory* of argument or argumentation. On the other hand, we should not go to the other extreme and act as if its normative character deprives it of any descriptive or empirical content, and of an interpretive aim. This point can be made more explicit by defining argumentation theory as the formulation, testing, systematization, clarification, and application of concepts and principles for the interpretation, evaluation, and practice of argument or reasoning.[3] This definition also makes more vivid the problem of distinguishing argument analysis from the theory of argumentation.

I have already said that the difference lies along the dimensions of generality, systematicity, and conceptual explicitness. In regard to generality, if one is engaged in the critical analysis of a particular argument and reaching a conclusion merely about that argument, then one is engaged in argument analysis; whereas if one is engaged in the critical analysis of a whole class of arguments or reaching a conclusion about arguments in

---

[2] Note that my reconstruction of this Galilean subargument implicitly contains an answer to the problem of how counterconsiderations or con reasons should be incorporated in a pro-and-con or conductive argument; cf. chapter 8. and 8.7 below.
[3] Finocchiaro 2005a, 21-64; Johnson and Blair 1985.

general, then one is doing argumentation theory. However, this dimension of generality makes it clear that the difference is a quantitative one of degree rather than a qualitative one of kind.

In regard to systematicity, what readily comes to mind is the Euclidean type of axiomatization. However, just as readily, I would want to add that it is both unrealistic and anachronistic to expect such systematicity in the theory of argument. But then the challenge becomes that of articulating a different kind of systematicity.

This problem may be expressed differently. Once one has formulated some principles of interpretation or evaluation, one may want to explore the logical relationships among them, logical in the sense of standard formal deductive logic. Certainly the argumentation theorist will want to do some of that. But is that the only kind of systematization one can undertake? If so, we would have the ironical situation that if the argumentation theorist wants to claim to do something above and beyond the argument analyst, he has to engage in deductive formal logic; if not, the question is what is this other kind of systematization.

At any rate, in regard to the difference between argument analysis and argumentation theory from the point of view of systematicity, it would seem that on the one hand systematicity is a matter of degree, but that on the other hand if a critical interpretation of an argument becomes systematic to any extent then one is taking a qualitative step (however small) away from argument analysis as such and toward argumentation theory.

Conceptual explicitness was the third distinguishing characteristic. It too is a matter of degree. Even the argument analyst can hardly avoid using such terms as argument, premise, conclusion, serial structure, intermediate propositions, linked versus independent reasons, interpretation, evaluation, criticism, objection, counterargument, fallaciousness, and so on. However, in the critical interpretation of a particular argument, many of the subtler ones of these concepts may be only implicit. On the other hand, I take it that part of the business of argumentation theory is to render more explicit concepts that are implicit in the practice of argument analysis.

There are probably other distinguishing characteristics, besides generality, systematicity, and conceptual explicitness; and even these three require more clarification and analysis than I have provided here. Nevertheless, given the aforementioned definitions of argument analysis and of argumentation theory, we may say that while they are both meta-argumentative activities concerned with the interpretation and evaluation of arguments, what distinguishes them is that argumentation theory has a higher degree of generality, systematicity, and conceptual explicitness. In terms of a slogan, we might say that argumentation theory is generalized or systematized argument analysis, and the latter is applied argumentation theory.

At this point, however, a few words need to be said about the relationship between argumentation theory and formal deductive logic. Elaborating and following up on our discussion above (chapter 1), it ought to come as no surprise that there is a large overlap between formal deductive logic and argumentation theory. This overlap is, however, only partial. My impression is that argumentation theory tends to be more empirical and interpretive oriented, whereas formal logic tends to me more *a priori* and normative oriented; but this does not mean that argumentation theory is, or can be, or ought to be completely *a posteriori* and value-free; similarly, formal deductive logic is not, should not, and cannot be completely *a priori* and evaluative; the difference is one of degree and of emphasis. Moreover, formal deductive logic tends to theorize about a special class of arguments (usually, mathematical proofs), about special interpretive aspects of their structure (such as the internal microstructural properties of propositions), and about a special kind of evaluation (namely, deductive validity or invalidity), whereas the theory of argument tends to have a more general scope. Finally, formal logic tends to me more

Basic Types of Meta-argumentation   41

systematic than argumentation theory.

As regards argumentation theory, it is beyond the scope of this chapter to give illustrations of theoretical meta-arguments, for such illustrations have an intrinsic interest in this investigation. Moreover, because of such pertinence, it would be odd and awkward to present the examples, without also advancing my interpretations and evaluation. Hence, the task of illustrating, interpreting, and evaluating the meta-arguments of various argumentation theorists will be left to the next several chapters, which make up part ii of this book.

3.6 Summary

In this chapter I have discussed several varieties of metareasoning and meta-argumentation. The prototypical case is argument analysis, namely the reasoned interpretation and/or evaluation of arguments. A second special case is that of self-reflective argumentation. A third special case is argumentation theory, or the theory of argument, conceived as the formulation, testing, systematization, clarification, and application of concepts and principles for the interpretation, evaluation, and practice of reasoning and argument. As compared with argument analysis, argumentation theory is more general, more systematic, and more conceptually explicitly, than argument analysis; whereas self-reflective argumentation is less so.

PART II
THEORETICAL META-ARGUMENTS

Chapter 4
Dialectical Definitions of Argument

The aim of this chapter is to examine the meta-argumentative aspects of work in logic and argumentation theory dealing with the problem of defining what an argument is. In particular, the theoretical position to be examined is the dialectical definition of argument. However, this examination will involve comparing and contrasting different varieties of that definition. Moreover, this examination will involve comparing and contrasting that dialectical conception with others that are different and use a different label (the illative definition), as well as with others that use the same label to mean something different (i.e., the dialogical approach).

4.1 Concepts of Dialectic

The concept of dialectic or dialectics is one of the most loaded and exploited notions in philosophy, the humanities, and the social sciences. Thus, it will be useful to mention briefly several concepts that are *not* pertinent here.

One of the first to be set aside is the meaning which this term has in the Hegelian and Marxist traditions of thought.[1] Although it is not to be excluded that the Hegelian concept might have some relevance, that would have to be revealed by some deeper level of analysis that presupposes the kind of investigation undertaken here.

Another meaning which is not our primary concern here is the one stemming from the Aristotelian tradition, which distinguishes in various ways what it calls dialectic from logic and from rhetoric. Such a contrast might turn out to be ultimately viable, but it is not part of our starting point here. Instead, in this context I am much more impressed by the fact that there are many argumentation theorists whose work combines fruitfully at least two of these three orientations.[2]

Finally, it is important to understand that the dialectical definition of argument to be studied here is not being equated with the dialogical definition. However, this particular exclusion needs some elaboration.

One of the best examples of the dialogical approach is a work entitled *From Axiom to Dialogue* by Else Barth and Erik Krabbe.[3] A critical examination of this work reveals that their achievement is not really to demonstrate the necessity to move from the axiomatic to the dialogical approach, by reducing the former to the latter; instead the structure of their proof is to demonstrate the equivalence of the methods of axiomatics, natural deduction, and formal semantics to the method of formal dialectics. However, as I have argued elsewhere (Finocchiaro 2005a, 231-45), the proof works both ways, so that the former methods acquire the merits of the latter, and the latter the limitations of the former; and the

---

[1] I have myself studied this other concept in Finocchiaro 1988.
[2] Here I am thinking of such works as Barth and Krabbe 1982, Blair and Johnson 1987, Woods and Walton 1989, Eemeren and Grootendorst (1992; 2004). For an attempt to sort out such interrelationships, see Johnson 2009.
[3] Barth and Krabbe 1982; cf. Barth and Martens 1982.

unintended consequence is that there is no logical difference between the axiomatic and the formal-dialectical method, and their difference will have to be located in some other domain.[4]

Another important result is due to James Freeman's (1991) work on *Dialectics and the Macrostructure of Arguments* and Francisca Snoeck Henkemans's (1992) work on *Analysing Complex Argumentation*. They have independently provided a dialectical analysis of complex argumentation, namely arguments where a conclusion is supported by more than just a single reason, either in the sense that two or more distinct reasons are given to support the conclusion, or in the sense that the reason which directly supports the conclusion is itself in turn supported by another reason. Their main accomplishment is to interpret arguments as the result of a hypothetical dialogue between a proponent or respondent and an opponent or challenger, a process during which the opponent asks various kinds of questions. However, as I have also argued previously (Finocchiaro 2005a, 244-45, 265-66), the questions asked are by and large evaluative questions, and so besides explicitly providing an illustration of the power of what might be called the informal-dialogical method, these authors have also implicitly suggested the evaluative dimension of complex argumentation. They may also be seen as having stressed the importance of complex argumentation and suggested that the usual emphasis on simple arguments is an undesirable oversimplification.[5]

A move in this direction (toward evaluation and complexity) has also been independently made by J. Anthony Blair. In a paper entitled "The Limits of the Dialogue Model of Argument" (Blair 1998), he has distinguished thirteen levels of complexity of dialogues depending on the complexity of the argument allowed at each turn of the dialogue; the thirteenth level is the one which is the norm in a scholarly paper or commentary. Blair also distinguishes between what he labels "solo" arguments and "duet" arguments: in solo arguments the respondent and audience are physically absent; their identity may not be known or fixed; and the norms of the discussion are not settled but open to dispute. Then he argues plausibly that to speak of dialogues for complex or solo arguments is metaphorical at best and probably distorting. Blair concludes with some theses that embody both a non-dialogical conception of the dialectical approach and of solo arguments. His words are worth quoting at length:

It would be nice if the term 'dialectical' were reserved to denote the properties of all arguments related to their involving doubts or disagreements with at least two sides, and the term 'dialogue' were reserved to denote turn-taking verbal exchanges between pairs of interlocutors. Then I could use this terminology to express the points that (1) all argumentation is dialectical, but by no means is all argumentation dialogical, and (2) the dialectical properties of dialogues, and the norms derived from the dialogue model, do not apply to non-dialogical argument exchanges, even though the latter are dialectical too. In other words, both duet arguments and solo arguments are dialectical, but only duet

---

[4] For a good, brief, and instructive example of translation of monological problems into dialogical terminology, see Krabbe 1998.
[5] I believe this double-edged nature of Freeman's and Snoeck Henkemans's work has been recognized by an exponent of the dialectical approach: Erik Krabbe has stated that the dialectical "obligation to handle objections can, in solo argument, be dealt with within the structure of a basic argument" (Krabbe 2000, 3), a basic argument being his (Walton and Krabbe 1995) label for what here I am calling complex argument. Similarly, he has suggested that "when studying more complex dialogues in which fallacy criticism is undertaken, not by an external evaluator, but by the participants themselves, profiles [i.e., sequences] of dialogue can again be used as a heuristic device" (Krabbe 2002, 155).

arguments are dialogues. [Blair 1998, 10][6]

One final explicitly critical contribution deserves mention. Chris Reed and Derek Long have stressed the importance and pervasiveness of what they call "persuasive monologue." A persuasive monologue is not merely a soliloquy, which is "a record of a chain of reasoning" (Reed and Long 1998, 2); nor an internal dialogue, "in which the speaker plays both roles" (ibid.); nor a "turn in dialogue" (ibid.). Instead persuasive monologue has two main characteristics: "firstly, the intuitive 'case building' of presenting arguments in support of the thesis" (Reed and Long 1998, 3); and "secondly, there is the more complex technique of presenting counterarguments to the thesis propounded, and then offering arguments which defeat those counterarguments" (ibid.). Although these authors' main interest seems to be the formal analysis and the computerized modeling of persuasive monologues, the point I would want to stress is that the second clause of their definition refers to replying to objections, and such criticism of counterarguments is a feature which many would not hesitate to call dialectical, in a sense of this word distinct from dialogical.

The upshot of these preliminary remarks is as follows. Several distinct concepts of dialectic exist even within argumentation theory, let alone in general: Hegelian, Aristotelian, dialogical, and (for lack of a better term) evaluative. Only the last one will be the focus of this investigation, and it involves not only evaluation, but also complexity, two-sidedness, and counterargumentation. However, it is noteworthy that some proponents of the dialogical approach to argumentation theory have produced results implicitly suggesting that dialogue may be dispensable (in favor of either deductive axiomatization or argument complexity and evaluation); and that some critics of the dialogical approach tend to stress monological argumentation, but in so doing they are quite sensitive to an aspect of argument which is dialectical in a sense other than the dialogical one.

4.2 Concepts of Argument

Let us now focus on what may be called the traditional conception of argument, or to be more precise, a version of the standard textbook definition. As many authors have done,[7] I too find it useful to quote Copi's definition: "An argument, in the logician's sense, is any group of propositions of which one is claimed to follow from the others, which are regarded as providing support or grounds for the truth of that one" (Copi and Cohen 1994, 5).[8]

However, although many of the same authors (e.g., Johnson 2000a, 148) take this to be a structural definition, I find it improper and misleading to speak of structure here because the structure involved is too insignificant to merit the name. The traditional concept does indeed define an argument as an ordered set of propositions, but the order introduced is simply that of designating one of the propositions as the conclusion; in other words, a distinction is made among all the propositions in the set, a distinction between the conclusion and the premises. However, such a single partition does not really yield a genuine structure, which for my sensibility would have to have at least two partitions; that

---

[6] Blair's clear distinction between (what he calls) the dialogue conception and (what I am calling) the evaluative conception of dialectics suggests the need to explore their relationship to other notions of dialectics, such as the classical Hegelian concept (cf. Finocchiaro 1988) and, more recently, Hilary "Putnam's dialectical thinking" (Cummings 2002).

[7] Walton 1990, 408-9; Johnson 2000a, 146; Hansen 2002, 264.

[8] This definition has remained essentially unchanged at least since the third edition of this classic textbook, which had "evidence" (Copi 1968, 7) in place of "support or grounds."

is, the minimal order I would want before calling it a structure is three propositions interrelated in such a way that one is supported by the second, which is in turn supported by the third. Instead of calling it structural, one might call this aspect of the traditional definition relational.

A second important feature of Copi's definition is the reference to the intention or purpose of the arguer. Again, although many commentators (e.g., Johnson 2000a, 148) have refused to attribute such a teleological character to Copi's definition, it seems obvious to me that when Copi says that the conclusion is claimed to follow from the premises, he is saying that the arguer claims this. And when he says that the premises are regarded as providing support or grounds for the conclusion, he is saying that the premises are so regarded by the arguer; that is, the arguer intends to use the premises to support the conclusion. In short, the purpose of the argument is to justify the conclusion by means of supporting reasons.

I am stressing that according to Copi's version of the traditional definition, an argument has function but no structure.[9] I believe there is a term that conveys both features, namely the term *illative*, which I adopt from Johnson (2000a, 150), who adopted it from Blair (1995). This traditional definition may thus be called the illative conception of argument. Illation is the special relationship that holds between premises or reasons and conclusion or thesis; it is not a purely abstract relation, but one that subsists in the mind of the arguer and of anyone trying to understand or evaluate the argument.

Two other versions of the traditional definition are worth mentioning, one more and one less abstract than the illative conception. The more abstract one avoids any reference to purpose and defines an argument simply as an ordered set of propositions partitioned into two subsets. For example, in *Choice and Chance,* Brian Skyrms stipulates that "an *argument* is a list of *sentences*, one of which is designated as the conclusion, and the rest of which are designated as premises" (Skyrms 1966, 1-2).[10] Those scholars who deny the teleological character of the traditional definition are probably thinking of this version, although of course it should not be equated with other versions such as Copi's.

The less abstract (or more concrete) version of the traditional definition adds a rhetorical condition to the illative conception, namely an element of persuasion. This conceives an argument as an attempt to persuade others that a conclusion is true by giving reasons in support of it. An example of such a definition comes from Michael Scriven's book *Reasoning*: "The simplest possible argument consists of a single premise, which is asserted as true, and a single conclusion, which is asserted as following from the premises, and hence also to be true. The *function* of the argument is to persuade you that since the premise is true, you must also accept the conclusion" (Scriven 1976, 55-56).[11]

These three versions of the traditional conception are importantly different, and

---

[9] This is almost the reverse of Johnson's (2000a) view, as it will emerge below.

[10] See also Kalish and Montague (1964, 13), quoted in Johnson (2000a, 123): "an argument, as we shall understand it, consists of two parts—first, a sequence of sentences called its premises and secondly, an additional sentence called its conclusion." See also Angeles (1981, 18), quoted in Walton (1990, 408): it defines argument as "a series of statements called *premises* logically related to a further statement called the *conclusion*."

[11] Cf. Epstein (2002, 5): "We're trying to define 'argument'. We said it was an attempt to convince someone, using language, that a claim is true ... An *argument* is a collection of claims, one of which is called the **conclusion** whose truth the argument is intended to establish; the others are called the **premises**, which are supposed to lead to, or support, or convince that the conclusion is true." Cf. also Govier 1989, 117.

constitute a sequence of increasingly more complex and narrow[12] definitions (as one moves from the purely abstract one through the illative one to the rhetorical). But they also share some very important features. All three lack any reference to a complex structure, or structure worthy of the name, as I have already mentioned. And all three lack any reference to dialectical matters, which will be our focus. Thus, let us now turn to what we may call the dialectical conception of argument.

The most natural version of the dialectical conception simply adds an element of criticism of objections to what I have called the rhetorical definition. We thus get that an argument is defined as an attempt to persuade someone that a conclusion is acceptable by giving reasons in support of it *and* defending it from objections. The best known example of this is the definition found in Johnson's book *Manifest Rationality*: "An argument is a type of discourse or text—the distillate of the practice of argumentation—in which the arguer seeks to persuade the Other(s) of the truth of a thesis by producing the reasons that support it. In addition to this illative core, an argument possesses a dialectical tier in which the arguer discharges his dialectical obligations" (Johnson 2000a, 168).

There is no time here for me to repeat or summarize the various clarifications that have been made to Johnson's definition by Johnson himself as well as by Trudy Govier, David Hitchcock, Hans Hansen, and others,[13] although I will say a little more later in the context of my analysis of Johnson's argument (section 4.6, below). However, I have already implicitly incorporated many of these clarifications when I gave my own formulation, before exemplifying it with Johnson's definition. In any case, a few remarks are in order and may be relatively novel.

One thing I would want to point out is that by calling illative *core* the set of conclusion and supporting reasons, Johnson suggests that the illative core is more fundamental than the dialectical tier. Now, this may very well be true; but it may not be. I would regard it as open question. Of this more presently. However, in order not to beg this question, I shall speak of the illative *tier* or *component* rather than *core*.

Another question I would want to ask is, why call *dialectical* tier or discharge of dialectical obligations such things as examination of alternative positions and reply to objections? What is the concept of dialectics inherent in such a terminological decision, and how is such a conception to be justified? Is it enough to do some hand waving in the direction of Plato's dialogues? Johnson's concept of dialectics is the one inherent in the following explicit statement: "that argumentation is dialectical means that the arguer agrees to let the feedback from the other affect her product. The arguer consents to take criticism and to take it seriously. Indeed, she not only agrees to take it when it comes, as it typically does; she may actually solicit it. In this sense argumentation is a (perhaps even *the*) dialectical process *par excellence*."[14]

Third, besides Johnson's references to written text, argumentative practice, purpose, persuasion, and truth, it is important to note the reference to both the illative and dialectical components or tiers. This implies that a text with an illative tier but without a dialectical one is not strictly speaking an argument (as some of Johnson's critics have pointed out, thus yielding an alleged *reductio ad absurdum* of his definition); but Johnson himself

---

[12] Although these three definitions are increasingly more complex and narrow, they are not necessarily increasingly more adequate, for as I shall argue below, the move from the "justification" of the illative definition to the "persuasion" of the rhetorical definition may not yield an increase in adequacy.

[13] Cf. Govier (1998; 1999, 223-40; 2000), Groarke 2002, Hansen 2002, Hitchcock 2002a, Johnson (2002a; 2003), Leff 2000, Rees 2001, Tindale 2002, Wyatt 2001.

[14] Johnson 1996, 107. Cf. Johnson 2000a, 161; Finocchiaro 2005a, 265-66.

prefers to say that such a text "does not fit the paradigm case of argument" (Johnson 2002a, 316).

In the present context, however, the point I want to stress is that there is a natural way to moderate Johnson's double requirement by disjoining the two conditions, in the sense of inclusive disjunction. We thus get the following conception: an argument is an attempt to persuade someone that a conclusion is true by giving reasons in support of it *or* defending it from objections. This is a weaker dialectical conception than Johnson's definition, but it still is importantly dialectical because it does call attention to the potential need to discharge one's own dialectical obligations, and because the inclusive disjunction obviously allows for cases where the argument contains both illative and dialectical tiers.

Such a more moderate dialectical conception has in fact been advanced by some scholars.[15] If I understand him correctly, I believe Alvin Goldman does this in his book *Knowledge in a Social World*. He explicitly allows for what he labels monological argumentation besides dialogical argumentation, as can be seen from this passage: "If a speaker presents an argument to an audience in which he asserts and defends the conclusion by appeal to the premises, I call this activity *argumentation*. More specifically, this counts as *monological* argumentation, a stretch of argumentation with a single speaker. ... I shall also discuss *dialogical* argumentation in which two or more speakers discourse with one another, taking opposite sides of the issue over the truth of the conclusion" (Goldman 1999, 131). And for Goldman, a crucial principle governing dialogical argumentation is this: "when there are existing or foreseeable criticisms of one's main argument, a speaker should embed that argument in an extended argumentative discourse that contains replies to as many of these (important) criticisms as is feasible" (Goldman 1999, 144).

We thus have two versions of the dialectical conception of argument, a stronger one exemplified by Johnson that regards the dialectical or critical tier as necessary for any argument, and a moderate one exemplified by Goldman that makes the dialectical or critical tier essential for one type of argument but not for all. Although these two versions of the dialectical conception are the most common and natural ones, there is actually a third version that deserves discussion and may be regarded as more strongly dialectical than Johnson's conjunctive version. This hyper dialectical conception would define an argument as an attempt to justify a conclusion by defending it from objections. According to this conception, replying to objections is both a sufficient and a necessary condition to have an argument; whereas for Johnson's strong dialectical conception, replying to objections is necessary but not sufficient; and for Goldman's moderate dialectical definition, replying to objections is sufficient but not necessary.

Unintuitive as it may sound, the hyper dialectical conception has been advanced by some scholars. In a 1980 book by the present author, entitled *Galileo and the Art of Reasoning*, in the context of a number of theoretical considerations, we find the following theoretical definition: "We may then say that an argument is a defense of its conclusion from actual or potential objections" (Finocchiaro 1980a, 419). More recently, in her review of Johnson's *Manifest Rationality*, Agnès van Rees has criticized his definition of argument for being insufficiently dialectical. Here are her revealing words: "According to this definition, producing reasons and discharging one's dialectical obligations are two different things. But in actual fact, if the notion of argument is indeed to be rooted in the dialectical practice of argumentation, the two should coincide. In a truly dialectical account, argument

---

[15] Besides Goldman, Reed (2000, 1) may be attributed this concept when he says, "The most fundamental problem facing the designer of an argument is premise availability: do there exist premises which can support a given conclusion or which can rebut or undercut some counterargument?"

per se would be defined as an attempt to meet the critical reactions of an antagonist, that is, to take away anticipated objections and doubt" (Rees 2001, 233).[16] And besides these two explicit formulations, the hyper dialectical definition has a memorable, emblematic, and brilliant illustration; that is, an argument by Alan M. Turing published in 1950 in the journal *Mind*, advocating that machines can think based primarily on a critique of nine objections to this conclusion.[17]

Once again, however, despite the differences among these three versions of the dialectical conception, my focus will be on what they have in common. Their common element is an emphasis on replying to objections or to criticism. It is such a dialectical component that provides an instructive contrast to the illative conception. And it is this dialectical tier that I want to understand better and evaluate. With such an aim, the next step will be to examine various arguments that have been advanced in favor of the dialectical conception of argument.

4.3 Goldman's Argument

What we are faced with now is an exercise in argument analysis: to identify, interpret, reconstruct, analyze, evaluate, and criticize the arguments for the dialectical conception of argument. In the rest of this chapter, I shall focus on arguments advanced by Johnson and Goldman. However, implicitly underlying my analysis will be my 1980 argument mentioned above and the pragma-dialectical argument, which will be explicitly examined later (chapter 5). Also implicitly underlying my analysis will be some arguments by John Stuart Mill, which will also be explicitly examined later, but not until chapter 10; they have already been injected into this discussion by Hansen (2002, 271), who speaks of Mill's "dialectical method" and quotes his striking claim that "when we turn to subjects [such as] morals, religion, politics, social relations, and the business of life, three-fourths of the arguments for every disputed opinion consist in dispelling the appearances which favor some opinion different from it" (Mill 1965, 286-87).

Let us begin with an argument which has been advanced by Goldman, or at least which I extract from Goldman; it deserves discussion because of its novelty. He intends to justify "a general thesis about critical argumentation and the probability of acquiring truth ... that lively and vigorous debate is a desirable thing"(Goldman 1999, 144), desirable in the sense that it "has positive veritistic properties" (Goldman 1999, 146). In other words, critical argumentation is likely to lead to the truth. The connection between this conclusion and the dialectical conception of argument may be elaborated as follows. Goldman (1999, 132) says that "critical argumentation is an attempt to defeat or undercut the proffered argument," and he contrasts it to "proponent argumentation [which] is a defense of the asserted conclusion by appeal to the cited premises" (Goldman 1999, 132). To this I add that if critical argumentation is a veritistically good thing, then it will also be desirable for the special case when the proffered argument is a critical argument, and so a reply to the critical argument is called for. Such a reply is precisely what the dialectical conception of argument stipulates.

Next, Goldman distinguishes at least three subtypes of critical argumentation, one that denies the truth of a premise, a second that questions the link between premises and

---

[16] She makes it clear that she is speaking from the "pragma-dialectical" point of view of the Amsterdam school of argumentation studies, and indeed one can find statements to this effect in such works as Eemeren and Grootendorst (1992, 73), Snoeck Henkemans (1992, 179), and Eemeren, Grootendorst, Jackson, and Jacobs (1993, 12, 14). For more details, see chapter 5 below.

[17] I first learned of this collector's piece from Reed and Long 1998, 3.

conclusion, and a third one which he calls "presenting a defeater" (Goldman 1999, 140). Then he formulates his argument for the special case of the latter, which is a notion he adopts from John Pollock[18] and may be explained as follows. Given an argument with premises P-1 through P-n and conclusion C, a defeater is a critical argument with the same premises plus one additional special premise D and conclusion not-C. A simple example given by Goldman himself is this: consider the argument that it will probably rain tonight because it was so stated last night in the local weather forecast of a reliable news medium; a defeater of this argument would be the counterargument that it will probably *not* rain tonight because although it was so stated last night in the weather forecast of a reliable local news media source, it was also stated this morning by the same source that the forecast had changed and the new prediction was fair weather tonight.

I will label Goldman's argument the "truth-in-evidence" argument because it is based on a premise which he himself calls the truth-in-evidence principle (TEP); it asserts the following: "a larger body of evidence is generally a better indicator of the truth-value of a hypothesis than a smaller, contained body of evidence, as long as all the evidence propositions are true and what they indicate is correctly interpreted" (Goldman 1999, 145). He points out that this principle is an epistemic version of the requirement of total evidence discussed in a methodological context by Rudolph Carnap (1950, 211) and Carl Hempel (1965, 64-67). Goldman also says that the principle is a generalization of Bayes' theorem. Although these are reasons of sorts, and although the principle has some inherent plausibility, Goldman (1999, 145-46) confesses that "I have no proof of this postulate."

But how does this principle support his conclusion that critical argumentation, or at least defeater presentation, is conducive to truth? Clearly, a defeater argument does encompass a larger body of evidence than the argument being criticized. So, Goldman's premise is indeed relevant.

But is it sufficient? He attempts to articulate such sufficiency by commenting on the two provisos incorporated into the principle. The condition of truth for the evidence propositions could be somewhat relaxed (he says) by requiring merely that they be justified. The other condition was that what the evidence propositions indicate be correctly interpreted; this seems to mean that it be correct to claim that the premises of the defeater argument imply or support the denial of the conclusion of the original argument. And Goldman himself suggests that in real-world cases such an implication would be itself controversial. The upshot of his articulation seems to be that in order for a defeater argument to have positive veritistic properties in accordance with the truth-in-evidence principle, the defeater has itself to be a good argument, namely have true or justified or acceptable premises, and these premises have to really support its own conclusion; but in realistic situations such goodness would be controversial.

Goldman seems to be aware of such difficulties, for despite his articulation and elucidation, what he claims about the premise-conclusion link of his argument is very modest. He says: "Suppose ... that (TEP) is correct. May we derive from it the veritistic desirability of engaging in defeater argumentation? In other words, does (TEP) imply that defeater argumentation usually has positive V-value? Although I shall not attempt to prove it, I suspect that this does follow" (Goldman 1999, 146).

Another difficulty that could be raised stems from the fact that Goldman intends his argument, which is specifically formulated in terms of defeater arguments, to apply to other cases of critical argumentation. However, its relevance to these other cases is questionable.

But to end on a more positive note, there are other valuable aspects of an argument besides such properties as truth of premises and validity of inference, or acceptability,

---

[18] Pollock 1986, 33-39; cf. Goldman 1999, 138-39, 139 n .11, 144-45.

relevance, and sufficiency of premises. Some of the additional values are what might be called suggestiveness or fruitfulness and novelty or originality.[19] And along these dimensions, I would rate Goldman's truth-in-evidence argument very highly.

4.4 Johnson's Argument: Its Illative Tier

Let us now examine Johnson's argument for the dialectical conception. As one would expect, his argument possesses an illative component as well as a dialectical tier, and the latter contains both replies to objections and criticism of alternative positions. The main alternative is what he calls the structural definition of argument, which (as suggested above) is not really structural and should rather be labeled the illative conception of argument.

One of Johnson's key supporting reasons I locate in passages where he makes statements such as the following: "Philosophers and others for whom argumentation is the principal methodology routinely include in their own arguments a section in which they voice and then deal with objections to their position ... If ... we look at the best practices of those who have the most at stake in this process, philosophers and logicians who have a vested interest in this practice, we will find that their arguments always take account of the standard objections ... Arguments with a dialectical tier are found in nonacademic discourse as well" (Johnson 2000a, 165-66).

Here we have an empirical argument, which some would call an induction by enumeration, and others more simply a generalization argument or an inductive generalization. Although in his book Johnson does not himself categorize this argument in this manner, I am encouraged to advance this interpretation by the fact that in one of his papers he does speak favorably of the "empirical turn" (Johnson 2000b, 14-15). Moreover, such a strand of argument is also found in other dialectically inclined authors. For example, Goldman, after defining an "extended argumentative discourse" as a series of nested arguments that present and answer objections, states that "in science, scholarship, law, and other polemical realms, extended argumentative discourses are the norm. Scholars are expected to report existing findings and literature that form the basis of predictable objections" (Goldman 1999, 144).

Also reminiscent of such an empirical argument is Mill's assertion, quoted earlier, that three-fourths of arguments involving human affairs consist of attempts to reply to objections.

If and to the extent that Johnson's case for the dialectical conception of argument includes an empirical inductive generalization, then its evaluation would have to deal with questions such as the following. For example, if indeed the best examples of arguments by philosophers have a dialectical tier, should our conclusion be that *good* philosophical arguments have a dialectical tier, or that *all* philosophical arguments have a dialectical tier? In other words, that all philosophical arguments *ought* to have a dialectical tier, or that they *do* have a dialectical tier? Moreover, can our conclusion be that good arguments *in general* (whether philosophical or not) have a dialectical tier; in other words, from a sample containing information about philosophical arguments, how can we reach a conclusion about arguments in general? Do we not also need data about the characteristics of arguments in other disciplines, such a science and the law, as Goldman indicates? Furthermore, if we gather such data about, for example, science, such scientific arguments

---

[19] This point about novelty is, of course, not novel. Johnson anticipates it to some degree (2000a, 336) and discusses it implicitly in replying to the objection that his own definition of argument is similar to Toulmin's notion of rebuttal (Johnson 2000a, 173-74). See also section 6 below.

that thus have a dialectical tier may happen to be those in special domains of scientific activity, such as in the context of peer discussion at the frontiers of research or in periods of scientific revolution; what are we then to say about other domains such as the context of scientific justification or the context of pedagogy or periods of normal science? And if indeed, as Mill states, three-fourths of arguments have a dialectical tier (indeed only this tier), should not the conclusion be qualified to make it a statistical rather than a universal generalization? And what are we to say about the other one-fourth of arguments? Finally, if we take seriously the possibility of an empirical inductive confirmation of the dialectical conception of argument, does not the above-mentioned evidence appear as merely anecdotal? Should we not attempt to devise more systematic and controlled tests or data gathering?

Some scholars have indeed undertaken such attempts. I am thinking of David Hitchcock's sampling of scholarly arguments to test his theory of inference; of David Perkins's studies of the difficulties in everyday reasoning; and of the present author's examination of arguments in Galileo's *Dialogue on the Two Chief World Systems*.[20] I believe that by and large the relevant parts of such inquiries do support the thesis of the dialectical nature of argument. In fact, it was in the context of such an empirical investigation that the present writer drew the conclusion mentioned earlier labeled the hyper dialectical conception of argument, namely "that an argument is a defense of its conclusion from actual or potential objections" (Finocchiaro 1980a, 419).

Although such an empirical approach has also been appreciated or advocated by other philosophers,[21] it is unlikely that most philosophers will have much interest in using such empirical argumentation to support their theories of argument; so let us go on to more theoretical and conceptual considerations, especially since we find such supporting reasons in Johnson's book.

A second supporting reason in Johnson's illative tier is a premise about the nature of the process of arguing. He states that "the process of arguing includes, by its very nature, feedback from the Other. Nor does the process of arguing end there. Also included as part of the process must be the response by the arguer to those objections and criticisms, as well as any revisions made by the arguer" (Johnson 2000a, 157). If we recall that for Johnson an argument is the so-called "distillate of the practice of argumentation" (Johnson 2000a, 168), then the relevance and sufficiency of this premise become obvious. What about its acceptability? Here we have to remember that Johnson is talking about one of several meanings of the word *argue* or *arguing*; and certainly for this particular meaning it is unobjectionable and true that the process of arguing does have such a dialectical component.

However, this claim about the nature of the process of arguing is almost analytically true. So an unfriendly evaluator[22] might at this point raise the possibility that Johnson's argument from the nature of the process of arguing begs the question. But a more friendly critic might point out that the function of this particular argument is to articulate a necessary connection between the process of arguing and the concept of argument, and the articulation of such analytic relations is a normal part of any theorizing; thus it is no defect

---

[20] Hitchcock 2002b; Perkins 1989; Perkins, Allen, and Hafner 1983; Finocchiaro (1980a; 2005a, 65-91).
[21] Barth 1985a; Krabbe 2000, 4; not to mention Toulmin 1958, concerning which see chapter 1 above.
[22] In this sentence and the next one, I am (for the sake of the argument) using Johnson's distinction between evaluation and criticism, although it seems to me that the concept of evaluation is broader than he allows and thus includes the concept of criticism as a special case. Cf. Johnson 2000a, 217-23, and chapter 2 above.

of a theory of argument that at some point it has to explain the links among various elements of its conceptual structure. However, such friendly criticism may have a consequence that suggests some possible revision by the arguer. That is, some arguments do not have rational persuasion as their telos, but rather the analytic elucidation of conceptual relations; and this is one of those arguments. But if we defend this argument in this manner from the criticism that it begs the question, then one has to revise the teleological rhetorical aspect of Johnson's definition. This could be done by stating the definition by saying that an argument is an attempt to justify a conclusion instead of saying that it is an attempt to persuade others that a conclusion is true.

Another way out might be to say that here both the friendly critic and the unfriendly evaluator are committing the fallacy of straw man, when they interpret these considerations about the nature of the process of argument as an argument in support of Johnson's definition; after all, he does not himself explicitly label them an argument. This denial of argumentative status would by itself be unproblematic, but it would begin to weigh if added to the previous unfriendly evaluation that Johnson's empirical inductive generalization was anecdotal. For one reply to that evaluation might have been that his empirical considerations should not be interpreted as a real or full-fledged empirical argument susceptible of evaluation in terms of the adequacy, variety, and representativeness of the sample used; and then we would be saying that neither the empirical considerations nor the conceptual ones were arguments. However, this consequence would not be fatal because there is at least one passage in Johnson's book that seems to advance a supporting reason as explicitly as the practice of argumentation allows. To this we now turn.

This is the passage where Johnson tries to ground the dialectical tier on the telos of rational persuasion. In his own words, "because the practice exists to achieve rational persuasion of the Other as a rational agent, the practice must also be dialectical" (Johnson 2000a, 160). To avoid straw-man problems, I quote this argument verbatim:

Because the arguer's purpose is rational persuasion, a second tier is required as well. Why? I have shown that the practice of argumentation presupposes a background of controversy. The first tier (the illative core) is meant to initiate the process of converting Others, winning them over to the arguer's position. But they will not easily be won over, nor should they be, if they are rational. The participants know that there will likely be objections to the arguer's premises. Indeed, the arguer must know this, so it is typical that the arguer will attempt to anticipate and defuse such objections within the course of the argument. If the arguer does not deal with the objections and criticisms, then to that degree, the argument is not going to satisfy the dictates of rationality; more precisely, to that very degree the argument falls short of what is required in terms of structure—never mind the content; that is, the adequacy of the response to those objections. For those at whom it is directed, those who know and care about the issue, will be aware that the argument is open to objections from those who disagree with its reasons, conclusion, and-or reasoning. Hence, if the arguer wishes to persuade Others rationally, the arguer is obligated to take account of those objections and opposing points of view. To ignore them, not to mention them, or to suppress them—these could hardly be considered the moves of someone engaged in the process of rational persuasion; thus, the process of persuasion must include a second—dialectical—tier in which objections and criticism are dealt with. [Johnson 2000a, 160; cf. 165]

This is a plausible argument, but I should like to point out that there are two, and not just one, final premises: the claim that the purpose of the argument is rational persuasion, and the claim that the process of argumentation occurs in a context of controversy. And the latter claim is both crucial and independent of the first. But as John Stuart Mill pointed out, in Euclidean geometry rational persuasion is achieved with just the illative tier, without any

Dialectical Definitions of Argument   53

need of dealing with objections.[23] Hence, the dialectical tier is not a consequence of just the telos of rational persuasion, but of this telos plus the controversial origin of argumentation.

In other words, Johnson gives the impression that the third reason of the illative tier of his argument for the dialectical conception of argument is the telos of rational persuasion. This impression is misleading because if this reason were the only premise it would be insufficient and because in fact Johnson himself combines it with the premise about controversial origin.

I suspect that Johnson's reaction to this criticism would be to insist that controversy is presupposed by all argumentation and to regard geometrical proofs as not arguments but mere inferences or entailments. This move would strike me as arbitrary insofar as Euclidean geometrical proofs are typically attempts to persuade oneself or others of the truth of the theorem in question by rational means. Moreover, the move would be questionable from the point of view of Johnson's own principle of vulnerability because the restriction of the domain to that of controversial situations tends to make the conclusion true by definition; that is, the dialectical tier becomes necessary as an immediate consequence of the controversial context. Then this third reason of Johnson's illative tier would basically reduce to his second reason.

I believe the way to remedy this difficulty is to revise the conclusion of Johnson's argument by weakening the requirement of the dialectical tier. For example, instead of being regarded as a necessary condition to have an argument, the dialectical tier might be regarded as a sufficient condition. This would amount to replacing the conjunctive version of the dialectical conception with the disjunctive one mentioned earlier.

4.5 Johnson's Argument: Its Criticism of Alternative Positions

So far we have examined the illative tier of Johnson's argument for his dialectical conception of argument. We have identified, analyzed, evaluated, and criticized three supporting reasons. Let us now go on to examine the dialectical tier of his argument. This tier consists of two parts, criticism of the alternative traditional conception of argument, and explicit replies to explicit objections to his definition. The criticism of alternatives can be easily subsumed under the notion of reply to objections, by regarding that criticism as a reply to the objection that the traditional conception of argument is adequate (and constitutes an alternative to the dialectical conception);[24] but it is useful to treat the two

---

[23] Mill, *On Liberty*, ch. 2, par. 23; 1965, 286; cf. chapter 10 below. This is not meant to deny that there are some parts of mathematics in which replying to objections is important, as argued by such authors as Lakatos 1976, Franklin 1987, Krabbe 1997, Mancosu 2005, Aberdein (2006, 2007), Dove (2007, 2009), and Finocchiaro (2010b, 97-120); but such controversial and dialectical aspects should not be exaggerated and regarded as the norm, rather they are refinements and second approximations; cf. the discussion in chapter 1.4 above.

[24] This only means that any alternative position generates an objection, not that any objection yields an alternative position. Indeed, as Govier (1999, 226-27) has argued, many objections do not involve alternative positions; for example, counterexamples to generalizations are objections but do not constitute alternatives. At times Johnson (2000a, 206-9) speaks of the dialectical tier as having a third part, namely dealing with undesirable consequences or implications of one's position; elsewhere he (Johnson 1998, 2) seems to accept Govier's (1998, 7-8) friendly critical revision that this is a special case of replying to objections; in still other places, Johnson (2002b, 3-4) speaks of *four* types of dialectical material, namely objections, alternative positions, criticism, and challenges. Such discussions suggest that more work is needed to clarify the concepts of objection and criticism, and indeed some scholars have undertaken this task; cf. Govier (1999, 229-32), Johnson (2000b; 2002b);

parts separately, as Johnson does explicitly in his book.

Let us begin with a criticism which I have implicitly already examined and so can be dealt with relatively briefly. According to Johnson, "argument has its structure (reasons in support of a thesis, or premises plus conclusion) because of the purpose it serves—rational persuasion. A significant limitation of the structural view is that it ignores this important aspect—purpose or function. The moral of the story is that if a satisfactory conceptualization of argument is to be developed, the purpose or function of the discourse must be referred to" (Johnson 2000a, 148, cf. 167). But earlier I pointed out that, although the purely abstract version of the traditional definition does lack any teleological aspect, the illative conception (such as Copi's) is teleological insofar as it does make the aim of argument the justification of the conclusion by means of supporting reasons. Although justification is indeed different from persuasion, this only means that the illative conception attributes to argument a purpose different from the purpose attributed to it by Johnson's dialectical conception. Now, this conception (and even the rhetorical one) may very well be correct that the purpose is persuasion, and the illative conception incorrect that it is justification, but it seems unfair to criticize the latter for conceiving argument as purely structural without function.

A second criticism advanced by Johnson (2000a, 147) is that "the traditional view ... must ultimately fail because it does not distinguish argument from other forms of reasoning." He discusses three problematic forms of reasoning: explaining, instructing, and making an excuse. This criticism would be relevant and strong if this claim of confusion were true. But it is not. An analysis of Johnson's supporting critical argument will show this.

Johnson (2000a, 146) says, "I offer reasons in support when I explain, 'The reason that your car won't start is that you have a dead battery, and also the starter is defective'. Here I am supporting one claim (your car won't start) by another (you have a dead battery) and another (your starter is defective)." This example and others given by Johnson involve a misconception of the notion of support used by the illative definition. In this example, the supporting reasons are not the ones mentioned but rather the observational reports (not mentioned) that after turning the ignition key nothing happened; the mentioned reasons are the causes offered to explain the observed fact. Correspondingly, in the quoted text, the claim that your car won't start does not function as a conclusion but rather as a presupposition or premise of an argument aiming to support the causal claim.

Besides this misunderstanding or misrepresentation of the notion of support used by the illative conception, Johnson's critical argument may involve an equivocation on the term *reason*. Admittedly a reason can mean a premise helping to establish a controversial conclusion in an argument, or it can mean a cause or explanation helping to account for why a given non-controversial claim is true. So it would be correct to say that one offers reasons when one explains, in one sense of *reasons*, but not that one offers reasons in support when one explains. Thus, here it may be the critic (and not the proponent of the illative view) who is failing to distinguish argument from explanation.

Johnson's remarks about the other forms of reasoning suffer from similar difficulties. He says, "I offer reasons when I instruct, 'If you want to get the best light for this shot, you are going to have to use a XDX-1000 filter combined with ...' Here I offer a reason (you are going to have to use a XDX-1000 filter) as support for the claim (if you want to get the best lighting), but the function of the discourse is not to persuade anyone that the claim is true" (Johnson 2000a, 146). Once again, there is no relationship of illative support between

---

Krabbe (2002, 160-62), Krabbe and van Laar 2011, van Laar and Krabbe 2012, and chapter 2 above. But much more remains to be done.

the two clauses, and there are not even two distinct claims; rather we have a single claim about a means-end connection between two things.

Johnson (2000a, 146) also says that "I offer reasons when I make an excuse, 'I can't go to the show tonight because I have to study for my exam tomorrow'. Here we have the structure of an argument as defined, but that is not sufficient to qualify it as an argument." I would counter that here we are in the domain of motivation, which is subsumable under the general concept of explanation, although it is also useful to treat it as an important special case. That is, one meaning of reason is that of motivation for an action. Clearly the quoted example is an explanation (of the speaker's not going), not an argument trying to persuade anybody that the speaker is not going. But this is an interpretation reached in the light of the illative conception, which clearly has the resources to say that the need to study is the motivating reason *why* one is not going, not the evidence proving *that* one is not going.

A third criticism advanced by Johnson is really a particular case of the second but deserves separate discussion and special attention. It claims that the illative conception presupposes an inadequate conception of a particular kind of reasoning, namely inference. This inadequacy has three overlapping aspects. First, the illative conception tends to conflate three forms of reasoning that ought to be distinguished, namely implication, inference, and argument (Johnson 2000a, 93-95). Second, it conceives the illative core in terms of a model that ought to be discarded, namely what Johnson calls "the (P+I) model, the view that an argument should be seen as consisting of a set of premises, plus an inference from them to the conclusion. The inference is typically represented as a bridge or a link from the premises to the conclusion" (Johnson 2000a, 166-67). Third, it fails to properly distinguish between argument and inference (Johnson 2000a, 177-78).

Much of this criticism is insightful and raises important issues. The key merit is to point out that the illative conception of argument has not been embedded in a wider theory of reasoning that would define, distinguish, and interrelate such concepts as reasoning, argument, inference, implication, and explanation. However, the effectiveness of this criticism is limited by the fact that the dialectical conception of argument, or more generally, Johnson's own pragmatic theory of argument, has also not been embedded within a more comprehensive theory of reasoning. He himself constantly reminds us that the aim of a theory of argument is narrower than that of a theory of reasoning, and that while the articulation of a theory of reasoning remains a desirable goal, it was not within the scope of his work on the theory of argument. Thus the crucial question is whether the theory of reasoning groped toward or adumbrated by Johnson's dialectical conception is more adequate than that presupposed by the traditional illative conception.

With this aim in mind, I would point out some inadequacies in Johnson's account, besides the ones discussed in connection with the previous criticism involving the relationship between argument, support, reasoning, and explanation (which would also be relevant to the present point). Johnson (2000a, 94) defines implication as "a logical relationship between statements or propositions, in which one follows necessarily from the others"; and he gives the following paradigm example: if it is true that if P then Q and that if Q then R, then it is true that if P then R. He gives this definition and illustration in a passage where he also defines and illustrates inference and argument, with the aim of showing how to distinguish and interrelate these three forms of reasoning. However, it seems to me that implication so defined is not a form of reasoning at all; it is an abstraction and not a form of mental activity. He is not defining implication as what might be called deductive reasoning, which instead is subsumed under his definition of inference as a special case; instead he is defining implication in a way that places it outside the domain of reasoning altogether.

In another passage, where he focuses merely on the distinction between inference and argument, he seems to come close to placing argument itself outside the domain of reasoning. In a summary extolling the advantages of the pragmatic approach, he says that one of these advantages is that "we can begin to get a handle on differentiating between arguments and inference. Arguments, as I have shown, are outcomes within the practice that are dialectical in nature and characterized by manifest rationality. What is an inference? In chapter 4, I presented inference as in one important sense something that happens in the mind, an activity perhaps spontaneous, perhaps calculated, by which the mind moves from one thought to another" (Johnson 2000a, 177-78). After pointing some similarities, he stresses the following difference: "an inference can be what it is while remaining within the mind of the inferrer; this is not true of argument. One way of drawing this contrast is to say that inferring is monolectical, whereas arguing is a dialectical process. Moreover, argument must be seen within the practice of argumentation, but no comparable requirement exists for inference" (Johnson 2000a, 177-78). The only point I want to make here is that by conceiving an argument as a dialectical process that subsists within the practice of argumentation and cannot remain within the mind of the arguer, Johnson seems to be saying that argument is not a form of reasoning; to be sure it originates in reasoning, but to become argument it has to metamorphosize into non-reasoning. Now, all this may be adequate, sound, and correct from various points of view, but it is clear that we need more than a theory of reasoning; the theory of argument would need to be embedded in something like a theory of action, or a theory of speech acts, or a theory of social interaction.

There is at least one other criticism which Johnson advances against the illative conception, namely its "failure to give an adequate representation of the dialectical character of argumentation" (Johnson 2000a, 165). I hesitate to include this point under the heading of criticism of alternatives, in the dialectical tier of Johnson's own argument; for there seems to be some circularity or question-begging in objecting to a position P by saying that it fails to describe the matter at hand the way its alternative Q does; in fact this failure is guaranteed by P's being an alternative to Q and is an immediate consequence of that fact. In short, to point out this sort of thing is part of the clarification or elucidation of the two positions. However, Johnson is explicit that he regards this as a "limitation" of the traditional conception of argument, and so it deserves some attention. Moreover, it will turn out that the examination of this criticism is useful from a methodological, meta-logical, or meta-theoretical point of view.

As one might expect, to justify this criticism Johnson elaborates an argument trying to show that arguments must have a dialectical tier. But this argument is and can only be one of the positive reasons supporting his dialectical conception; that is, one of the elements of the illative tier of his argument. In fact, he repeats the third reason discussed above, namely the one trying to establish that the telos rational persuasion implies the necessity of the dialectical tier. Clearly, here I need not repeat my own criticism of that reason. Instead, I want to elaborate a meta-theoretical point.

I want to stress that a positive reason of the illative tier has become a critical reason of the dialectical tier. I believe this could always happen. In fact, consider the illative tier R-1 through R-n of an argument whose conclusion is P; and consider that part of the dialectical tier consisting of various criticisms C-1 through C-n of alternative Q. Now, if and insofar as any one supporting reason is relevant and sufficient, then it would be more or less true to say that, for example, if R-n is true then P is also true. Moreover, if and insofar as the various criticisms are cogent, then it would be approximately correct to claim that, for example, if C-n is true then Q is false; indeed any particular criticism of Q can be phrased in this manner, for that is what makes it a criticism. Next, note that the fact that P and Q are alternatives means that they are at least contraries (though they need not be contradictories), so that, for example, if P is true then Q is false. Finally, putting together the three claims expressed in my last three sentences, we get that if R-n is true then Q is false, and hence that R-n is a criticism of Q; but R-n was one of the supporting reasons in the illative core, and so any such reason can generate a criticism in the dialectical tier.

My final comment here is that not all criticisms of alternatives are or need be dialectical rephrasings of illative supporting reasons. For example, Johnson's first three criticisms are not like that. Thus in discharging one's own dialectical obligations, it seems important to distinguish between criticisms that are independent of the illative tier and criticisms that are not. Only the independent criticisms would seem to add anything new to the argument, whereas the dependent ones may be useful rhetorically or pedagogically but add little to the logical strength of the argument. And this was the methodological, meta-theoretical lesson I wanted to elaborate.

4.6 Johnson's Argument: Replies to Objections

I have been analyzing Johnson's argument in favor of the dialectical conception in light of his own definition. To complete the analysis there remains to examine the second part of its dialectical tier, consisting of his explicit replies to objections. In his book *Manifest Rationality*, he lists, numbers, and discusses five objections.

The first objection (Johnson 2000a, 169-71) is that the definition is too restrictive because it disqualifies from the category of arguments discourses that lack a dialectical tier. Johnson's reply is that his definition is indeed restrictive insofar as it does imply this disqualification, but it is not excessively restrictive because this restriction is quite proper. The restriction is proper because it focuses attention on the paradigm and central instances of argument, rather than on the derivative cases that might be called "proto-arguments."

My criticism of this reply is that Johnson's definition is indeed too restrictive because the desired redirection of focus can be accomplished equally well by the moderate, disjunctive version of the dialectical conception. Recall that that conception states that an argument is an attempt to justify a conclusion that gives reasons in support of it and/or defends it from objections.

The second objection (Johnson 2000a, 171-73) is that the dialectical tier is unnecessary in the definition of argument because the work it does could be accomplished in other ways: for example, one could make the dialectical tier part of the normative requirements of a good argument;[25] or part of the definition of such more complex discourses as extended arguments, or cases, or supplementary arguments, or full-fledged arguments, and the like, as distinct from mere arguments or ordinary arguments. Johnson's reply is that the first criticism would have difficulties distinguishing between arguments that are bad insofar as they lack a dialectical tier, and arguments that are bad insofar as their dialectical tier fails to reply effectively to objections. Similarly, the second criticism would lead to the question, when is it enough to give an ordinary argument and when is it necessary to present an extended argument?

I agree that it is important to distinguish between the factual existence of a dialectical tier and the evaluative adequacy of it, and so it may be impossible to go through with the suggestion that the question of the dialectical tier belongs wholly to the theory of assessment, and can be removed from the theoretical problem of definition. However,

---

[25] Johnson (2000a, 171 n. 20) attributes this objection to Blair in Blair and Johnson 1987; one can also find it in Govier 1998. This objection was also implicitly raised above in my criticism of Johnson's inductive generalization argument, when I asked whether the conclusion should be formulated as saying that all arguments have a dialectical tier, or that all good arguments have it.

regarding extended, supplementary, or full-fledged arguments, I do not see any difficulty, and that part of this objection seems to me to reinforce the moderate, disjunctive definition.

The third objection is that the definition is circular because it defines argument in terms of argumentation, among other things.[26] Johnson replies by admitting that there is a slight circularity, but claiming that the circularity is not vicious or objectionable. I agree with him, in part because I would add that the reference to argumentation is not necessary. In fact, my formulation of the three versions of the dialectical conception avoids such reference, without I believe any loss of generality, at least from the point of view of the contrast between the illative and the dialectical conceptions.

However, another comment is in order. Even if we eliminate mention of argumentation in the wording of the dialectical conception of argument, this eliminates only a potential internal circularity; but this does not avoid what might be called a potential external circularity, that is a circularity in the justification of the definitional claim. In fact, we saw earlier that one of Johnson's supporting reasons in the illative tier of his argument was the one based on the nature of the process of arguing; that that argument could be criticized as begging the question; that that criticism could be answered only by making a revision in the definition; and that the revision was to broaden the concept to include justifications that did not aim at rational persuasion but at conceptual clarification.

The fourth objection (Johnson 2000a, 173-74) is that the definition is not novel because it is similar to that advanced by Stephen Toulmin (1958, 101), who used the notion of rebuttal. Johnson replies that his definition is indeed similar to Toulmin's, but has some novelty insofar as it is a generalization of it. I agree that Johnson's definition has considerable originality, vis-à-vis Toulmin's. Thus, we might say that the objection is false. However, I believe another issue needs to be raised here.

That is, is this objection relevant? And if it is, why is it relevant? Neither the dialectical nor the illative conception of argument says anything about novelty. Nor do the traditional or Johnson's theories of assessment. In particular, his definition speaks of the truth of the conclusion, and it is the conclusion's truth that needs to be supported with reasons and dialectically defended from objections; it is not the conclusion's novelty or originality. In short, if this fourth objection were true, namely, if Johnson's definitional conclusion were not novel, why would that be a problem? I do not have an answer to this question, although my intuition tells me that such an objection is relevant, that novelty is important, and so that this is an important question.

The fifth and last objection discussed in Johnson's book is that the problem addressed by the definition is one of "just semantics" (Johnson 2000a, 174-75) because the word argument has many meanings, and so it is arbitrary to choose or invent one particular meaning. Johnson replies that although the problem motivating the definition is in a sense semantical, it is not "just semantics" in the sense of being unimportant. In the course of his reply, Johnson probably admits too much when he concedes some force to this objection and declares that his conception is a stipulative definition. For as Hansen (2002, 272-73) has argued and as Johnson himself (2002a, 313-14) later admitted, Johnson's thesis is really a theoretical definition; that is, a claim that is part of a theory of argument aiming to provide concepts and principles for the identification, understanding, interpretation, analysis, evaluation, and criticism of arguments.

Besides these replies to these five objections, contained in the book *Manifest Rationality*, since the book was published several scholars have advanced other objections

---

[26] Johnson 2000a, 173; cf. Hitchcock 2002a, 289.

and Johnson has replied to them.[27] Thus a complete account or an extended discussion of Johnson's argument would have to include these additional objections and replies. However, my aim here is theoretical rather than historical, and so my discussion of them here will be guided by their relevance to the problem under discussion, namely the adequacy of the dialectical definition of argument and its relative merits vis-à-vis the illative definition.

To begin with, some of the objections raised after publication had been anticipated by Johnson. This applies to the objections that his definition is too restrictive, that the dialectical tier is unnecessary, and that the definition is circular. However, needless to say, after the book's publication, new nuances and clarifications have emerged from these discussions.

In regard to other objections, I wish to reiterate something I stated at the beginning of this chapter. That is, when I introduced the dialectical conception of argument (section 2), I first gave my own formulation, and then I quoted Johnson's definition as an illustration. I also claimed that in giving my own formulation, I had attempted to incorporate the most important clarifications and most telling objections that had emerged from post-publication discussions. One example will have to suffice.

Both Hansen (2002, 269-70) and Hitchcock (Hitchcock 2002a, 289) have independently objected that Johnson's definition, as he words it, states or implies that to be an argument (at least in the paradigm sense of the concept) a discourse must contain *all* (and not just some of) the reasons that support the conclusion; plus *only* those reasons that *actually* support it, as contrasted to those that are intended to support it; plus replies to *all* objections and criticisms, and not just to some; and *only actual* replies, rather that attempted replies. Thus, they have suggested that when Johnson says that an argument presents "the reasons that support it [the conclusion]" (Johnson 2000a, 168) the definition should instead speak of "reasons in support of the conclusion" or "reasons that attempt to support the conclusion." And when he says that in the argument's dialectical tier "the arguer discharges his dialectical obligations" (Johnson 2000a, 168), the definition should instead say that the arguer attempts to discharge some dialectical obligations.

These revisions also take care of the infinite-regress objection advanced by Govier (1998, 8; 1999b, 232-37). This is the objection that if all arguments must have a dialectical tier, then a reply to an objection must also have a dialectical tier, since such a reply is or should be an argument; thus, a reply to an objection to the original conclusion must contain replies to the objections to the reply, and so on *ad infinitum*. Johnson's best reply to this objection involves "pointing out the parallel between the illative core and the dialectical tier. That is, the same line of reasoning that prevents an infinite regress in the illative core can also be deployed to prevent the exfoliation of the dialectical tier" (Johnson 2003, 562).

Johnson does not elaborate. I suppose what he has in mind is the following traditional difficulty: if an argument must have an illative tier containing "the" reasons that support the conclusion, it must contain not only all the reasons that directly support it (which is already an indefinitely long process), but it must also contain all the lower-level reasons that directly support those direct reasons and so indirectly support the original conclusion; but such indirect reasons must themselves be supported by further reasons, and so on *ad infinitum*. This well-known infinite regress is usually stopped by saying that the illative tier need contain only those reasons that seem appropriate in the given context.

The situation with the dialectical tier is analogous: one replies only to those objections that seem appropriate in the context. At the level of the formulation of the definition of

---

[27] Govier (1998; 1999, 223-40; 2000), Groarke 2002, Hansen 2002, Hitchcock 2002a, Leff 2000, Tindale 2002, Rees 2001, Wyatt 2001. Cf. Johnson 2002a; 2003.

argument, elimination of the definite article I believe does the trick: if we say that an argument is an attempt at justification which gives reasons in support of a conclusion or defends it from objections, this clearly means that it gives one or more reasons in support of the conclusion and/or defends it from one or more objections.

4.7 Conclusion: A Moderately Dialectical Conception

To recapitulate, Johnson's argument for the dialectical conception of argument is complex and multi-faceted. It has an illative tier that advances at least three supporting reasons: empirical support reminiscent of an inductive generalization; the argument from the nature of the process of arguing; and the argument from the telos of rational persuasion. Johnson's argument also has a dialectical tier consisting of two main parts, criticism of alternatives and replies to objections. He criticizes the traditional conception of argument in at least four ways: that it conceives argument as having structure but no function; that it fails to distinguish argument from other forms of reasoning, especially explanation; that it presupposes an inadequate conception of inference; and that it fails to give an adequate account of the dialectical nature of argumentation. Furthermore, he replies to at least six objections that allege the following charges: excessive narrowness of scope, dispensability of the dialectical tier, vicious circularity, lack of novelty vis-à-vis Toulmin, triviality beyond semantical issues, and infinite regress.

In light of this interpretation, reconstruction, and analysis of Johnson's argument, it is obvious that it satisfies its own definition. And given the stringent requirements of this definition, this satisfaction represents a considerable merit. In the course of my discussion, I have also assessed, evaluated, and criticized that argument. It would be too tedious to recapitulate these assessments, but it is important to point out that they have been partly negative, unfavorable, and destructive, and partly positive, favorable, and constructive. That is, they have been partly critical (in the ordinary sense of criticism connoting negativity), and partly critical in Johnson's (2000a, 217-23) own sense (connoting fruitful constructiveness). To add a further dimension to such constructiveness and to try to provide a synthetic overview of the forest after our long journey through its trees, I would suggest that the upshot of my assessments is that Johnson's argument is cogent insofar as it justifies the following thesis, and implausible otherwise. The thesis is the claim that an argument is an attempt to justify a conclusion by giving reasons in support of it, or defending it from objections, or both.

I further claim that this is a moderately dialectical conception and that I have provided an argument in favor of this conception. A question now arises. What kind of argument have I provided? Is it self-referentially consistent? That is, does my argument fit my own definition? Is my argument an instance of its own conclusion? I believe it is. It may be viewed primarily as a defense of this moderately dialectical conception by means of criticism of Johnson's alternative strongly dialectical conception. Such a defense would suffice to make it an argument (in the sense of the moderate definition), even though it is obvious that I have not explicitly defended the moderate conception from the other alternatives, namely the hyper dialectical, the illative, the rhetorical, and the purely abstract conception. But I have presumed that in the present context no such defense was needed. If this is correct, this point would further reinforce the moderate conception. On the other hand, my initial remarks about the dialogical model of argument may be seen as an explicit, if summative and sketchy, defense of the same moderate conclusion from the dialogue model.

Moreover, although my moderately dialectical conception of argument does not require every argument to possess an illative tier, my argument may be taken to have such a tier, consisting of two supporting reasons. One is provided by my interpretation, reconstruction, and appropriation of Goldman's truth-in-evidence argument; the other consists of my suggested replacement for Johnson's empirical argument, namely my more systematic version based on the data from Galileo's *Dialogue*.

Finally, does my argument include replies to objections? I have already pointed out that the criticism of an alternative can be conceived as a reply to the objection that there is no reason to prefer the given conclusion to the alternative. This link is, of course, what enables my previous considerations to instantiate the moderately dialectical conception of argument. Furthermore, incidentally and in passing, these considerations also contain replies to possible objections. I would have to admit, however, that so far I have not presented explicit replies to explicit objections. I can also say that I welcome objections, although I do so with some hesitation. In fact, such welcoming leads to a paradox.[28]

The arguer's welcoming of objections is certainly important. And Johnson (2000a, 161, 165) has written eloquent words in this regard. But if this open-mindedness is to be more than a desirable psychological trait, one would have to say that a good argument *should* elicit objections; indeed that an argument is good (in part) if and insofar as it generates objections. An argument should not fall on deaf ears; if it does, more than being a sign of its conclusiveness, it is probably a sign of its sterility. Of course, to be really good, an argument should also have the resources to answer or refute subsequent objections. So it is not really the existence of objections or the possibility of generating them that adds value to an argument. It is its ability to elicit refutable, implausible, or invalid objections. It is these that I welcome with open arms.

To encourage this process, I end by volunteering some of these objections myself. The first one may be formulated as follows. It is undeniable that there is a difference between the conjunctive and the disjunctive dialectical conceptions: it is one thing to say that an argument is an attempt to make a conclusion acceptable by means of both reasons in support of it and replies to objections; and it is another to say that an argument is an attempt to make a conclusion acceptable by means of either reasons in support of it *or* replies to objections. But this is a very small difference: *sub specie aeternitatis*, they are both dialectical conceptions; and even in the less Olympic earlier presentation of various conceptions of argument, both of these conceptions were treated as special cases of the dialectical conception and were regarded as having much more in common than they had differences, especially vis-à-vis the various versions of the traditional conception. So it is unclear what all the fuss is about; the difference is so minor as to approach triviality.

My reply to this objection is that even the eternal gods who view these matters from Olympus need to cultivate their powers of discrimination and their ability to make fine distinctions. So there is no good reason to ignore the difference between conjunction and inclusive disjunction. The important point is that differences should not be magnified or exaggerated, but the other side of this coin is that they should not be minimized or underplayed. However, such balance has been precisely what I have stressed in this chapter, by beginning to point out that these two definitions were versions of the dialectical approach, and by ending with the conclusion that the disjunctive conception seems to be preferable to the conjunctive one.

A second objection to my argument involves Johnson's notion of manifest rationality, by which he means the attempt to not only be rational, but also to look and appear rational. The objection would allege that in my reconstruction of his argument I have ignored the

---

[28] This paradox is a version of the problem discussed by Johnson (2000a, 223-36) in connection with his principle of vulnerability and his argument that no argument is conclusive. Mill (1965, 293-95 = *On Liberty*, ch. 2, par. 31-33) also discusses a version of this problem. Cf. chapter 10 below.

following element of its illative tier: that an argument must have a dialectical tier because "argumentation [is] more than just an exercise in rationality" (Johnson 2000a, 163), it is also an exercise in manifest rationality; and "manifest rationality is why the arguer is obligated to respond to objections and criticisms from others" (Johnson 2000a, 163-64). This argument is hard to miss since it is being referred to in the title of Johnson's book; since it is given in the body of the work when the idea of manifest rationality is explicitly discussed (Johnson 2000a, 163-64); and since elsewhere (Johnson 2000b, 3) he explicitly presents it as an additional line of justification, besides the argument from the telos of rational persuasion. Moreover, the argument is apparently important because if it is cogent it would justify the strongly dialectical conception of argument, but not the moderate definition.

My response to this objection starts with a criticism of the argument from manifest rationality. In regard to its premise that argument is an exercise in manifest rationality, it may be deemed acceptable, but I do wonder whether its acceptability exceeds that of the conclusion. Moreover, I agree that the premise is more or less relevant, but I question whether it is sufficient. One reason for questioning its sufficiency is that, as we saw in the case of the argument from the telos of rational persuasion, an additional linked premise is needed, namely a proposition about the controversial origin of argumentation; otherwise, as proofs in Euclidean geometry suggest, rationality can be achieved without replying to objections.

A more specific and important reason for questioning the sufficiency of the present premise of manifest rationality is that it is unclear that it really adds anything to the argument from the telos of rational persuasion. Johnson's key point here seems to be that whereas rationality as such might be taken to require that one answer only objections that are really relevant, manifest rationality requires that one answer objections that appear to be relevant, even if in reality they are not. But I would point out that the inclusion of apparent, as distinct from real objections, is required by the telos of rational persuasion, for it would not be persuasive to neglect objections that are believed (even if incorrectly) to be forceful. In other words, in Johnson's own account the two operative notions are manifest rationality and rational persuasion, and these seem to me to be two sides of the same coin, rather than two distinct concepts. Although similar considerations have led Hitchcock and Hansen[29] to conclude that Johnson's notion of manifest rationality is rhetorical after all, my own conclusion here is that this argument from manifest rationality has no force above and beyond the argument from the telos of rational persuasion.

There is another conclusion I would want to draw. I originally did not include this supporting reason in my reconstruction of the illative tier of Johnson's argument because I judged it to be devoid of the additional force just mentioned. In doing so, I was operating, I believe, from the point of view of strict rationality, as distinct from manifest rationality. I was telling myself that, appearances to the contrary, and despite Johnson's own explicit statements to the contrary, he really had no distinct argument; so also using the principle of charity, I decided it was better to neglect these considerations, rather than interpreting them as an argument and then criticizing the argument as worthless. In that sense and to that extent, my previous neglect was justified, and hence the present objection has no force.

However, in discussing this objection now, I was taking the point of view of manifest rationality at the metalevel. That is, given all the appearances (at the ground level) that manifest rationality is one of Johnson's reasons in support of the dialectical tier, I explored whether these appearances correspond to reality; whether this reason, besides being meant to support his conclusion, does really support it. If and insofar as my doing so has added to

---

[29] Hitchcock 2002a, 7; Hansen 2002, 273-74; cf. the reply in Johnson 2002a, 327-29.

the persuasive force of my own argument, then perhaps I have come closer to achieving my present aim of rationally persuading you that the moderate conception is preferable. Furthermore, such rational persuasion has perhaps been enhanced by the fact that I explicitly included at least one objection which for me did not have even prima facie plausibility, but which might be plausibly advanced by Johnson or anyone taking his point of view; and this in turn may be taken to enhance the practical value of manifest rationality, practical in the sense of the practice of argumentation, even if it remains true, as I would hold, that theoretically speaking manifest rationality is an aspect of rational persuasion.

A third objection now comes to mind. It is one from the point of view of the illative conception,[30] and it is based on my criticism of Johnson's criticism of this conception. That is, if it is inaccurate to object, as Johnson does, that the illative conception lacks a teleological aspect and fails to attribute a telos to argument; and if it is inaccurate or unfair to object that the illative conception fails to properly distinguish between argument and other forms of reasoning such as explanation and inference; and if it is circular to object that the illative conception ignores the dialectical nature of argument; then does it not seem that the illative conception can survive Johnson's criticism, and is perhaps adequate? In other words, even if my critique of Johnson's illative tier and of his explicit replies to explicit objections shows that my moderate conception is preferable to his strong one, it does not show that the moderate conception is superior to the illative definition, especially when we recall my criticism of Johnson's criticism of the illative conception. Does not my argument need one other component, namely a criticism of the alternative illative position?

Part of my answer to this objection lies in stressing two arguments discussed earlier: my reconstruction or appropriation of what I have called Goldman's truth-in-evidence argument and of what I have called Johnson's empirical inductive generalization. Insofar as they support my moderate version of the dialectical conception, they do not support the illative conception. I would also want to exploit my claim about the symmetry of the illative and dialectical tiers; insofar as that claim is correct, it suggests that even if the illative conception were otherwise acceptable, the moderately dialectical re-description of the situation would be more encompassing and therefore better.

And this suggests that perhaps the best line of defense here is to question whether the illative definition of argument is really an alternative to the moderately dialectical conception. One reason to see that they are not really alternatives is to stress that the moderately dialectical definition as I phrased it and am defending it here at the end of my discussion speaks of justification rather than persuasion. This is a revision of the dialectical conception required by the difficulties that emerged when I discussed Johnson's argument from the nature of the process of arguing; that argument could not be regarded as a successful attempt at persuasion (since from the point of view of persuasion it may be

---

[30] Analogously, another objection could be raised from the point of view of that I have called the hyper dialectical conception. This is the view that an argument is a defense of its conclusion from actual or potential objections. Accordingly, one could object that the moderate definition is a step in a direction opposite to what is needed: presumably one should further strengthen Johnson's strong conception rather than weakening it into the moderate conception. And the important reason to take a step toward the hyper conception is the following: given any claim that has been asserted, one could always raise the question, what reasons if any there are in support of the claim; this question may be regarded as the prime or minimal objection to any claim; if one anticipates it, one constructs the illative tier and gives the supporting reasons even before the objection has actually been raised; or one can wait until after the objection has been explicitly raised; in either case, the illative component can be interpreted as a part of the dialectical tier. The answer to this objection will be given when I discuss the hyper dialectical definition (in chapter 5), the key point being the symmetry between the dialectical and illative tiers.

begging the question), but it was better regarded as a justification (insofar as it is an analytical conceptual elucidation in the context of theorizing). This move from persuasion to justification raises the question of the precise relationship between these two notions, but this issue cannot be elaborated here; suffice it to say that I conceive justification partly as a weakening of the notion of proof, and partly as a requirement for rational persuasion. Thus my moderate conception incorporates an element of the illative definition.

To be sure, they are not identical, but they are not inconsistent either. In a sense, the illative conception entails the moderately dialectical one, since the latter is a disjunction of which the former is the first disjunct. But although this entailment ensures their formal consistency, it perhaps points in a misleading direction because it suggests that the moderate conception is a special case of the illative one, whereas the opposite is more nearly correct. That is, the moderate conception is more general than the illative one, in the sense that it can subsume under itself all the particular instances that the illative conception subsumes, plus others that the illative conception does not subsume. If then, the proper relationship between the moderately dialectical and the illative conceptions is not one of competition or inconsistency, but rather compatibility or species and genus, then in arguing for the moderately dialectical conception one need not reject the illative definition, but one should rather incorporate it.

In short, the moderately dialectical conception of argument may be seen as a synthesis of the illative conception and of the strongly dialectical definition. And if the latter two alternatives are dubbed thesis and antithesis, then I may be allowed to end my analysis on this dialectical note, in the Hegelian sense of dialectic. And this ending adds suggestiveness to my initial remark (in section 4.1) that, at a higher level of approximation or deeper level of analysis, the Hegelian notion may turn out to be not altogether irrelevant.[31]

4.8 Summary

This chapter began by distinguishing several distinct notions of dialectic. Then it proceeded to a comparison and contrast of the illative and the dialectical definitions of argument. I distinguished three versions of the latter: a moderate conception for which the dialectical tier is sufficient but not necessary; a strong conception for which the dialectical tier is necessary but not sufficient; and an hyper conception for which the dialectical tier is necessary and sufficient. Then I reconstructed, analyzed, and evaluated Goldman's argument for the moderate conception based on the truth-in-evidence principle and Johnson's multifaceted argument for the strong conception. Johnson's argument contains an illative tier of three supporting reasons, and a dialectical tier consisting of four criticisms of the illative conception and replies to six objections. The result of my critical analysis is an argument for the conclusion that the moderate conception is correct (i.e., that an argument is an attempt to justify a conclusion by supporting it with reasons, or defending it from objections, or both); my argument contains supporting reasons appropriated from the acceptable parts of Goldman's and Johnson's arguments, and criticism of the alternative strong conception. I ended by defending the moderate conception from three objections.

---

[31] I am of course referring to the Hegelian view of dialectic as the synthesis of a thesis and an antithesis. My reference is made partly in jest, for I am aware that it is questionable whether this triadic interpretation of Hegelian dialectic is anything more than a vulgar oversimplification and has anything more to do with the dialectical philosophy of Hegel than the *terza rima* has anything to do with the poetry and art of Dante's *Divine Comedy*. Cf. Findlay 1964, 353; Finocchiaro 1988, 183.

Chapter 5
The Hyper Dialectical Definition of Argument

In the last chapter, the hyper dialectical conception of argument emerged as one out of four possible theoretical definitions of argument that can be formulated in terms of the presence or absence of supporting reasons (illative tier) and replies to objections (dialectical tier). The focus there was the comparative evaluation of the strongly and moderately dialectical definitions, that is to say, the meta-arguments for the strong definition in the context of those for the moderate conception. It is now time to examine the hyper dialectical definition by doing something analogous in its regard.

5.1 Eemeren's Hyper Dialectical Conception

In various works, Frans van Eemeren and his followers in the pragma-dialectical school of argumentation studies have advanced a conception of argument that deserves more explicit discussion than it has received. One formulation of this conception asserts that "in principle, every single argumentative move serves to remove some form of doubt that the antagonist may have with regard to the standpoint" (Eemeren and Grootendorst 1992, 73). Similarly, another member of the pragma-dialectical school has stated that "in a dialogical approach, every argument is regarded as a means to overcome some form of doubt or criticism. Sometimes this doubt or criticism is left implicit by both parties so that it must be inferred from the arguments that are advanced. But if the arguer explicitly mentions the criticism to which he is reacting in his arguments, the analyst has much more to go on" (Snoeck Henkemans 1992, 179).

These formulations embody a very strong claim, as is suggested by their syntactical form of universal generalizations. Indeed, the claim is even stronger than that; for the first formulation adds the qualification that this is so as a matter of principle, while the second formulation adds the qualification that the doubt or criticism can be implicit as well as explicit. So the claim seems to be that the removal of doubt or criticism is a property of all arguments that does not just happen to belong to them, but that must belong to them as a matter of definitional necessity.

The definitional status of this claim has been made explicit by a third representative of the pragma-dialectical school, van Rees (2001). The context of her discussion was that of a review of Johnson's (2000a) theory of argument. One of her criticisms is directed at Johnson's definition of argument, which reads: "an argument is a type of discourse or text—the distillate of the practice of argumentation—in which the arguer seeks to persuade the Other(s) of the truth of a thesis by producing the reasons that support it. In addition to this illative core, an argument possesses a dialectical tier in which the arguer discharges his dialectical obligations" (Johnson 2000a, 168). Van Rees's objection is that "according to this definition, producing reasons and discharging one's dialectical obligations are two different things. But in actual fact, if the notion of argument is indeed to be rooted in the dialectical practice of argumentation, the two should coincide. In a truly dialectical account, argument *per se* would be defined as an attempt to meet the critical reactions of an antagonist, that is, to take away anticipated objections and doubt" (Rees 2001, 233).

I do *not* interpret van Rees's objection as amounting to the criticism that Johnson's definition is not "really" dialectical because the only "really" dialectical definition is the pragma-dialectical one. Rather I interpret it as the criticism that Johnson's definition is insufficiently dialectical and that her proposed definition is more highly or deeply dialectical. For Johnson defines an argument in terms of both an illative component of reasons supporting the conclusion and a dialectical tier of replies to objections; so for him the dialectical component is just part of an argument, a necessary part to be sure, but only one of two parts, the other part being the illative tier. Whereas Van Rees is proposing that an argument be defined in terms of just the dialectical tier; that is, the dialectical tier is not only necessary, but also sufficient, to have an argument.

Now, it is important to note that Johnson's definition is already more dialectical than some others, i.e., that he gives the dialectical tier more importance than other definitions do. In fact, as we saw in the last chapter, there are other definitions that give the dialectical tier a role, without making it necessary. For example, according to Goldman (1999, 131), "if a speaker presents an argument to an audience in which he asserts and defends the conclusion by appeal to the premises, I call this activity *argumentation*. More specifically, this counts as *monological* argumentation, a stretch of argumentation with a single speaker. … I shall also discuss *dialogical* argumentation in which two or more speakers discourse with one another, taking opposite sides of the issue over the truth of the conclusion." This amounts to claiming that answering doubts or criticisms is a sufficient but not necessary condition for an argument, and that the illative component is another sufficient but not necessary condition. Clearly, Johnson's definition is more strongly dialectical that Goldman's, and so I find it inappropriate to deny dialectical status to it. Similarly, Goldman's definition should not be denied dialectical status either, because it does take into account the dialectical tier and may be contrasted to other definitions that do not. These are the purely illative definitions, such as the following one advanced by Copi and Cohen (1994, 5): "an argument, in the logician's sense, is any group of propositions of which one is claimed to follow from the others, which are regarded as providing support or grounds for the truth of that one."

In my framework, the situation is as follows. One type of definition conceives an argument as just an attempt to support a conclusion with reasons. Besides Copi, this conception corresponds to the definitions of many other authors. Adopting a term from Johnson (2000a, 150), who adopted it from Blair (1995), we may call the supporting reasons the *illative tier* (not the illative *core*), and then we may call this the purely illative definition of argument, or the illative definition, for short; here the illative tier is a necessary and sufficient condition to have an argument.

Another type of conception defines an argument as an attempt to justify a claim by supporting it with reasons, defending it from objections, or both. Besides Goldman, such a definition has also been advanced by Reed (2000, 1). In the last chapter, I elaborated and justified it. And it may also be attributed to John Stuart Mill (1997, 68-69; *On Liberty*, ch. ii, paragraph 23), as I shall argue later (chapter 10). If we call the defense from objections the dialectical tier, then we may regard this as a *moderately* dialectical definition, in which the dialectical tier as well as the illative tier are each a sufficient but not necessary condition to have an argument.

A third type conceives an argument as an attempt to justify a claim by supporting it with reasons *and* defending it from objections. Johnson's definition is the best known example of this type. Here the two tiers are individually necessary and jointly sufficient conditions for an argument, but neither is individually sufficient; this may be called the strongly dialectical definition of argument.

Finally, there is a fourth type, of which Eemeren's pragma-dialectical definition is the best known example, according to which an argument is just an attempt to justify a claim by defending it from objections. Here the dialectical tier is both necessary and sufficient to have an argument, whereas the illative tier is neither necessary nor sufficient; a natural label I used before (chapter 4, above) is the *hyper* dialectical definition of argument.

5.2 The Pragma-dialectical Argument

The next question I want to discuss is, why should we define an argument the way Eemeren does? How is the hyper dialectical definition justified? What arguments does the pragma-dialectical school of argumentation give for their hyper dialectical definition?

Such arguments are not easy to find. For example, let us search in Eemeren and Grootendorst's (1992) work. If we begin with the first quotation above which I took as a pragma-dialectical formulation of the hyper dialectical definition (Eemeren and Grootendorst 1992, 73), we find that the nearby passages contain no argument for this claim. In that context, the claim is part of a recapitulation of previous theoretical claims for the purpose of tackling a new topic, that of complex argumentation.

In fact, if we examine other parts of the same work, we do find other formulations of the hyper dialectical definition. For example, in the first chapter we are told that "dialectification is achieved by treating argumentation as a rational means to convince a critical opponent and not as mere persuasion. The dispute should not just be terminated, no matter how, but resolved by methodically overcoming the doubts of a rational judge in a well-regulated critical discussion" (Eemeren and Grootendorst 1992, 10-11). However, there is no argument. Here, this claim is just one of four that articulate various aspects of the pragma-dialectical approach, the other three being claims about externalization, functionalization, and socialization.

If we examine their later work, we find that "in ... argumentation ... protagonists advance their arguments for their standpoints that are intended to systematically overcome the antagonist's doubts or to refute the critical reactions given by the antagonist" (Eemeren and Grootendorst 2004, 61; cf. 1992, 35). This is a definition of the "argumentation stage," and so again we have a variant of the hyper dialectical definition. But, again there is no justifying argument in sight. Instead, this is one of four claims, the other three being about the confrontation, opening, and concluding stages of a critical discussion.

Thus it may appear that pragma-dialectical theorists do not argue for their definition of argument; that they do not engage in argumentation when it comes to defending their own conception of argument. This would be ironic, for it would mean that, for all their emphasis on argumentation as the subject matter of their theorizing, they do not themselves engage in argumentation in their own theorizing. And in turn this would suggest that pragma-dialectical theorists do not practice what they preach. This would be serious criticism if it were accurate, *pace* the fallacy theorists who still believe that to criticize someone for incoherence in the theory-practice relationship is a variant of the *ad hominem* fallacy. But is the criticism accurate? That it is not completely accurate may be seen by focusing on a clue contained in what we have already examined.

We have seen that in each of the three places where we located a formulation of the hyper dialectical definition there was no justifying argument, but that the last two passages occurred in contexts where this definition was being stated together with several other theoretical claims. Now, it is a well-known phenomenon in the epistemology of scientific theorizing that normally theoretical claims cannot be confirmed individually or in isolation, but only collectively or holistically or in conjunction with others. For example, consider such a scientific achievement as Isaac Newton's *Mathematical Principles of Natural Philosophy* (1687). One of its basic theoretical claims is the so-called first law of motion or law of inertia. If we were to examine this work to locate arguments directly justifying this principle, we would be as disappointed as for the case of the hyper dialectical definition in Eemeren's and Grootendorst's works. Instead, we would find that the law of inertia must be combined with the law of force (the second law of motion), with the law of action and reaction (third law), and with the law of universal gravitation in order to yields those explanations that justify them, that is explanations of such established facts as Kepler's laws of planetary motion, Galileo's law of falling bodies, and the tides. Similarly, for the case of the pragma-dialectical theory, we should not expect that the hyper dialectical definition can be justified by means of simple direct arguments; perhaps we should look for holistic indirect arguments in which that definition is combined with other theoretical claims and then the whole system used to yield analyses of various facts of argumentative practice.

With these considerations in mind, let us see what we can find. At the end of chapter four of Eemeren and Grootendorst's (1992) work, they assert that

argumentative discourse can, in principle, always be dialectically analyzed, even if it concerns a discursive text that, at first sight, appears to be a monologue. The monologue is then, at least partially reconstructed as a critical discussion. The argumentative parts are identified as belonging to the argumentation stage and other parts as belonging to the confrontation, the opening or the concluding stage. Usually, this reconstruction is not so much of a problem as it may seem. A speaker or writer who wants to convince his audience should, after all, make it clear ... that he intends to offer ... arguments in favor of his standpoint that overcome all critical doubt ... [Eemeren and Grootendorst 1992, 42-43]

Here we do have, along with a restatement of the hyper dialectical claim, what I would call an argument-sketch: the programmatic claim that if we apply the various principles of the pragma-dialectical theory, we can show that all argumentative discourse consists of attempts to overcome critical doubts; namely that, appearances to the contrary, all actual arguments can be shown to conform to the hyper dialectical conception. This is a programmatic claim in the sense that it describes a research program for argumentation theorists to conduct this kind of analysis in regard to actually occurring argumentation. Such research would presumably yield the empirical claim that all actual arguments are indeed attempts to remove the doubts and criticism of real or potential opponents; and so what we have here is an empirical argument-sketch. Finally, the argument is also indirect, theoretical, and holistic because such an empirical claim would be the result of the application of the hyper dialectical claim together with others, such as the pragma-dialectical principles of externalization, functionalization, and socialization, and the definitions of confrontation, opening, and concluding stages.

In accordance with this theoretical-empirical research program, i.e., in an attempt to give content and substance to this argument-sketch, the pragma-dialectical school has been, and continues to be, working hard to produce the relevant analyses. Even when the empirical material analyzed is relatively uninteresting, it is undeniable that the pragma-dialectical analysis provides some support in favor of their theory (e.g., Eemeren, Grootendorst, Jackson, and Jacobs 1993). But when the material studied is intrinsically interesting and important, then such analyses are doubly welcome. For example, there are indications that Eemeren and Houtlosser (1999b; 2000; 2003b) have been working on the historically important case study involving the apologia of William of Orange (1581) against the proclamation of Philip II, King of Spain, banning him (William) from The Netherlands. More generally, the pragma-dialectical analysis of fallacies (Eemeren and Grootendorst 2004, 162-86; 1992, 93-207) could be regarded as another instance of providing content and substance to the argument-sketch under discussion. Another important such instantiation is that provided by Snoeck Henkemans's (1992) analysis of complex argumentation.

However, at this point the discussion becomes more complex in another dimension. That is, on the one hand, we have seen that the perceived paucity of (direct) argumentation for the hyper dialectical definition that we find in the pragma-dialectical theory can be corrected first by the realization that here we are dealing with a systematic theory whose arguments are necessarily indirect and holistic, and second by the undeniable fact that pragma-dialectical theorists have tried and are trying to provide the relevant analyses of actual, empirical argumentative practice. On the other hand, my point now is that the cogency of such a theoretical, indirect, empirical argument is dependent on more than just producing such analyses, or even on showing that they are good and enlightening analyses. The additional requirement is that the pragma-dialectical analyses of such phenomena should be better and more enlightening than other analyses of the same phenomena. For the problem at hand, this issue reduces to the question of whether and why the hyper dialectical analysis is better that the strongly dialectical analysis, or the moderately dialectical one, or the illative one; for these are the four alternatives in this context.

In other words, it may very well be that the reconstruction of argumentative discourse accomplished in the project of Eemeren, Grootendorst, Jackson, and Jacobs (1993) provides empirical support for the pragma-dialectical theory, including the hyper dialectical definition. It may also be that more empirical support is provided by Eemeren and Houtlosser's analysis of William of Orange's argument against the King of Spain; by Eemeren and Grootendorst's analysis of fallacies; and by Snoeck Henkemans's analysis of complex argumentation. Now, from the point of view of the illative definition of argument (perhaps with some holistic qualifications) such empirical support generates supporting reasons for the hyper dialectical definition, and hence an argument in its favor. The same holds from the viewpoint of the moderately dialectical definition. However, for the strongly dialectical definition, and even more so for the hyper dialectical definition, such empirical support per se is not even an argument, let alone a good argument; it is at best an incomplete argument. From the point of view of the hyper dialectical definition, what is required is at least to "translate" such empirical support into something that looks like the removal of doubts, criticism, or objections by opponents. That is, such empirical support must be given a "dialectical garb," to use Barth and Krabbe's (1982) expression. In short, by their own criteria, the analyses which pragma-dialectical theorists have advanced in favor of the hyper dialectical definition are either not arguments, or not good arguments, or nor complete arguments.

5.3 The Symmetry of the Dialectical and Illative Tiers

At this point, one could say that the answer to this latest difficulty is easily forthcoming, indeed that it is already implicit in the progression of this very chapter. For the support provided by the pragma-dialectical analyses of empirical material has emerged as an answer to the questions, doubts, criticism, and misgivings which this writer was expressing toward the hyper dialectical definition of argument. And so such pragma-dialectical analyses, although they may *appear* to be illative monological exercises, can be easily reconstructed as being dialectical replies to the objections of an antagonist.

What's more, this is true in general. As suggested earlier (chapter 4.7), given any claim, one could always raise the question, what reasons if any there are in support of the claim. This question may be regarded as the prime or minimal objection to any claim. If one anticipates it, one constructs the illative tier and gives the supporting reasons even before the objection has actually been raised. Or one can wait until after the objection has been explicitly raised. In either case, the illative component can be interpreted as a part of the dialectical tier.

However, I do not think this establishes the exclusivity, or even the primacy, of the dialectical tier. For in a way analogous to how a dialectical theorist attempts to interpret the illative component in terms of the dialectical tier, a proponent of the illative definition can perhaps try to do the reverse and reinterpret the dialectical tier in terms of the illative component. The general argument would be the following.

Consider an argument whose illative component consists of premises P-1 through P-n and conclusion C. And suppose the argument also has a dialectical tier with objections O-1 through O-k, respectively answered by replies R-1 through R-k. Now consider the conjunction of an objection and its corresponding reply, (O-j & R-j), or some appropriately reworded phrasing of it that might be needed for grammatical propriety. It seems to me that such a conjunction would constitute a reason supporting the conclusion C. It would be like saying that one reason for accepting the conclusion is that if one objects to it in such and such a way, such an objection would be incorrect;[1] or collectively considered, one reason for accepting conclusion C is that all objections against it fail, i.e. that there are no objections to it. In other words, an objection to a conclusion C may be seen as a reason against it, a reason for claiming not-C; and if a reason R for not-C is a bad reason, then the claim that R is a bad reason for not-C, may be seen as a reason for C. Of course, such a reason would not be a conclusive reason, and to claim such conclusiveness would be to commit a damaging version of the fallacy from ignorance. But we are clearly dealing with reasons that, however strong, fall short of conclusiveness, and for such cases the explicit refutation of an explicit objection may be viewed as a supporting reason.

The upshot of these considerations is that while the presentation of supporting reasons may be regarded as a reply to a weak or minimal objection, the refutation of objections may be regarded as a weak or minimal supporting reason. There thus seems to be a symmetry between the illative and the dialectical tiers. Applied to the present issue, this symmetry implies that although the illative empirical support for the hyper dialectical definition of argument can be interpreted to constitute a dialectical defense as well, and thus to satisfy this very definition, the hyper dialectical definition is itself conceptually indistinguishable from the purely illative definition; their differences are either purely linguistic or two sides of the same coin.

More generally, this symmetry also strengthens the moderately dialectical definition of argument, since it answers the objection (formulated and left unanswered in chapter 4.7) that the illative tier of an argument can be interpreted as part of the dialectical tier. The answer now is that the dialectical tier can also be interpreted as part of the illative tier. In other words, the symmetry also provides an additional reason for giving the illative and the dialectical tiers distinct and equal roles, as the moderate definition does with its inclusive disjunction. And we also have another reason for dropping the talk of illative *core* and speaking of illative *tier*, as suggested earlier (in chapter 4.2).

5.4 Another Argument

Although, as indicated above, the problem of the formulation and justification of the hyper dialectical definition of argument emerges naturally within the pragma-dialectical theory, it also emerges in other ways in other scholarly contexts. Especially after the impasse just reached, and while waiting for a possible reply from the pragma-dialectical school, it will be valuable to examine some of those other contexts.

In a previous work, the present author has also formulated and defended the hyper dialectical definition of argument. One of the arguments is worth quoting verbatim:

---

[1] An instance of this is the subargument from Q and R to F in the extended example discussed in chapter 3.4 above. This issue also relates to the problem, discussed in chapter 8 below, of how to conceive counterconsiderations or con reasons in a conductive argument.

The basic purpose of an argument is to make a proposition more acceptable than it would be in the absence of the argument. This means that without the argument its conclusion would be less acceptable. But how could the acceptability of a proposition be less, when there is no supporting argument? Only if there are present certain objections to the proposition; the argument is then an attempt to remove such objections. We may then say that an argument is a defense of its conclusion from actual or potential objections. No argument would be needed if there were no need to make the conclusion more acceptable, i.e. if there were no previous objections. Thus the objections are prior, and objections are nothing but critical arguments. [Finocchiaro 1980a, 419]

The fifth sentence in this passage is a clear and explicit statement of the hyper dialectical definition of argument. It may be regarded as the final conclusion of the argument contained in this passage. A final, unsupported premise is a proposition (expressed in the first sentence) which is being adopted from Hamblin (1970, 245); this embodies a conception of the purpose of argument in terms of increasing the acceptability of the conclusion. An intermediate proposition here is unstated in this passage, it being an immediate and obvious consequence of the last sentence, namely that critical arguments are prior to constructive ones. This proposition is being adopted from Johnstone (1959, 82); here, by critical argument is meant one whose conclusion explicitly expresses a negative evaluation of some proposition or argument, whereas a constructive argument is one whose conclusion is explicitly about some event or object in the world. Johnstone's point is that apparently "constructive" arguments are really critical, i.e., justify their conclusion by criticizing objections and counterarguments. Thus, historically and "dialectically" speaking, this argument was also an attempt to combine the ideas of these two pioneers of recent argumentation theory.

Now besides this argument, there is a second, much longer, and very different one advanced in that previous work of mine. This second argument is essentially an inductive one, indeed an inductive generalization. The sample of arguments consists of those in Galileo's *Dialogue on the Two Chief World Systems, Ptolemaic and Copernican.* Galileo's arguments are examined in great detail with an eye toward deriving or testing useful claims in the theory of argument. It turns out that in the sample, by far "most arguments are critical, rather than constructive" (Finocchiaro 1980a, 418; cf. 2005a, 43-44); this provides empirical support for the above-mentioned Johnstonian thesis. Then from this thesis, via the argument quoted above, one arrives at the hyper dialectical definition of argument.

Now, modesty and potential embarrassment prevent me from evaluating these arguments. Instead it will be more instructive to try to locate other possible arguments for the hyper dialectical conception.

5.5 Mill's Hyper Dialectical Argumentative Practice?

One source that could be explored is John Stuart Mill. I believe this exploration will be instructive despite the fact that he appears to subscribe to what I call a moderately dialectical, rather than strongly or hyper dialectical definition (cf. chapter 10, below). Nevertheless, I would begin by saying that, however "moderately dialectical" he may be according to my terminology used here, it is undeniable that he preaches and practices a dialectical approach. My second consideration is that it is generally acknowledged that Mill is pursuing certain ideas quite single-mindedly, passionately, coherently, so much so that we would expect him to articulate a "hyper" position on the subject. Third, he is the author of a very revealing and emblematic remark, made in his essay *On Liberty*, that "when we turn to subjects [such as] morals, religion, politics, social relations, and the business of life, three-fourths of the arguments for every disputed opinion consist in dispelling the appearances which favor some opinion different from it" (Mill 1997, 69). Now, although this claim, even if true, would not prove the hyper dialectical definition without further ado, the hyper dialectical implications of it are obvious: it would show that what I have found to be the case for the arguments in Galileo's *Dialogue* is relatively common; in other words, if the hyper dialectical definition applies in 75% of the cases, then it is more promising than it appears at first. In line with this remark, we could then examine an instructive example of such an explicitly hyper dialectical argument mentioned by Reed and Long (1998, 3), namely Turing's (1950) attempt to show that machines can think based primarily on a critique of nine objections to this conclusion.

However, an even better example may be Mill's own argument in the first chapter of *The Subjection of Women*, if Hansen's (2005) suggestion is correct that Mill's argumentative practice there corresponds to the hyper dialectical conception. According to Hansen (2005, 7),

in *The Subjection of Women* ... chapter 1 serves double duty: it is first a general introduction to the work, an essay which aims to show that 'the legal subordination of one sex to the other' is wrong ... Secondly, ... Mill imagines a presumption will be claimed in favour of the existing social relationship between the sexes because it is the *status quo*; this means that the burden of proof will fall on anyone who wishes to change the system ... Mill ... accepts the burden ... The way the argumentation unfolds in the heart of this chapter is that Mill is proposing the thesis that the present arrangement between the sexes is unjustified, and he answers objections to this ... In this chapter Mill does not present any particular argument in favour of his thesis. What he has done is to consider a number of objections to the thesis, and then replied to them, one by one.

Before testing this interpretive hypothesis, let us note that in this last sentence Hansen seems to betray a commitment to either the illative or the strongly dialectical conception of argument. We need not share this presupposition, for from the point of view of the moderately or hyper dialectical conceptions, if Mill considers and then replies to several objections to his thesis, even without offering any constructive, illative, supporting reasons, he is thereby advancing an argument for his thesis.

However, I believe that an examination of Mill's *Subjection of Women* reveals that neither his argumentative practice nor his meta-argumentative reflections correspond to the hyper dialectical conception. Mill's book tries to justify the claim that "the principle which regulates the existing social relations between the two sexes—the legal subordination of one sex to the other—is wrong in itself, and now one of the chief hindrances of human improvement; and that it ought to be replaced by a principle of perfect equality, admitting no power or privilege on the one side, nor disability on the other" (Mill 1997, 133). He states that for a satisfactory justification of this claim, he has to do the following:

I should be expected not only to answer all that has ever been said by those who take the other side of the question, but to imagine all that could be said by them ... and besides refuting all the arguments for the affirmative, I shall be called upon for invincible positive arguments to prove a negative. And even if I could do all this, and leave the opposite party with a host of unanswered arguments against them, and not a single unrefuted argument on their side, I should have thought to have done little; for a cause supported on the one hand by universal usage, and on the other by so great a preponderance of popular sentiment, is supposed to have a presumption in its favor, superior to any conviction which an appeal to reason has power to produce in any intellects but those of a high class. [Mill 1997, 134]

Mill sees himself as holding the negative side of the question whether the legal subordination of women to men is right; but what he holds can also be stated affirmatively, at least grammatically speaking, namely that the subjection of women is wrong. He feels bound to abide by what he would regard as the usual requirement; that is to refute or criticize all arguments against or objections to his own claim, and to provide unanswerable supporting reasons in its favor. And in most of the essay he does that. For example, in the last part of chapter 1 (Mill 1997, 149-55), he tries to refute the objection that the subordination of women is not wrong because it corresponds to the nature of men and women. Then in chapter 2, he tries to show that the subordination of women is wrong because it produces considerable evil and harm in marriage. In chapter 3, he argues that it is wrong because of its harmful effect in the public sphere of employment and citizenship. In chapter 4, he gives reasons for thinking that it is wrong because its abolition would result in considerable benefits and advantages in marriage, social relations, and the psychological well being of women. So far, Mill is being moderately dialectical in regard to the *definition* of argument, but quite inclusive in regard to what's needed to *justify* one's conclusion. Note for example, that he thinks it is important to imagine or invent counterarguments or objections to one's one conclusion. However, in this case something more radical is needed.

In this case all these things are insufficient because, for Mill, the belief in the subjection of women is not the result of argumentation but rather of non-rational causes. Mill mentions the two chief ones, custom and emotion. In such cases, an opinion "gains rather than loses in stability by having a preponderating weight of argument against it" (Mill 1997, 133), and so what is also needed is to undermine those non-rational causes. Part of this undermining involves exposing them. But perhaps more is needed after exposure, partly rational criticism of such causes, and partly counter-acting or counter-balancing such causes by others. In fact, most of chapter 1 (Mill 1997, 135-49) deals with this task of exposure and counter-balancing. The major counter-balancing factor discussed by Mill is the modern progressive tendency for greater and greater equality.

This is an extremely interesting and provocative idea, but a fuller discussion is beyond the scope of this chapter and will have to wait until chapter 11. In part Mill is pointing out a limitation of rationality and argumentation. In part he is proposing a way to deal with this fact of cultural life. His proposal requires that in some cases, besides rational argumentation we need what might be called causal explanation and causal intervention. So he is making the task harder for the practitioner and theorist of argumentation. In that sense he is advocating the kind of radical and uncompromising position he is famous for. Although such a position could be called hyper according to the general dictionary sense of this word, it is not hyper in the sense defined in this chapter. That is, although he does not seem to hold a hyper dialectical definition of argument, he seems to hold a very stringent principle for the justification of conclusions, so laden with what might be called "dialectical" obligations that it would not be improper to label it a hyper dialectical principle of justification.

5.6 Summary

To summarize, the hyper dialectical definition of argument claims that an argument is an attempt to remove the doubts, criticisms, or objections of an opponent. This claim can be attributed to Frans van Eemeren and the pragma-dialectical school generally. This theoretical definition can be contrasted to other conceptions, such as Johnson's strongly dialectical, Goldman's moderately dialectical, and Copi's purely illative definitions. I then discussed the pragma-dialectical argument for the hyper dialectical definition; although difficult to identify, such an argument can be reconstructed. This pragma-dialectical argument was defended from one possible criticism, but also seemed to face the insuperable objection that the various analyses which pragma-dialectical theorists advance to support their definition do not show it is preferable to all alternatives. A general argument was then advanced for the unique superiority of the hyper definition over the others, but apparently it fails because of the symmetry between supporting reasons and replies to objections. Another argument was then presented, retrieved from a 1980 work by the present author, based on theses advanced by Hamblin and Johnstone and on an empirical study Galileo's arguments in his *Dialogue*. Finally, I examined Mill's argumentative practice and meta-argumentative reflections in *The Subjection of Women*, to see whether they conform to the hyper dialectical definition; it turns out they do not.

Chapter 6
Common Methods of Argument Criticism

We have already seen on several occasions that the evaluation of arguments can be done seriously only if one gives reasons supporting the evaluative claim. Such a reasoned evaluation is obviously an argument, and since the subject matter is the original argument, the evaluation is clearly a meta-argument. The aim of this chapter is to reconstruct in explicitly meta-argumentative terms relevant insights by various scholars that deal with several common methods of criticizing arguments.

6.1 Krabbe's Formal-Fallacy Criticism

An important type of meta-argument occurs when a ground-level argument is criticized for having committed a fallacy. As Krabbe (2002, 162) has stated, "in fallacy criticism it is upon the critic to show why an alleged move in critical discussion is so completely wrong that is cannot even prima facie be accepted as a serious contribution to the discussion. Thus fallacy criticism leads to a critical discussion on a second level, a discussion about the permissibility of a move in the ground level discussion."

This meta-argumentative interpretation of fallacy criticism is presented by Krabbe (1995, 338; 2003, 642) as a generalization of a thesis advanced by Hamblin (1970, 283-303). In his seminal work, Hamblin argues that it is important to distinguish "points of order" from "topic points" in a critical discussion: the former raise metacognitive questions, whereas the latter raise substantive issues; this is Hamblin's way of distinguishing between meta-level and ground-level. Furthermore, he argues that equivocation criticism essentially involves meta-arguments and metadialogues, writing in the last sentences of the book: "the road to an understanding of equivocation, then, is the understanding of *charges* of equivocation. For this, the development of a theory of charges, objections or points of order is a first essential" (Hamblin 1970, 303). More recently, the metadialogical ramifications of this thesis have been systematically explored by Laar (2002; 2003a; 2003b).

Krabbe's thesis about fallacy criticism is also presented by him as a solution to the problem of the asymmetry between favorable and unfavorable evaluations of arguments. In several challenging papers, Massey (1975a; 1975b; 1981) had asked and answered negatively the question, "Are there any good arguments that bad arguments are bad?" By contrast, Krabbe (1995) asks and answers affirmatively the question, "Can we ever pin one down to a formal fallacy?" Despite the terminological variance, and the opposition of their respective conclusions, the metadialogical dimension of the discussion is obvious. What is being discussed is the nature and cogency of meta-arguments to the effect that some ground-level argument is bad, fallacious, or invalid. Let us reconstruct Krabbe's own argument (a third-level meta-argument!) that it is possible to construct cogent (second-level) meta-arguments to the effect that some ground-level argument is a formal fallacy.

First, what is a formal fallacy? For Krabbe (1995, 336), "*a formal fallacy, in dialogue, is committed as soon as a party presents a formally invalid (i.e., not formally valid) argument that violates the code of conduct of the dialogue.*" Here it is important to note that, besides formal invalidity, there is a second element in this definition—code violation; i.e., a violation of some rule agreed upon by the two interlocutors, or arguably relevant in the context of that discussion. Although it is unrealistic to expect prior or explicit agreement about the rules of a particular discussion, learning the contextual relevance of various types of arguments and criticism is a normal part of the education designed to achieve mastery of a given field. For example, historians often argue for chronological theses by means of arguments which, however strong, are formally invalid; and the same happens in the experimental branches of empirical science when one gives evidence to support some empirical generalization. But everybody knows, or ought to know, that in these contexts such formally invalid argument do not violate the rules of the game. My point is simply to underscore the fact that, following Krabbe, there are at least two things and not just one that must done to prove a formal fallacy;[1] and since these two things embody different claims, two distinct meta-arguments must be advanced in effective formal-fallacy criticism.

Next, what is formal invalidity? Or equivalently, what is formal validity? And more fundamentally and generally, what are validity and invalidity? Again, I follow mostly the spirit and occasionally the letter of Krabbe's (1995) discussion. An argument is *valid* iff there is no "situation, actual or fictitious (a possible world, if one wishes) such that in that situation all the premises are true and the conclusion is false" (Krabbe 1995, 335-36); i.e., iff it is impossible for the premises to be true while the conclusion is false; i.e., iff "there is no *counterexample* to it" (Krabbe 1995, 336). Such a counterexample to an argument should not be confused with a counterexample to an argument-form, which is an argument instantiating the form and having true premises and false conclusion; thus to be clearer, we may speak of *counterexample-situations* (to arguments) and of *counterexample-arguments* (to forms). Finally, an argument is *invalid* iff it is not valid.

Formal validity is a special case of validity. An argument is *formally valid* iff "it can be correctly paraphrased ... such that its schema (or form) is valid" (Krabbe 1995, 336); i.e., iff it instantiates a valid argument form; i.e., iff it instantiates a form that has no counterexample arguments. And an argument is *formally invalid* iff it is not formally valid; i.e., iff it does not instantiate any valid argument form. Note that this is not equivalent to instantiating an invalid form. Thus, validity is more general than formal validity: all formally valid arguments are valid, but not all valid arguments are formally valid; and all invalid arguments are formally invalid, but not all formally invalid arguments are invalid.

Given these definitions, Krabbe discusses several methods of proving formal invalidity, i.e., several types of meta-argument concluding that a ground-level argument is formally invalid.

One method is what Krabbe, following Massey, calls "the trivial logic-indifferent method."[2] This amounts to proving that the argument's premises are true and the conclusion is false. By the definition of validity, such a proof shows that it is possible for the premises to be true while the conclusion is false, and hence that the ground-level argument is not valid. Then from the relationship that formal validity is a special case of validity, we get that the ground-level argument is not formally valid, i.e., is formally invalid. Krabbe does not deny the correctness of this method but agrees with Massey that it does not go very far.

For example, suppose someone, perhaps in a context of learning geography, thought that: (1) Reno is the capital of Nevada, because (11) Las Vegas is not, and (12) if Reno is the capital of Nevada then Las Vegas is not. Based on empirical or archival research one can easily show that Las Vegas is indeed not the capital of Nevada, since the capital is Carson City; but it remains true that if Reno were the capital then Las Vegas would not be, since American states have only one capital; and it is false that Reno is the capital, again

---

[1] Actually, there are several other requirements, each corresponding to the various distinct aspects of the traditional definition of a fallacy; that is, a common type of argument that appears correct but is not. Cf. Finocchiaro (2005a, 109-47; 2013a); Woods 2013, especially chapter 4.

[2] Krabbe 1995, 341. Cf. Massey 1975a, 64; 1981, 494.

because the capital is Carson City.

I agree with Krabbe and Massey that here we have triviality and little if any logic. However, I would stress two things: we do have, inevitably, argumentation, indeed a meta-argument; and the proof is indirect in the sense that the meta-argument shows formal invalidity without appealing to anything "formal," but rather by showing (simple) invalidity, and using the principle that all formally valid arguments are valid.

The same indirect proof is used in another method, which Krabbe discusses at greater length. He calls it "the method of *counterexample*. This is the royal road of showing invalidity" (Krabbe 1995, 340). Krabbe clarifies that "counterexample" is commonly used with several different meanings, but that here he is using it in the sense defined above, namely a *situation* in which the premises are true and the conclusion is false. The correctness of this method is grounded on the definition of validity (to intermediately conclude invalidity), and on the relationship between validity and formal validity (to finally conclude formal invalidity). Krabbe also gives the following insightful description:

In general, the method of counterexample works as follows: find an obviously consistent set of logically simple and perspicuous sentences that together demonstrably entail the truth of the premises and the falsity of the conclusion, i.e., find the description of a counterexample. A counterexample may be fictitious, therefore it is not required that these sentences be true. To find the required set, logical analysis may be very helpful. E.W. Beth's method of semantic tableaux, in particular, is an effective instrument for the discovery of counterexamples. But in order to convince one's opponent, one need not expound the techniques used in the discovery of the counterexample. It suffices to convince her that these sentences describe a possible situation and then derive the required truth values from the premises and conclusion of the original argument. [Krabbe 1995, 342-43]

For example, regarding the argument above about the capital of Nevada, without doing any empirical research or knowing whether Las Vegas or Reno is the capital, we can simply imagine a situation in which neither Reno nor Las Vegas is the capital. If one is acquainted with the names of other Nevada cities (e.g., North Las Vegas, East Reno, Virginia City), one could simply imagine that one of them, say East Reno, was the capital. It would then follow that Las Vegas is not, and so the first premise is true; the second premise would still be true, by the rules of states' administration; but it would also follow that Reno is not, and so the conclusion is false. Here is then a situation in which the premises are true and the conclusion false. Therefore, by the definition of validity, the argument is not valid. Therefore, formal validity being a special case of validity, the argument is formally invalid.

From the general description of the method of counterexample-situation, and from this example, the meta-argumentative nature of the process is obvious.

Krabbe (1995, 341, 343-44) admits that because of the indirectness of such proofs of formal invalidity, it might be preferable to reserve the label "formal fallacy" to cases where one proves formal invalidity more directly by exploiting logical forms. This he calls the method of formal paraphrase (Krabbe 1995, 340). Such a preference would amount to modifying the definition of formal fallacy to read: a formal fallacy is an argument (1) which is formally invalid, (2) which violates some rule of critical discussion, and (3) whose formal invalidity is shown by the method of formal paraphrase. This is an interesting suggestion, but Krabbe does not explicitly commit himself to it; so we too shall leave it merely as a possible terminological convention.

More importantly, the method of formal paraphrase appeals explicitly and directly to the definition of formal validity. The ground-level argument is paraphrased in some more or less formal logical system, and "the reason that the argument is [formally] invalid is expressed as follows: 'this paraphrase captures the gist of your argument (meaning: the ground of its presumed validity), and this paraphrase constitutes an invalid logical form" (Krabbe 1995, 340). It is crucial to understand that there are three things which the meta-argument must try to prove: (1) that the ground-level argument instantiates a particular argument form; (2) that this argument form is invalid; and (3) that that this argument form captures "the gist of the argument," or "the ground of its presumed validity," or all logically important features of the argument. The third clause is especially important; if it is ignored, one would conclude that a ground-level argument is formally invalid simply because it instantiates an invalid argument form, even though it might also instantiate another form that is valid, thus committing "the fallacy behind fallacies" exposed by Massey (1981).

For example, consider again the argument about the capital of Nevada. Firstly, one could claim that it is of the form: (2) R because (21) not-L and (22) if R then non-L; indeed this is the well-known form "affirming the consequent." Secondly, one could point out that this form is commonly known to be invalid; if need be, this invalidity could be exhibited by assigning the truth value falsity to both R and L, or by constructing this counterexample-argument: (3) New York is the capital of the USA, because (31) Boston is not the capital of the USA, and (32) if New York is the capital then Boston is not. Thirdly, one would have to argue that affirming the consequent is all that is happening in the original argument; that is, that the form affirming the consequent does indeed capture the gist of the argument. To better grasp that this third point is needed in this case, let us contrast it to another case in which the claim would not hold.

Consider this argument, devised for this purpose by Massey (1981, 492): (41) if something has been created by God then everything has been created by God; (42) everything has been created by God; therefore, (4) something has been created by God. This argument instantiates affirming the consequent: if S then E; E; so, S. However, this form ignores another crucial feature of the argument, namely the relationship between the second premise and the conclusion; the conclusion is a special case of the second premise; indeed the conclusion follows from the second premise alone, by the rule of universal specification.[3] Hence affirming the consequent per se is an improper paraphrase of the argument (4), and the third clause of the method of formal paraphrase rules out this paraphrase.

There is a fourth method briefly mentioned by Krabbe (1995, 340), the method of logical analogy. He does not elaborate. However, other authors have. Let us therefore go on to examine this other type of meta-argument.

6.2 Govier's Refutation by Logical Analogy

An important technique for criticizing arguments is "refutation by logical analogy." Without claiming to give a complete history of this notion, it will be useful to review some literature.[4]

One of the first to explicitly use this label was Oliver (1967, 469), who describes the technique as follows: "To prove the invalidity of any argument it suffices to formulate another argument which (*a*) has the same form as the first, and (*b*) has true premises and

---

[3] It could be objected that Massey's example works only if we understand the universal quantifier to have existential import. However, the existential import that is being presupposed is the principle that '(x)Φx' implies 'Φa', and hence 'there is an x such that Φx'; not the principle that '(x)(Fx → Gx)' implies 'there is an x such that (Fx & Gx)'.

[4] Besides the works discussed below, other useful contributions has been made by Juthe (2009), who uses the label "refutation by parallel argument," and by Whaley (1998) and Whaley and Holloway (1996), who speak of "rebuttal analogy."

false conclusion. This method is based upon the fact that validity and invalidity are purely *formal* characteristics of arguments, which is to say that any two arguments having the same form are both valid or both invalid, regardless of any differences in the subject matter with which they are concerned."

Oliver goes on to dismiss this method as incorrect; for he interprets it as a version of the flawed utilization of formal paraphrase, that is the oversimplified version of the method of formal paraphrase that ignores the third requirement that the form in question should capture the gist of the argument. But we have seen that there is no reason to oversimplify the method of formal paraphrase in this manner. Moreover, there is no reason to equate this method even with the properly nuanced method of formal paraphrase, because to say that one argument has the same form as another is not equivalent to saying that each has a unique form and that these two forms are identical.[5] Furthermore, as the label suggests, here we are talking about two arguments being logically *analogous*, rather than of their having the *same* form; and logical analogy should be understood as similarity of reasoning,[6] or as some kind of one-to-one correspondence from the formal point of view.[7] This may be the reason why in his brief description, Krabbe (1995, 340) says that "this technique consists of drawing up another, *formally analogous*, argument such that it can be shown … that its premises are true, whereas its conclusion is false" (italics added).

Besides being important for its terminological priority, Oliver's account is notable because its description makes it clear that this method is more widespread than the term "refutation by logical analogy" suggests. In fact, some scholars give a definition of what they call the "method of counterexample" that is identical to Oliver's definition of refutation by logical analogy. For example, W. Salmon (1984, 21) states that "a common way of exposing a fallacious argument is to compare it with another argument of the same form in which the premises are true but the conclusion is false. We shall call this method of proving invalidity the *method of counterexample*." Obviously this is *not* equivalent to Krabbe's method of counterexample, which we have seen involves a counterexample-situation; instead Salmon's "method of counterexample" involves a counterexample-argument, and so it rather corresponds to Krabbe's "method of logical analogy" (as one can infer from his brief description quoted in the last paragraph).

Govier, too, speaks explicitly of refutation by logical analogy. However, she gives a slightly different definition than Oliver's: "the technique of refuting arguments by constructing logically parallel ones … is based on a perception that the argument refuted has a structure which is general. If that structure is shown flawed by the presentation of another argument which has the structure and is flawed, then the original argument is refuted" (Govier 1985a, 27).

Moreover, unlike Oliver's unfavorable evaluation, she claims that the method is in principle correct: "the technique of logical analogy can in some cases provide a conclusive refutation of an argument" (Govier 1985a, 30). As suggested by her language of "flawed" arguments, by contrast to Oliver's talk of invalidity, she argues that this technique "seems to be applicable to nondeductive arguments as well as deductive ones" (Govier 1985a, 27). Furthermore, Govier's account is less formalist than Oliver's, partly because her notion of

---

[5] My argument here would be analogous to Quine's move about meaning and synonymy: sameness of meaning need not presuppose the existence of mysterious entities called meanings which words have, but may be conceived as a relationship of pairs of linguistic expressions; this relationship should be labeled synonymy in order to avoid being misled; see Quine 1961, 11-12, 22, 48.
[6] For an interpretation along these lines, see Guarini 2002, 3-8.
[7] Juthe (2009, 153-55) deserves credit for showing awareness of this problem, for advancing a version of this thesis, and for attempting to formulate some supporting arguments.

"parallelism" involves structural considerations but not multiple instantiation of the same unique form: "we construct a parallel argument in which the central features of the original are preserved while its incidental features may be varied. In this case we do not formalize in order to reveal the structure of the argument. Rather, we make structure appear by presenting a logical analogy. The structure or 'form' is repeated in the parallel argument. We 'see' it as we see sameness of shape in a blue circle and a red circle. The shape is common to both and can be seen as such without appearing as a separate structure" (Govier 1985a, 30).

Finally, Govier's "parallelism" involves generality, but not formalism: "the technique of logical analogy illustrates the fact that connections may be general without being, in the standard logician's sense, formal" (Govier 1985a, 30-31).

In 1986, Copi included a discussion of refutation by logical analogy in both the seventh edition of *Introduction to Logic* and the first edition of *Informal Logic*. Like Govier, Copi gives a favorable evaluation of the method. And like Govier, Copi views the technique as aiming to prove not only invalidity, but also, more generally, other logical errors. In fact, he is more explicit than Govier that "the method of refutation by logical analogy can be used with (almost) equally telling effect in criticizing an inductive argument" (Copi 1986b, 423). Moreover, Copi's examples are so incisive that they deserve quotation. One example comes from a 1952 opinion of the U.S. Supreme Court written by Justice Clark: "It is urged that motion pictures do not fall within the First Amendment aegis because their production, distribution, and exhibition is a large-scale business conducted for private profit. We cannot agree. That books, newspapers, and magazines are published and sold for profit does not prevent them from being a form of expression whose liberty is safeguarded by the First Amendment. We fail to see why operation for profit should have any different effect in the case of motion pictures" (Copi 1986b, 423).

Another example comes from a campaign speech by Abraham Lincoln on 2 March 1860:

The South were threatening to destroy the Union in the event of the election of a republican President, and were telling us that the great crime of having destroyed it will be upon us. This is cool. A highwayman holds a pistol to my ear, with "stand and deliver, or I shall kill you, and then you will be a murderer." To be sure the money which he demands is my own, and I have a clear right to keep it, but it is no more so than my vote, and the threat of death to extort my money, and the threat of destruction to the Union to extort my vote, can scarcely be distinguished in principle. [Copi 1986a, 189]

Finally, Copi's account contains a novel claim that deserves special attention. That is, refutations by logical analogy are themselves inductive arguments, specifically arguments by analogy. Although Copi does not give an explicit elaboration of this point, the claim is implicit in the fact that the section discussing this topic (Copi 1986b, 421-23) is in chapter 12, on "Analogy and Probable Inference," which in turn is a chapter of part III, on "Induction." However, this claim is partly explicit when Copi (1986b, 421) introduces the topic by saying "there is a special kind of argument that uses an analogy to prove that another argument is wrong, or mistaken." The proof involves "constructing a refuting analogy. A refuting analogy for a given argument is an argument of exactly the same form or pattern as the given argument, but whose premises are known to be true and whose conclusion is known to be false" (Copi 1986b, 422). What Copi here calls a "refuting analogy" corresponds to what Salmon calls a counterexample, and what I have called a counterexample-argument. Although a better label might have been "refuting analogue," Copi's label indicates that he views this technique as involving analogical reasoning about

Common Methods of Argument Criticism   81

two analogous arguments, namely a meta-argument by analogy.

The question of the inductive nature of refutations by logical analogy should not be confused with the question whether this technique applies to inductive as well as deductive arguments. The former question is about the meta-argument, the latter is about the ground-level arguments. Even if Copi's suggestion here were incorrect, the inclusiveness of the technique might still hold; in that case we would have deductive meta-arguments about deductive or inductive ground-level arguments. Copi does not ask these questions, let alone answer them. But in his brief mention of this technique, Krabbe refers to a contribution by Woods and Hudak (1989) that discusses these issues. To this we now turn.

6.3 Woods's Parity of Reasoning

In an article ostensibly about arguments by analogy, but more revealingly entitled "By Parity of Reasoning," Woods and Hudak (1989) focus on arguments which I will call *arguments by parity of reasoning*. These are such that "they argue that two or more target arguments stand or fall together and that they do so because they are relevantly at parity, that they possess similar deep structures by virtue of which they coincide in their logical form" (Woods and Hudak 1989, 127). This brief description is elaborated when they say that these arguments have a *"basic structure* ... somewhat as follows: 1. Argument A possesses a deep structure whose logical form provides that the premises of A bear relation R to its conclusion. 2. Argument B shares with A the same deep structure. 3. Therefore, B possesses a deep structure whose logical form provides that its premises likewise bear R to its conclusion. 4. Hence, ... A and B are [both] good or [both] bad arguments, by parity of reasoning, so called" (Woods and Hudak 1989, 127).

The two "target" arguments A and B can sometimes be usefully distinguished from each other insofar as one is the "original" argument and the other is a "comparison" argument. That is, an argument by parity of reasoning is "an argument to the effect ... that another argument—let's call it a 'comparison' argument—shares an identical form with the original argument. Thus the ... [meta]argument both makes an argument and presents a (comparison) argument. The argument it presents ... is not the argument it makes. The argument it makes ... holds that the comparison argument is identical or—at a minimum—relevantly similar in form with the original, and therefore that the original stands or falls with it" (Woods and Hudak 1989, 128).

Finally, there is a fourth explanation. An argument by parity of reasoning is

an argument to the effect (schematically represented) that since argument
A: 1. p / 2. q / ... / n. Therefore, w
and another argument
B: 1. s / 2. t / ... / n. Therefore, u
both instantiate (or are cases of) an argument
Q: 1. S_1 / 2. S_2 / ... / n. [Therefore,] S_n
and, furthermore, since B draws an assessment-verdict, V, by virtue of its relationship to Q, so too should A draw down the same verdict. [Woods and Hudak 1989, 132]

It should be noted, that by contrast with refutations by logical analogy, arguments by parity of reasoning do not always advance a negative assessment of the original argument, but sometimes advance a favorable judgment; the key point is that they advance the same assessment of the original argument as of the comparison argument. Moreover, in arguments by parity of reasoning, the notion of parity suggests equality, identity, or sameness,[8] whereas in refutations by logical analogy, the notion of analogy suggests similarity; and similarity is not identity.

In the just-quoted passages, I have studiously avoided and edited out any talk of analogy, which I found confusing and confused. Insofar as I could understand such language, Woods and Hudak seem to be claiming that arguments by analogy (in the ordinary sense) are arguments by parity of reasoning (in their sense). This I found to be an untenable claim.

So far, I have also avoided Woods and Hudak's language of meta-argument or metadialogue, but here my motivation was to avoid a reconstruction in which it is true by definition that arguments by parity of reasoning are meta-arguments. This is a second important claim they advance, for if they are first taken to define arguments by parity of reasoning as indicated in the above quotations, then this second claim is an immediate consequence of those definitions. These authors explicitly advance this claim, for example when they say that "arguments by parity of reasoning ... are arguments about arguments, *meta-arguments*" (Woods and Hudak 1989, 127). I believe it is better to regard the meta-argumentative nature of arguments by parity of reasoning as a consequence, rather than a part, of the definition also because then we can appreciate better their metadialogical character. This emerges when Woods and Hudak (1989, 128) point out that arguments by parity of reasoning typically occur when one is attempting a "dialectical breakout from a stand-off," where "a stand-off is a kind of dialectical black hole," namely when a critical discussion has reached an impasse because of extremely deep disagreements; the point is that in such situations a change of level into a meta-discussion is a natural step to take and perhaps the only thing that can help the discussion.[9]

In fact, their paradigm example of an argument by parity of reasoning is Judith Thomson's (1971) celebrated "argument designed to show that the termination of a rape-induced pregnancy is morally justified" (Woods and Hudak's 1989, 127). In this argument, the conclusion is defended by imagining a situation in which a violinist has been connected to my body, without my knowledge or consent, in order to use my kidneys to process his blood, which his own diseased kidneys cannot do. Thus, a third valuable claim I would attribute to Woods and Hudak is that Thomson's violinist-abortion argument is an argument by parity of reasoning.

Fourthly, Woods and Hudak seem to claim that arguments by parity of reasoning (as defined) are valid. For in arguments by parity of reasoning, the target arguments "share a deep structure by virtue of which they stand or fall as arguments. Deep structure deserves the name of logical form when it binds logical appraisal in such ways. Of course, not every appraisal of an argument is determined by its deep structure; in simple cases, validity is settled thus, but not soundness" (Woods and Hudak 1989, 134).

Here they seem to be saying that arguments by parity of reasoning are deductively valid because the target arguments share the same logical form, and the concept of logical form implies that arguments with the same logical form share the same logical appraisal. This is a plausible (meta)argument, but it may be questioned by questioning whether parity

---

[8] One might object that in their third explanation quoted above, Woods and Hudak seem to be denying this suggestion of mine when they say that "the comparison argument is identical or—at a minimum—relevantly similar in form with the original" (1989, 128). However, I think that this property of being "at a minimum, relevantly similar" amounts to being "identical in the reasoning"; the point is not that the two arguments are identical simpliciter, but rather that their reasoning is identical.

[9] For more on this issue, on deep disagreements in general, and on what Woods calls standoffs of force five, see chapter 7 below.

Common Methods of Argument Criticism 83

of reasoning can be reduced to identity of logical form and what exactly the notion of logical form entails.[10] In any case, the point is not true of all argument appraisal; for example, it does not apply to truth of premises. However, it is not restricted to deductive appraisal, for "whatever verdict—whether of deductive validity, inductive strength or what not—that is conferred upon a given argument by virtue of the logical form of its deep structure is also conferred upon any argument sharing that structure" (Woods and Hudak 1989, 126).

In summary, Woods and Hudak have defined an important class of arguments, called arguments by parity of reasoning. These are meta-arguments that conclude that some original argument should receive the same logical assessment as some comparison argument because these two ground-level arguments share the same logical form. Thomson's argument about abortion and the violinist is a significant example of such a meta-argument by parity of reasoning. Such meta-arguments by parity of reasoning may be regarded as deductively valid. But to say that arguments by analogy (as ordinarily defined) are meta-arguments by parity of reasoning is a problematic claim.

6.4 Conclusion: Common Types of Critical Meta-arguments

We have seen that the trivial logic-indifferent method of proving formal invalidity argues that (M1) an argument A is formally invalid because (M111) its premises are true and its conclusion is false, and hence (M11) A is invalid. The method of counterexample-situation proves that (M2) an argument A is formally invalid because (M211) there exists some situation in which the premises are true and the conclusion is false, and hence (M21) A is invalid. The method of formal paraphrase is the meta-argument that (M3) argument A is formally invalid because (M31) A instantiates some argument form F, (M32) F is an invalid argument form, and (M33) F captures the gist of A. A refutation by logical analogy is the meta-argument that (M4) argument A is flawed in the sense F because (M41) A is logically analogous to argument B, given that (M411) there is a one-to-one correspondence between their respective structures and contents, and (M42) B is flawed is the sense F. Arguments by parity of reasoning show that (M5) argument A receives an evaluation E because (M51) A has the same logical form as argument B, given that (M511) both arguments instantiate argument form F, and (M52) B receives evaluation E. Finally, note that we have also discussed formal-fallacy criticism, i.e., meta-arguments concluding that (M6) some argument A commits a formal fallacy because (M61) it is formally invalid and (M62) it violates the agreed-upon rule to use only formally valid arguments.

These are obviously definitions of types of meta-argument since their subject matter is arguments and each definition stipulates what type of conclusion and what type of premises the meta-argument has. These six types are not, nor are they meant to be, exhaustive. For example, a special and interesting class of meta-arguments consists of arguments whose conclusion is the denial of the conclusion of another argument; such arguments correspond to those which chapter 2 called counterarguments or conclusion-refuting criticism. Moreover, here the names have been adapted *ad hoc* from the literature, but by further reflection we might also devise a more systematic way of naming meta-arguments.

Additionally, interesting questions arise about the inter-relationships among these six meta-arguments. For example, what are the implications of the fact that any of the first three can be combined with the sixth one, insofar as (M61) is identical to (M1), (M2), and (M3)? Is a meta-argument by parity of reasoning really different from the method of formal

---

[10] For a criticisms along these lines, see Guarini 2002; for a response, see Woods 2002; for a related criticism, see Guarini 2004.

paraphrase? Is it really different from a refutation by logical analogy? Are parity of reasoning and logical analogy really different? What is the ground for the correctness of refutations by logical analogy (when they are correct)?

Such meta-argumentative reflections have implications regarding metadialogues, at least as understood by scholars who follow a dialogical approach. To see this, first, it should be noted that there is a considerable overlap between dialogue and argument, insofar as a significant subclass of dialogues (called persuasion dialogues or critical discussions) consist of arguments (Walton and Krabbe 1995, 65-85). Moreover, Barth and Krabbe (1982) have famously proved the equivalence between the axiomatic and dialogical methods; and this proof may be taken to suggest (see Finocchiaro, 2005, pp. 231–245) not only that the monolectical way of talking about arguments can be translated into a dialogical way of talking, but also that the reverse is the case. Third, an important strand of this chapter can be regarded as a translation of Krabbe's dialogical account of formal-fallacy criticism into a monolectical framework. It follows that a metadialogical theorist could now undertake to translate into a dialectical framework the meta-argumentation of logical analogy and of parity of reasoning sketched above.

Finally, more empirical or historical-textual analyses are desirable. This should be done partly to find significant illustrations of the various meta-arguments described here, and thus give concrete content to these relatively abstract conceptualizations. However, the reverse methodological possibility should also be left open, namely that perhaps the historical-textual study of meta-arguments will lead to the discovery of other types and principles of meta-argumentation. These will be guiding ideas in the historical case studies undertaken in later chapters, in the third part of this book.

6.5 Summary

The aim of this chapter has been to reconstruct in explicit meta-argumentative terms some relevant insights by various scholars that deal with common methods of argument criticism. Krabbe (1995) showed that formal-fallacy criticism (and more generally, fallacy criticism) consists of metadialogues, and that such metadialogues can be profiled in ways that lead to their proper termination or resolution. Here I reconstructed Krabbe's metadialogical account into monolectical, meta-argumentative terminology by describing three-types of meta-arguments corresponding to the three ways of proving formal invalidity which he studied: the trivial logic-indifferent method; the method of counterexample situation; and the method of formal paraphrase. A fourth type of meta-argument corresponds to what Oliver (1967), Govier (1985a), and Copi (1986a; 1966b) call refutation by logical analogy. A fifth type of meta-argument represents my reconstruction of arguments by parity of reasoning studied by Woods and Hudak (1989).

Chapter 7
Deep Disagreements, Fierce Standoffs, Etc.

One area that is suffused with meta-arguments is the following cluster of cognitive phenomena that have been discussed under various labels. Fogelin (1985; 2005) has called them "deep disagreements." Friemann (2002; 2003; 2005) has studied so-called "intractable quarrels." Woods (1992; 1996) has analyzed them as "standoffs of force five," here dubbed "fierce" standoffs. And Johnstone (1954; 1959; 1978) has been concerned with "fundamental philosophical" controversies or disagreements. The aim of this chapter is to study this cluster of topics from the point of view of meta-argumentation.

7.1 Fogelin on Deep Disagreements

The meta-argumentative character of Fogelin's argument is obvious from his own summary: "deep disagreements cannot be resolved through the use of argument, for they undercut the conditions essential to arguing" (2005, 8). It is important, however, to understand its full complexity. Fogelin's meta-argument may be reconstructed as follows:

[F12111]   normal argumentation "takes place within a context of ... shared beliefs and preferences ... [and] procedures for resolving disagreements" (6);[1]
[F12112]   deep disagreements are those when the contending parties do not share any relevant beliefs, preferences, or resolution procedures (7);
[F1211]   therefore, deep disagreements lack the conditions of normal argumentation, i.e., they make normal argumentation impossible (7), i.e., they "undercut the conditions essential to [normal] arguing" (8);
[F121]   therefore, "deep disagreements cannot be resolved through the use of argument" (8);
[F12a]   therefore, deep disagreements "are not subject to rational resolution" (11);
[F12b]   they are amenable only to "persuasion" (9) and "persuasion techniques" (11);
[F11]   but there are important disagreements that are deep;
[F1111]   for example, the controversy over abortion reduces to the issue over the moral status of the fetus, and there are no common grounds about that,
[F111]   so it is a deep disagreement (8-9); and
[F1121]   the controversy over affirmative action reduces to the question the existence of groups rights, above and beyond individual rights, and there are no common grounds about that,
[F112]   so it is a deep disagreement (8, 10, 11);
[F1]   therefore, there are important disagreements that are not subject to rational resolution, but only to rhetorical persuasion (11).

The structure of this argument should be apparent from the usual reasoning indicators explicitly used, and can be made completely clear from the standard labeling I have used to name its various propositions. That is, the numbering system for the propositions in my reconstruction is the one I discussed earlier (chapter 2), which in turn was a variation of the system presented by various authors when they discuss the representation of complex arguments by means of structure diagrams in the shape of either tree branches or tree

---

[1] In this section, numerals in parenthesis are references to the page numbers in Fogelin 2005.

roots.[2] The key idea is that if a given claim within an argument is labeled *n*, then the premises that directly support it are labeled *n1*, *n2*, *n3*, etc.; and if proposition *nm* is also part of some subargument, then the premises directly supporting it are labeled *nm1*, *nm2*, *nm3*, etc. The structure of Fogelin's argument may thus be represented in the diagram in Figure 7.1.

Figure 7.1

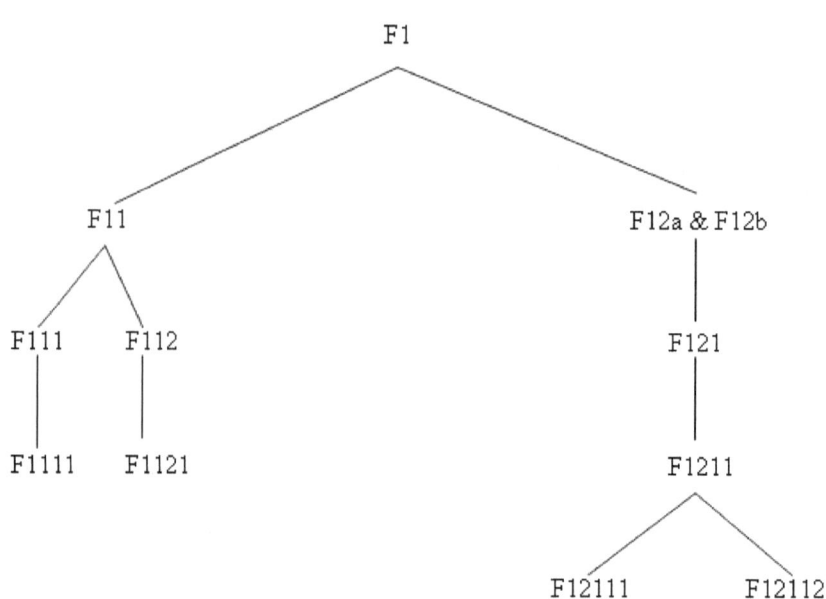

Besides clearly portraying the logical structure of Fogelin's argument, another advantage of such a reconstruction and diagram is that they enable us to find our way in the maze of the many comments and criticisms which Fogelin's argument has received. To them I now turn.[3]

7.2 Critiques of Fogelin's Argument

7.2.1 The Dynamics of Rationality (Lugg 1986)

In 1986, Andrew Lugg was the first to advance a criticism of Fogelin. It may be reconstructed as follows: [1111] "Reason may not be sufficient to decide a particular issue here and now but it may still contribute significantly to its resolution later on" (Lugg 1986, 48). This can happen in several ways. [11111] "Individuals who disagree deeply may still be able to narrow the distance between themselves by dint of argument, debate, inquiry, and research ... [11112] Individuals can also bring about a shift in one another's allegiances by demonstrating hidden strengths of their own views and by eliciting hidden weaknesses of alternative views. [11113] Furthermore, they may find themselves having to

---

[2] Cf., e.g., Angell 1964, 369-93; Scriven 1976, 41-43; Finocchiaro (1980a, 311-31; 2005a, 39-41); Eemeren and Grootendorst 1984, 87-93; Eemeren, Grootendorst, and Kruiger 1984, 17-36; Freeman 1991; Snoeck Henkemans 1992; and chapter 2 above.
[3] Besides those analyzed in detail here, one should also mention the following critiques of Fogelin's argument: Freeman 2012, Kraus 2012, Naess 1992, and Zarefsky 2012.

shift ground as a result of their discovering things *wrong* with the views that they accept and things *right* with the ones that they reject" (Lugg 1986, 48). It follows that [111] even if deep disagreements are not subject to rational resolution *immediately*, this does not mean that they are not subject to rational resolution *later*. [11] Fogelin fails to distinguish immediate resolution and later resolution, and this failure amounts to [1] presupposing a "static" conception of rationality and failing to appreciate a "dynamic" conception.

Lugg's criticism amounts to a meta-argument that focuses on Fogelin's propositions F121, F12a, and F1. Lugg argues that Fogelin's argument embodies an ambiguity, such that these propositions may be justified in one sense, but are not justified in the other sense. This criticism strikes me as being cogent.

7.2.2 The Power of Internal Criticism (Davson-Galle 1992)

In 1992, Davson-Galle criticized some of Lugg's criticism and expressed some appreciation of Fogelin. That is, "contra Lugg and in support of Fogelin, [2] rational discussion might be futile and futile in more cases of disagreement than even Fogelin seems to allow" (Davson-Galle 1992, 154). [21] One type of case involves "putting the other cognitive agent into a situation where a primitive epistemic act occurs" (Davson-Galle 1992, 150), for example, [211] "persuading someone to accept that it's raining by opening a blind so that he can see that it is so" (Davson-Galle 1992, 150).

While I do not see that this affects Lugg's point about the dynamics of rationality, it does appear to strengthen Fogelin's proposition F12a. Davson-Galle's point is that some disagreements are not subject to rational resolution, meaning resolution by reasoning and argument, because they must be resolved through simple observation. This may be so, but if a disagreement can be resolved by simple observation, then it is not a deep disagreement by Fogelin's definition (proposition F12112), which explicitly excludes procedures for resolutions among the common grounds. Thus, appearances to the contrary, there is no reinforcement of proposition F12a.

At any rate, Davson-Galle also raises an objection to Fogelin: "yet rational persuasion might be more powerful in other ways than Fogelin allows … one might rationally persuade someone of the error of his/her ways by tracing distasteful consequences (to him/her but not to you)" (Davson-Galle 1992, 154). This point may be interpreted as the meta-argument that [3] proposition F121 does not follow from proposition F1211; for [31] even if deep disagreements cannot be resolved through what Fogelin calls "normal" argumentation (i.e., even if proposition F1211 is true), it may happen that they *can* be resolved through some special kind of argument (i.e., proposition F121 may be false), such as [311] showing that the opponent's thesis leads to consequences not acceptable to the opponent.

This is an extremely important point, and the criticism is essentially correct.[4] Although Davson-Galle does not elaborate or give illustrations, the point has been elaborated in other contexts by others. In fact, this other special kind of argumentation corresponds to "*ad hominem* argument" according to a conception that goes back to Galileo and Locke, passes through Whately, and culminates with Henry Johnstone Jr.[5] This conception should not, of course, be confused with the *ad hominem* fallacy of ordinary language and contemporary textbooks; it states that an *ad hominem* argument is one which criticizes a thesis by arguing

---

[4] A similar point has been made later by Zarefsky (2012, 80-82, 84-85), under the labels of charges of hypocrisy, circumstantial *ad hominem*, and interfield borrowing.
[5] See Johnstone (1952; 1959; 1978); Finocchiaro (2005a, 277-91, 329-39); Eemeren and Grootendorst 1993; and the discussion in section 7.6 below.

that it implies consequences not acceptable to its proponent. I shall elaborate later, when I discuss Johnstone's position in more detail.

7.2.3 The Rationality of Suspending Judgment (Feldman 2005)

Next, let us review the critiques found in the special issue of *Informal Logic* published in 2005 and devoted to the discussion of Fogelin's original essay.

Feldman (2005) objects that one kind of rational resolution is the one that results in suspension of judgment after a critical analysis of the arguments and counterarguments, and that deep disagreements are often rationally resolvable in this manner, such being the case in Fogelin's own example about affirmative action.

Here, Feldman is pointing out an ambiguity in the notion of rational resolution. One kind of rational resolution occurs when it is possible to show that the arguments favoring one side of the two disagreeing parties are better that the arguments on the other side, and so the disagreement can be resolved in favor of the former and against the latter. But another kind of rational resolution occurs when it is not possible to show that the arguments on one side are better than those on the other; this means that the arguments on the two sides are equally good or equally bad; and such an evaluation implies that one ought to suspend judgment.

The meta-argument here is this: [41] Fogelin's argument embodies an ambiguity with regard to the notion of resolution, for [411] this term has two different possible meanings that might be labeled the strong and the weak sense of resolution, and [412] this ambiguity is present in propositions F121, F12a, and F1. [42] Even if these propositions were justified in the strong sense, they would not be justified in the weak sense of "resolution." Indeed, [43] when a disagreement cannot be rationally resolved in the strong sense, this failure implies that it *is* rationally resolvable in the weak sense. Therefore, [4] in that sense the argument for F1 is self-contradictory.

This meta-argument strikes me as being essentially correct.

7.2.4 The Power of Complex Argument (Turner and Wright 2005; Campolo 2005)

The next objection in this sequence is advanced by Turner and Wright (2005). They criticize Fogelin by stressing that there is a difference "between his relatively modest claim that some disputes cannot be resolved through argument and his more radical claim that such disputes are beyond rational resolution" (Turner and Wright 2005, 25). They seem to strengthen the modest claim, insofar as they point out that much argumentation is "epigrammatic" (Turner and Wright 2005, 28) reason-giving, and certainly some disputes involve much more than that. However, "we often change our minds about something as the result of education and experience the significance of which cannot be captured in a sentence or two ... [and] to stigmatize ... [this] as irrational demeans the concept of rationality" (Turner and Wright 2005, 33).[6]

I interpret this criticism as an objection to Fogelin's inference from proposition F121 to F12a. Turner and Wright are arguing that [5] this inference is incorrect. Their meta-argument is that on the one hand [51] proposition F121 is essentially true, because [511] arguing often consist of brief reason-giving, [512] which indeed cannot resolve deep disagreements; on the other hand [52] this inference assumes that brief "epigrammatic"

---

[6] For elaborations of such an approach to reasoning and argumentation (stressing skills, experience, and understanding, above and beyond simple reason-giving), see Wright (1995; 1999; 2001a; 2001b; 2002) and Campolo and Turner 2002; cf. also Turner 2005.

reason-giving is all there is to rationality, but [53] this assumption is false since [531] rationality often depends on time-consuming education and long experience.

This meta-argument is largely acceptable, but with some provisos. Turner and Wright are focusing on what may be called simple argumentation, admitting its limitations with regard to deep disagreements, and advocating that rationality should not be reduced to simple argumentation. So far so good. However, their own argument suggests that not all argumentation is, or need be, epigrammatic reason-giving consisting of one or two sentences. Sometimes argumentation is complex, consisting of multiple or long chains of constructive reasoning, as well as equally complex chains of destructive or critical objection-answering.[7] What they call education, experience, and reflection essentially involve such complex argumentation. Thus, although we may agree with Fogelin, Turner, and Wright that deep disagreements cannot be resolved by means of simple argumentation, this is not to say that they cannot be resolved by means of complex argumentation; indeed, complex argumentation is often the key to the rational resolution of the dispute.

A point of view similar to Turner and Wright is expressed by Campolo (2005). However, his view of argumentation and rationality is applied primarily to reinforce Fogelin's argument-limitation thesis, claim F121, rather than refuting his rationality-limitation thesis, proposition F12a.

Campolo's argument can be reconstructed as follows: [6] Deep disagreements, indeed, cannot be resolved through the use of argument because [61] argumentation is essentially "reasoning together," that is, reason-giving and reason-assessing; and [62] reasoning together "depends on a variety of resources and skills" (Campolo 2005, 38); "but [63] the path to expertise, competence, and intersubjectivity is paved with training, practice, study, apprenticeship, immersion in a tradition or way of doing something. [64] Reasoning together, on its own, cannot bring about any of this" (Campolo 2005, 45). [641] It is only "a way of repairing small gaps in intersubjectivity" (Campolo 2005, 47).

My criticism of Campolo's meta-argument is similar to my criticism of Turner and Wright's. That is, besides the simple and brief reasoning and argumentation of which Campolo speaks, there is long and complex argumentation. The training, practice, study, apprenticeship, and immersion in tradition of which he speaks can be seen as ways of learning and mastering complex argumentation.

7.2.5 The Importance of Being Open-minded and Fair-minded (Friemann 2005)

The next critique is prima facie very different since it is an attempt to inject into the discussion ideas from psychotherapy and its treatment of "intractable quarrels." This has been done by Friemann (2002; 2003; 2005) both in his contribution to the special issue of *Informal Logic* and other related papers. Intractable quarrels are emotionally charged disagreements typically between married persons over such issues as money, sex, and children. Such quarrels usually cannot be resolved by the persons involved, but are helped by psychotherapists. In so doing, psychotherapists act as mediators or third parties and try to instill in the quarreling persons the skills and attitudes of openness, empathy, and

---

[7] At its simplest, a complex argument may be defined as one which has at least one intermediate proposition, which is the conclusion of one subargument and the premise of another subargument. Next, one could define a complex argument as one with at least one intermediate proposition and/or at least two independent reasons. Next one could add the further conditions "and/or an acknowledgment of contrary reasons." Thus, as should be expected, the complexity of argumentation can quickly become a very complicated business. Cf., for example, Angell 1964, 369-93; Finocchiaro 1980a, 313-31; Freeman 1991; Snoeck Henkemans 1992; and chapter 2 above and 8 below.

identification.[8] The meaning of these notions seems to be this. Openness is the willingness to listen to the views, complaints, and reasons of the other party. Empathy is the readiness to show some appreciation for and sympathize with those views, complaints, and reasons. And identification is the willingness to accept the correctness of those views, complaints, and reasons. Despite the fact that strong emotions usually accompany intractable quarrels, but need not accompany deep disagreements, Friemann defends the similarity between intractable quarrels and deep disagreements on the grounds that "feeling is not really distinct from thinking" (Friemann 2005, 53, 59).

Friemann's position can be reconstructed as the following argument: [711] Deep disagreements are analogous to intractable quarrels because [7111] "feeling is not really distinct from thinking" (Friemann 2005, 53, 59). [712] Intractable quarrels are resolved with the help of psychotherapists insofar as the contending parties learn to practice openness, empathy, and identification. Therefore, [71] deep disagreements can be resolved insofar as the contending parties practice openness, empathy, and identification. [72] Openness, empathy, and identification are ideals that are part of rationality. Hence, [7] deep disagreements can be rationally resolved, i.e., proposition F12a is false.

The first step of this argument (subargument 711, 712, 71) is, of course, an argument from analogy, and could be questioned in the usual manner by questioning the strength of the analogy on the basis of the significance of the similarities and the dissimilarities. The second step (71, 72, 7) is a slight amplification that connects Friemann's argument more explicitly with Fogelin's. Furthermore, if we relate this argument to Fogelin's proposition F121, it seems that Friemann is essentially agreeing that deep disagreements and intractable quarrels cannot be resolved through normal argumentation, namely argumentation *not* guided by the ideals of openness, empathy, and identification.

In my view, the idea of having argumentation guided by various principles of rationality is extremely important. As we shall see later, other scholars (e.g., Woods 1992; 1996) have discussed it. For now, two points deserve mention. First, we should examine carefully what is involved in such guidance. My hunch is that it is essentially meta-argumentation, that is, argumentation about whether the ground-level arguments that define the deep disagreement or intractable quarrel are in accordance with some principles of rationality. Second, the identification and formulation of such principles is a challenging task And here is it worth raising the question whether identification belongs to such a list. In fact, I do not think it does.

In a commentary on one Friemann's papers, Nosich (2002) has argued that although openness and empathy are important, identification has no place, even in a therapeutic context. I agree with this claim, but my own argument is slightly different. I would argue as follows. I would begin by stressing an argument-theoretical conception of openness and empathy. I would define openness, or open-mindedness, as the willingness and ability to learn from and listen to the arguments favoring the other side, i.e., the arguments against one's own view. Empathy would be the willingness and ability to understand the details and appreciate the strength of these contrary arguments; this notion corresponds to what various informal logicians have called charity in the sense of the principle of charity, or fair-mindedness,[9] which are the terms I would use. Then I would stress the value of open-mindedness and fair-mindedness, both on empirical historical grounds and theoretical

---

[8] Memedi (2007; 2011) has also introduced the idea of a third party, but does not say that this idea had been elaborated earlier by Friemann (2002; 2003; 2005), although Memedi does cite Friemann 2005 in a general way; cf. also Campolo 2007.

[9] See Ennis 1996, 171; Fisher 1991; Fisher and Scriven 1997, 90-91, 137-43; Paul 1990, 110, 111, 198; and Scriven 1976, 71-73, 166-67.

philosophical grounds.[10] In this context, identification would be the acceptance by one party of the other party's evaluation of the arguments. Of course, this will resolve the disagreement if the ideal is practiced by only one side; but if both were to try to comply with it, the attempt would be merely the joint switching of sides, and the disagreement would remain. In short, identification is an incoherent ideal: whereas one has the duty to know, understand, and appreciate the arguments advanced by the opposite side, one can have no duty to share the opposite side's evaluations of those arguments.

7.2.6 The Metacognitive Awareness of Deep Disagreements (Adams 2005)

Let us now go on to the last critique in that special issue of *Informal Logic*. In an essay revealingly entitled "Knowing When Disagreements Are Deep," Adams (2005) advances a beautiful argument that deserves extended quotation:

> the logic of deep disagreements makes it impossible to specify *a priori* conditions such that, for any disagreement, satisfaction of just those conditions would be necessary and sufficient epistemically to conclude that the disagreement is deep. The only way for the parties to know whether such a state of affairs obtains is by continuing to work through an attempt at rational discourse, and this because the question of whether a given disagreement is deep can only be settled by exhausting the possible resources of normal discourse. All of this means that the only way for the parties to establish that their disagreement is deep is to reject the very path of non-rational persuasion recommended by Fogelin and concentrate instead on their collective efforts at mutual persuasion by reasons. The only way, in other words, to come to know whether discourse is normal is to proceed as if it is. [Adams 2005, 76]

That is: [81] there is no *a priori* way of knowing when a disagreement is deep; [82] the only way of finding out is by a process of argumentation yielding the result that every argument advanced by one side fails to convince the other side, or is criticizable by the other side; therefore, [8] argumentation is essential in a deep disagreement. Or: [91] either the disagreement is deep, or it is not; [92] if not, then the two parties should engage in argumentation in order to resolve it; [93] if the disagreement is deep, the two parties should engage in argumentation in order to find out that the disagreement is deep; therefore, [9] the two parties should engage in argumentation in any case.

This is an elegant argument, and its most striking feature is its *ad hominem* character, in Johnstone's sense of the term. From plausible premises that are or should be acceptable to Fogelin, Adams reaches a conclusion (proposition 8 or 9) not acceptable to Fogelin. In particular, Adams's conclusion clearly contradicts Fogelin's claim about persuasive techniques, F12b. It does not conflict with Fogelin's denial of rational resolution (F12a) or of resolution by argument (F121), because it implies nothing about resolution, but only about discussion, i.e., that deep disagreements are rationally discussable and amenable to argument.

Adams's conclusion also appears to conflict with Fogelin's undercutting thesis, F1211. But here we must be careful. This undercutting thesis does seem to follow from the premises of its subargument, the theses about the conditions of normal argumentation (F12111) and of deep disagreements (F12112), and these premises appear to be true. And

---

[10] By empirical historical arguments I mean arguments based on the practice and accomplishments of past great thinkers, e.g., Galileo; in this regard, see Galilei 1997, 339-56, 376, 388, and Finocchiaro 1980a, 114-15, 134-35, 177. By theoretical philosophical grounds I mean arguments based on general conceptual considerations, such as those found in chapter 2 of Mill's *On Liberty*; see, for example, Mill 1997, 52-84, and chapter 10 below.

yet, what Adams seems to be doing is to start with those same premises, add the other considerations of his own argument, which are plausible and unobjectionable, and arrive validly at the thesis of the necessity of argument (proposition 8 or 9).

Perhaps the way out of this apparent paradox is to distinguish ground-level argumentation and meta-argumentation. In all three propositions of that subargument, Fogelin is obviously talking about ground-level argumentation. But Adams is talking about meta-argumentation since a key claim in his argument is that every argument advanced by one side fails to convince the other side, and this is obviously a meta-argumentative claim. On the other hand, such meta-argumentation is dependent on ground-level argumentation, and so the necessity of meta-argumentation requires the presence of the ground-level argumentation. Thus, it is not only meta-argumentation that is essential in deep disagreements, but ground-level argumentation is also essential, and so deep disagreements cannot be undercutting the conditions of arguing, as claimed by Fogelin. It follows that his proposition F1211 remains problematic.

7.2.7 In-principle vs. In-practice Resolution (Phillips 2008)

After these 2005 critiques of Fogelin, Phillips (2008) has published a more recent contribution that advances a new criticism and reinforces some already discussed.

One of Phillips's criticisms is that "we cannot measure the success of an argument merely by whether the issue is rationally resolved on the spot. Argument can also prove effective by beginning a process of reflection leading to eventual resolution, even if the argument's advancement is thwarted in the short-term" (Phillips 2008, 88). This is essentially identical to Lugg's (1986) point, discussed above, that we must distinguish between static and dynamic rationality, and that a deep disagreement may have a rational resolution from the point of view of the latter, even if it does not have a resolution from the point of view of the former.

Phillips (2008, 97) also objects that "the possibility of rationally resolving a disagreement does not, in theory, require any antecedent common beliefs about the topic at hand. It does, however, require certain joint procedural commitments and competencies with respect to the argumentative exchange itself. For instance ... the willingness of arguers to subject challenged beliefs to rational appraisal is a crucial requirement for the successful progression of argument." This corresponds to the criticism which earlier I extracted from Friemann (2002; 2003; 2005), to the effect that in deep disagreements appealing to evidence is insufficient, but appealing to methodological principles is necessary. Here Phillips is mentioning a principle that is more basic than Friemann's principles of open-mindedness and fair-mindedness, and he attributes this principle to Eemeren and Grootendorst (1988; 2004, 191). I would call it the principle of rational-mindedness (cf. Finocchiaro 2010b, xli, 132-34). I would add that, although rational-mindedness may be necessary and more fundamental than open-mindedness and fair-mindedness, by itself it is not sufficient and not especially helpful for the resolution of a deep disagreement. On the other hand, open-mindedness and fair-mindedness are more helpful, since they include rational-mindedness and go beyond. Such interrelationships will be further clarified below, in our reconstruction of Woods's standoffs of force five.

Phillip's main criticism is based on a distinction between two senses in which a disagreement may or may not be subject to rational resolution: in theory and in practice; i.e., in principle and *de facto*. He argues that Fogelin's own arguments and examples (and those of his followers) do not show that deep disagreements are not rationally resolvable in principle, but rather that they are very difficult to resolve in practice: "what makes productive argument seem impossible in many of the examples given by Fogelin and his supporters, then, is merely a problem of execution. Interlocutors are frequently unwilling or unable to rationally assess certain beliefs, to communicate sincerely, or to recognize where their differences lie" (Phillips 2008, 101). I believe Phillips's distinction between resolution in principle and resolution in practice is very important, and it will re-emerge later in our criticism of Woods's argument.

Fourthly, Phillips advances a positive constructive thesis, to the effect that the practical difficulty affecting deep disagreements actually enhances the importance of argumentation: "being conscious of those barriers and working to overcome them where possible is perhaps the most crucial step to enhancing the power of argument in our world" (Phillips 2008, 101). This claim is analogous to the one I extracted from Adams (2005) earlier: there the claim was that argumentation is essential in deep disagreements insofar as only argumentation can reveal if and when a disagreement is deep; here the claim is that argumentation is essential insofar as only argumentation can enable us to overcome the undeniable practical difficulties of deep disagreements, which difficulties we know can in principle be overcome.

From these four claims in Phillips's critique, one could elaborate a more explicit reconstruction of his meta-argument. The procedure would involve doing things such as the following: for each critical claim, determining which proposition or inferential step of Fogelin's reconstructed argument is affected and how; determining what reasons Phillips gives or would give for each of his critical claims; determining whether any one of the four claims supports any one of the others; and determining whether the four claims or some subset of them could be combined to yield some more general or overarching criticisms of Fogelin's argument. However, except for Phillips's third criticism, there is no need in this case to undertake such an explicit reconstruction of his meta-argument, and the task may be left as an exercise for the reader. But that third criticism may be reconstructed as follows: [{10}1][11] Fogelin's argument, specifically proposition F12a, embodies an ambiguity with regard to the notion of resolution, for [{10}11] there are two senses in which a disagreement may or may not be subject to rational resolution: in theory and in practice, i.e., in principle and de facto; [{10}2] Fogelin's own arguments and examples do not show that deep disagreements are not rationally resolvable in principle, but [{10}3] rather that they are very difficult to resolve in practice; it follows that [{10}] the in-principle resolution of deep disagreements remains an open question.

7.2.8 Conceptual Disagreements and Rational Persuasion (Godden and Brenner 2010)

More recently still, Godden and Brenner (2010) have discussed the relationship between "Wittgenstein and the Logic of Deep Disagreements," elaborating a Wittgensteinian perspective, partly motivated by Fogelin's remarks that his own account was a development of Wittgenstein's views. Although the Wittgenstein connection is beyond the scope of the present inquiry,[12] one principal strand of Godden and Brenner's critique raises a novel and valuable point.

To appreciate that strand, we should begin with their argument that "not all differences are disagreements. Disagreement is the contrary of agreement. Thus, it would seem that

---

[11] Here and elsewhere braces are used to treat numbers larger than single digits as a unit rather as an ordered pair resulting from the application of the standard labeling system introduced in chapter 2 above. That is, here '10' designates the initial label for Phillips's third critical conclusion.

[12] See Johnson 2010 for an exhaustive discussion of the possible connections between Wittgenstein and informal logic, and a negative conclusion regarding the Wittgensteinian character of Fogelin's view of deep disagreements.

disagreement is only possible where agreement is also possible. Yet, agreement is only possible where understanding is possible, and understanding, being the result of successful communication, is only possible where communication is possible. So, it would seem that there are a number of important preconditions to what might be called *meaningful disagreement*" (Godden and Brenner 2010, 46-47). It follows further that "it is no failure of rational argumentation that it cannot resolve differences between parties incapable of communicating with each other" (47).[13]

This leads the authors to focus on the existence and role of "differences in conceptual schemes …disagreement in concepts" (48), or, as I shall call them, conceptual disagreements. When these are taken seriously, then we can see that "deep disagreements occur when there is a partial but incomplete accordance in the disputants' use of an expression as well as a partial but significant variation. The depth of a disagreement is due to fact that some aspects of the use of an expression are either indeterminate or incongruous (e.g., disputants might disagree about what should count as evidence for the claim). The meaningfulness of a disagreement is due to the similarities in the use of an expression (e.g., disputants might agree about the consequences of the acceptability of the claim). Thus, deep disagreements tend to occur on the fringes of understanding. Importantly, these are the very features that make deep disagreements impervious to the normal operations of reasons and evidence" (49-50).

However, it is equally important to realize that reasoning can continue to operate in abnormal or exceptional conditions. Indeed, as Wittgenstein (1969, 612) eloquently put it, "at the end of reasons comes persuasion. (Think what happens when missionaries convert natives.)" That is, for Godden and Brenner, "the settling of how this is to be done needn't be either irrational or nonrational. Instead, it involves a kind of 'persuasion' … as a form of rhetoric in the service of concept-formation. While the type of reasoning and argumentation involved here is dialectical rather than demonstrative, amorphous rather than uniform, indeterminate rather than binary, it is neither fraudulent nor relativistic nor arbitrary. To be 'won over' through such persuasion involves accepting a certain picture of the world; it involves learning to apply concepts in a way to which one was, perhaps, not initially inclined, and then 'recognizing' (understanding, judging, appreciating) that this use of concepts is befitting of one's projects … The resultant conceptual shift will involve a new understanding of things; it will be holistic rather than singular—commonly it will involve broad-reaching changes in one's activities (including judgments, inferences, explanations and attitudes). Similarly, it will be made for holistic rather than individual 'reasons'" (76-77).

All these considerations may be reconstructed as an argument, indeed the following meta-argument. [{11}11] It is indeed true that, as Fogelin (proposition F12111) claimed, normal argumentation presupposes shared beliefs, preferences, and procedures. [{11}12] But normal argumentation also involved shared *concepts*. On the other hand, [{11}13] the key feature of deep disagreements is not the absence of shared beliefs, preferences, and procedures (Fogelin's proposition F12112), but rather the presence of partial agreements *and* partial disagreement about concepts; for [{11}1311] deep disagreements must be disagreements in the first place, and so [{11}131] they presuppose communication and understanding; but [{11}132] if communication and understanding were total, argumentation would be normal and disagreement would not be deep. Now, [{11}14] when there is partial agreement and partial disagreement about concepts, rational persuasion is possible. Therefore, [{11}1] deep disagreements are subject to rational persuasion.

---

[13] Starting here in this subsection, references to Godden and Brenner 2010 will be given by means of just page numbers in parenthesis.

Therefore, [{11}] deep disagreements are subject to rational resolution.

This is an argument whose final conclusion contradicts one of Fogelin's principal conclusions (proposition F12a), but whose starting point is an acceptance of Fogelin's final premises (propositions F12111 and F12112). Or to be more precise, there is a qualified acceptance of these final premises. The other considerations this meta-argument adds are claims that are basically Wittgensteinian, and so they should be largely acceptable to Fogelin. In short, this is an *ad hominem* argument, in Johnstone's sense. One interesting feature of this meta-argument is that it apparently reverses Fogelin's relationship between persuasion and reasoning. That is, he seems to be saying that deep disagreements are amenable only to persuasion techniques (F12b) because they are not amenable to rational resolution (F12a). On the other hand, Godden and Brenner are saying that deep disagreements are amenable to rational resolution ({11}) because they are amenable to appropriate persuasion techniques ({11}1).

This interpretation and reconstruction of the Godden-Brenner critique correspond, I believe, to the one which they themselves hint at. In fact, at one point they summarize their critique as follows: "(i) Fogelin, following Wittgenstein, highlights a kind of disagreement that he calls "deep"; (ii) he ascribes to reason a stereotypically different role in deep versus normal disagreements; and (iii) because of this, 'deep disagreements' thus defined allow only for nonrational persuasion in their resolution. We proceeded to argue that (iii) neither follows from (i) and (ii) nor represents Wittgenstein's position. The following section illustrates, through a series of examples, mostly inspired by Wittgenstein, various ways that 'rational persuasion' (as we call it) can operate in disagreements having depth." (56). And I believe my reconstruction only makes this summary more explicit as an argument and more pertinent to Fogelin's own argument.

As reconstructed, Godden and Brenner's meta-argument is largely cogent and plausible. It amounts to stressing the aspect of deep disagreements that involves differences about concepts, as distinct from differences about beliefs (or judgments), preferences (or values), and procedures, which are the only differences Fogelin seems to consider. Their meta-argument is also a plea for a more adequate understanding of persuasion techniques, according to which such techniques can be both good and bad, and when they are good they may be deemed to be rational and to instantiate argumentation; that is, more adequate than apparently allowed by Fogelin, who seems to have an oversimplified view of persuasion techniques, relegating persuasion automatically to the realm of the irrational or nonrational. We will see in our discussion of Woods that his account of standoffs of force five also shows a similarly sounder appreciation of rhetorical persuasion than Fogelin.

My only reservation about the Godden-Brenner meta-argument involves their conception of concepts. That is, once we start taking seriously what concepts are, we may discover much more than the comparison frequently mentioned and exploited by Godden and Brenner, namely that "*concepts : judgments :: measures : measurements*" (71). This comparison is acceptable as far as it goes, but it does not go very far. However, it may turn out that the nature of a concept is to be analyzed in terms of its role in reasoning, or inferential role, that is, in terms of the inferences that would lead one to express a judgment applying the concept, as well as the further inferences one would make from a given application of the concept. Such an inferentialist analysis of concepts has been elaborated and defended with great subtlety, skill, and originality by Robert Brandom (1994; 2000, 1, 10-11, 36-37, 50, 76), and it is not my aim here to elaborate or even utilize it. Rather the only thing I want to note is that, if Brandom's inferentialist conception of concepts is correct, then even conceptual disagreements would be instances of *normal* reasoning, and this would enhance the power of argumentation and rationality for the resolution of deep disagreements. Thus, the exploration of such a possible connection would seem to be a very fruitful project for future investigation.

7.2.9 Proving the Existence of Deep Disagreements

Finally, this is perhaps the proper place to raise a difficulty about another part of Fogelin's argument, the part supporting the claim (proposition F11) that there are (or have been) in fact deep disagreements. He gives the examples of the controversies over abortion and over affirmative action. The structure of this subargument was reconstructed above. This difficulty is indirectly addressed by Adams (2005), who in the course of his general argument about how to show that a disagreement is deep examines the depth of some actual controversies, such as the Terri Schiavo case over life support for comatose patients. And at least one other author mentions this difficulty in passing when, with regard to the controversy over affirmative action, he writes: "even granting that the debate reduces to a fundamental clash of views concerning the cogency of appeals to the rights of groups (as Fogelin plausibly argues), there remains the possibility of settling the issue one way or the other by mounting arguments that are neutral with regard to the question of group rights" (Lugg 1986, 48).

My criticism starts by focusing on Fogelin's claim that the controversy over affirmative action reduces to the issue over the existence of group rights (in proposition F1121). The term *reduces* hints at the fact that in this controversy there are other arguments that are advanced based on other considerations, as indeed is the case. The reduction claim is asserting that all these apparently different arguments can be reformulated, without loss of meaning, as arguments about group rights. Or perhaps the claim is asserting that those other arguments are correct if and only if the group-rights argument is correct. Now, these assertions are generalizations about a class of arguments. To justify them, one must examine each such argument and show that and how it is equivalent to some argument hinging on group rights. I wonder whether this has ever been done by anyone. But even if it had been done, such a justification would be meta-argument about this class of arguments and their relationship to group-rights arguments. Obviously the reduction claim cannot be just asserted without justification, nor can it be proved by merely formulating and defending a group-rights argument. Thus, Fogelin has not adequately shown that the affirmative-action controversy is indeed a deep disagreement.

The same applies to the abortion controversy. My conclusion is two-fold. First, Fogelin's existential claim (proposition F11) has not been adequately justified. Second, meta-argumentation would be required in order to adequately justify it.

Let us now summarize. Fogelin famously argued that deep disagreements are incapable of rational resolution, because they lack the conditions of normal argumentation (shared beliefs, preferences, and procedures) and hence cannot be resolved by argument; and since some important disagreements are deep, it follows that some important disagreements are incapable of rational resolution. His critics have advanced the following objections. First, rationality is dynamic rather than static, and so even if Fogelin's skeptical conclusions (F12a, F1) were true at a particular time of cultural evolution and intellectual history, they might be falsified at a later stage. Second, rationality ought to be aware of its own limitations, and so when a deep disagreement cannot be rationally resolved in favor of one side and against the other, the rational thing to do may be to suspend judgment. Third, rationality cannot be equated with argumentation, but includes other activities and mental attitudes, such as education, experience, training, practice, study, apprenticeship, immersion, open-mindedness, and fair-mindedness; thus, when a deep disagreement is not resolvable through argument alone, it is often resolvable through argument combined with some of these elements of rationality. Fourth, argumentation should not be limited to the normal case of brief, epigrammatic, reason-giving based on shared or common grounds about the subject matter being disputed, instead other kinds of arguments may be necessary: for example, *ad hominem* arguments (in Johnstone's sense) may be effective, but do not rely on common grounds; complex argumentation is a way of incorporating the results of long experience and practice, and it may work even if simple argumentation does not; and meta-argumentation may accomplish what ground-level argumentation cannot, by appealing to such principles as open-mindedness and fair-mindedness and determining whether the disagreement is a deep one in the first place. Fifth, techniques of persuasion may be necessary to help resolve deep disagreements, but not all persuasive techniques are irrational or nonrational, rather some are rational; in this case we can speak of rational persuasion, and indeed of persuasive argumentation, and so we would have rational and/or argumentative resolution of a deep disagreement. Finally, although it seems undeniable that some disagreements are deep, Fogelin has not shown that any existing disagreements, e.g., the controversies over affirmative action and abortion, are deep in his sense; and this could not be shown merely based on the existence of some argument that does not resolve a given controversy and the unproven assertion that all other arguments in the controversy reduce to it.

7.3 Woods on Standoffs of Force Five

7.3.1 A Limitation Theorem

We have already seen that some scholars have discussed under other labels some of the issues examined by Fogelin under the heading of "deep disagreements." For example, as described above, Friemann (2005) has studied "intractable quarrels" and argued that there are similarities between them and deep disagreements, such that we may extrapolate to the latter some of the lessons from the former. And there have been others, who have spoken of "enduring dissensus" in practical argumentation (Kock 2007c; cf. 2003; 2007a; 2007b); of "world-view dissensus" (Freeman 2012, 73); of "polemic" and "culture sensitive" arguments (Kraus 2009; 2012); and of "intractable disputes" (Memedi 2011). The best example of such similarity of content or problem under different terminology or framework is found in Woods's (1992; 1996) account of "standoffs of force five," or fierce standoffs for short. Its insightfulness and instructiveness will become obvious in the course of the analysis below.

Both the similarities and the differences between Fogelin's deep disagreements and Woods's standoffs of force five are immediately apparent from the preliminary remarks with which Woods introduces his discussion: "Standoffs are described in that branch of logical theory which deals with the rational adjudication of conflict, itself a branch of dialectic. Dialectic is, among other things, the logic of negotiated rational acceptance. It lies open to a limitation theorem roughly as follows. Disagreements sometimes yield to ascending strata of intractability and so approach a limit at which they go into a dialectical black hole, to speak voluptuously, whereupon conditions for further negotiations lapse" (Woods 1992, 98). And his concluding remarks show the same identity in difference, while displaying the same inimitable eloquence: "It cannot be denied that the dominant elites' handling of the Canadian abortion mess is dialectically malodorous in all sorts of ways. But it must not be said of these dialectical felonies that they commit the further crime of displacing intellectually pure procedures for rational consensus. They are not capable of doing that. For recall, we have a limitation theorem. No such procedures are available" (Woods 1992, 108).

7.3.2 Degrees of Depth

An important improvement which Woods introduces into the discussion is that he distinguishes five different kinds of deep disagreements depending on the degree of their depth. He calls them standoffs of force one through five, which is meant to be reminiscent of the classification of hurricanes into categories one through five, depending on rain intensity, wind speed, and destructiveness. Woods's definition is cumulative in the sense that a standoff of force $n+1$ is defined as a standoff of force n when an additional condition $n+1$ is present. Thus the definition of a force-five standoff has five conditions, each of which corresponds to one kind of standoff in the cumulative progression just described. That is, a standoff of force five is a disagreement such that:

"(ia)   no consensus exists ... which settles the matter with regard to [the disputed claim] S; and
(ib)   no consensus exists ... about procedures that would or might establish such a consensus" (Woods 1992, 98);
"(ii)   there is no honorifically procedural consensus to let things drop" (Woods 1992, 98);
"(iii) there is no explicit agreement to send the dispute to third-party determination" (Woods 1992, 99);
(iv)   there is no "presumption of prior consent to a legislated solution by a lawfully constituted government, recognized as such" (Woods 1992, 103); and
(v)   the participants do not "acknowledge that the opposite opinions are 'real possibilities'" (Woods 1992, 103), and in that sense are "closed-minded" (Woods 1992, 104).

And as previously suggested, a standoff of force four is one that satisfies conditions (i)-(iv); a standoff of force three is one that satisfies conditions (i)-(iii); a standoff of force two, conditions (i)-(ii); and a standoff for force one, condition (i), which of course has two parts, (ia) and (ib). Several other points are worth noting about this conceptual framework.

Condition (i) amounts to the absence of common grounds regarding factual claims, normative preferences, and resolution procedures, and so it is essentially equivalent to the definiens of Fogelin's definition of a deep disagreement; it follows that what Fogelin calls deep disagreements correspond to what Woods calls standoffs of force one.[14] In fact, each additional condition amounts to defining an additional potential difference that increases the depth of a disagreement or the force of a standoff; for example, condition (iii) explicitly excludes the third-party idea which Friemann claims to be one of the main remedies to intractable quarrels. Thirdly, each additional higher-force condition could be reformulated in more cognitive sounding terminology; for example, condition (ii) is equivalent not only to saying that there is no agreement to disagree, but also to saying that there is no agreement about the significance or insignificance of the dispute. Similarly, condition (iv) is equivalent to saying that there is disagreement about whether the controversy is political or moral; here, a political dispute is one that can be properly resolved through the legislative making, executive application, or judicial interpretation of laws by the legitimate authorities; and a moral dispute is one that includes the question of whether positive public law has any role and if so what.[15]

---

[14] Despite what is suggested by Woods 1992, 104 n. 20.
[15] Here I am adapting Woods's (1992, 103) own distinction between a political and a moral standoff.

Condition (v) deserves special attention. In fact, Woods himself goes to great lengths in elaborating it. My own interpretation of his complex analysis is this. To begin with, at the level of nominal definition, some connections are clear. Open-mindedness is the acknowledgment that the position of the opposite party is a "real possibility." Closed-mindedness is the unwillingness to acknowledge that the opposite position is a real possibility, i.e., the denial that the opposite position is a real possibility, i.e., the claim that the opposite position is *not* a real possibility. I believe these definitions imply not that open-mindedness and closed-minded are contradictories, but rather that they are contraries, in the technical sense of contradiction and contrariety. That is, two contradictories can be neither both true nor both false, but must be one true and the other false; whereas contraries cannot be both true, but can be both false. In fact, if we do not lose sight of the fact that we are in the context of a controversy and that both open-mindedness and closed-mindedness are attitudes or dispositions of one side toward the other, then we can see that both have something in common. Both the open-minded and the closed-minded disputants reject the opposite position, i.e., regard it as false; but at the same time they regard it as a possibility, i.e., admit that it is logically possible for it to be true. In short, in both open-mindedness and closed-mindedness, one regards the opposite position as false, but not necessarily false. Their difference comes with regard to whether or not the possibility is a "real" or serious one.

Woods (1992, 104) starts with an explicit definition of the negative case, when a position is *not* regarded as a real possibility. His definition is mostly in terms of the notion of knowledge and belief, which I will try to rephrase into argumentation terminology. The first step of my reformulation would yield that to claim that the opposite position is not a real possibility is to claim that (a) the position is false; (b) there are good arguments against it; and although (c) it is possible that these arguments are bad, (d) this possibility is not a good reason to think that these arguments are bad. Although this makes some sense as it stands, it is not clear that this captures the notion of closed-mindedness; moreover, the non-satisfaction of these four conditions would yield nothing approaching the notion of open-mindedness.

To remedy this, we can start by dropping condition (a), leaving it implicit, it being understood, as mentioned before, that both closed-minded and open-minded proponents of one side think that the other side's position is false. Next, let us rephrase conditions (c) and (d) in terms of what is likely to be their underlying rationale; (c) can be rephrased as the claim that there are arguments in favor of the opposite position, and (d) as the claim that these (contrary) arguments are not (necessarily) good. Combining the two modified conditions, we get that there are no good arguments in favor of the opposite position. In other words, to think that the opposite position is not a real possibility is to think that there are good arguments against it and no good arguments in favor. Then, to think that the opposite position is a real possibility would be to think that either there are no good arguments against it or that there are some good arguments in favor of it. However, the latter still does not capture the notion of open-mindedness, in the sense of an open-minded opponent of a position; rather it captures the notion of a *proponent* of a controversial position.

To get at the root of this difficulty, I believe we need to admit that there is a third property shared by closed-mindedness and open-mindedness. This is in addition to the two mentioned earlier, namely that both closed-minded and open-minded disputants think that the opposite position is false, but not necessarily false. This third property is that they both think that there are good arguments against the opposite position, i.e., good arguments in favor of their own position. Given that they share these three characteristics, one possible difference might be that the open-minded disputant admits, and the closed-minded disputant denies, that there are good arguments in favor of the opposite position. But this is still insufficiently precise because although it will work as a conception of closed-mindedness, the implication for the definition of open-mindedness is implausible. The implication would be that open-mindedness means to think that there are good arguments against the opposite position and there are good arguments in favor. However, the more plausible formulation would be to admit that there are some good arguments in favor of the opposite position, but to claim that there are better (or much better) arguments against it. In other words, it would be preferable to define closed-mindedness as claiming that there are good arguments against the opposite position and no good arguments in favor, i.e., good arguments in favor of one's own position and no good arguments against it; whereas open-mindedness would be claiming that there are some good arguments in favor of the opposite position and better arguments against it, i.e., some good arguments against one's own position and better ones in favor of it.

However, although these definitions would be consistent and viable, open-mindedness so defined would include what I earlier called fair-mindedness, but it is useful to distinguish these two cognitive traits. To do so, we can start with the closed-minded disputant's characteristic claim that there are no good arguments in favor of the opposite position (i.e., against his own position). This obviously includes two possibilities: that there are no contrary arguments, and/or that the contrary arguments are worthless. Similarly, in the other definition there are two parts in the clause that there are some good arguments in favor of the opposite position (i.e., against one's own position): that there are some such contrary arguments, and that some of these contrary arguments have some worth. Now, an important and common intermediate attitude is to claim that there are contrary arguments, but that they are completely worthless; and this is an attitude that is not closed-minded because of the admission of the existence of contrary arguments, but which is not open-minded (in the Woodsian sense extracted above). What label should we give to this intermediate disposition? I prefer to call it open-mindedness (in a different sense than the Woodsian meaning), and relabel fair-mindedness the Woodsian notion just elaborated.

These considerations yield the following framework, which I believe is in the spirit (not the letter, to be sure) of Woods's discussion, as well as in accordance with the intuitions which I expressed apropos of Friemann's critique. In controversies of various kinds, deep disagreements, intractable quarrels, and standoffs of various forces,[16] normally the advocates of one side claim that the position of the other side is false, refutable by good arguments against it, albeit not necessarily false but logically possible. With this in the background, then there are three distinct cognitive stances. To be closed-minded means that one also claims that there are no contrary arguments, i.e., arguments in favor of the opposite position or against one's own. To be open-minded means that one also acknowledges that there are contrary arguments. To be fair-minded means that one acknowledges not only that there are contrary arguments, but also that some of them have some worth (although less than the good arguments favoring one's own position or against the opposite one). It is obviously possible to be open-minded but not fair-minded, although if one is fair-minded one is already open-minded; that is, fair-mindedness implies (includes) open-mindedness, but not conversely. Moreover, closed-mindedness (as now defined) is the contradictory ("the opposite") of open-mindedness, but not of fair-mindedness, for obviously open-mindedness affirms and closed-mindedness denies that

---

[16] But note that similar remarks apply to similar phenomena that have been studied under other labels, such as political debates and enduring dissensus; so it is not surprising that in such similar contexts scholars have elaborated a notion of open-mindedness and a need for a metacognitive stance. See, for example, Adler 2004a, Dryzek and Niemeyer 2006, and Kock (2007a; 2007c).

there are any contrary arguments, whereas both closed-mindedness and fair-mindedness could be absent if and when one claimed that there are contrary arguments but that they are worthless. On the other hand, the relationship between closed-mindedness and fair-mindedness is one of contrariety, for if either one of them is present then other one is not.

7.3.3 Appealing to Methodological Principles

Be that as it may, another important improvement which Woods introduces into the discussion is the explicit consideration of appeals to methodological principles to resolve disagreements or standoffs. That is, when it appears as if one is involved in an intractable disagreement and argumentation seems powerless to resolve it, a worthwhile possible remedy is to "look for a heretofore unrecognized 'tie-breaking' methodological principle, one on which the protagonists might be expected to agree, and apply it" (Woods 1992, 100). The principles he considers make up a rich collection pregnant with meaning.

One principle is called Ramsey's Maxim. In fact, Woods (1992, 100) quotes it from Ramsey's book on *The Foundations of Mathematics*: "in such cases it is a heuristic maxim that the truth lies not in one of the two disputed views, but in some third possibility which has not yet been thought of, which we can only discover by rejecting something assumed as obvious by both disputants" (Ramsey 1931, 115-16).[17]

One example suggested by Woods is the following. In the abortion controversy, pro-choice advocates tend to formulate their position as the claim that the choice whether or not to abort should always be available to a woman (i.e., should never be curtailed), and that pro-life advocates tend to formulate theirs as the claim that abortion is always wrong (i.e., never permissible). Now, it is probably the case that a common assumption of both sides is that the issues over abortion, choice, and life should be resolved by means of absolute, categorical, universally valid principles speaking of what holds always or never. If this is so, then one could apply Ramsey's Maxim by rejecting this assumption and claiming that one should look for qualified and nuanced claims about what is sometimes the case and try to explore and formulate appropriate conditions under which life ought to prevail over choice and other conditions under which choice ought to prevail. Of course, this may not work, but it would cast the debate in a new light.

This example is also an illustration of another methodological principle, "a second maxim, the Maxim of Moderation: Avoid extreme answers" (Woods 1992, 101). In this case, extremism would be the tendency to formulate one's position by means of universal generalizations asserting what is always or never the case. And moderation would be the focus on the nuances allowed by the existential quantifier "sometimes." Woods is clear that the Maxim of Moderation and Ramsey's Maxim are not equivalent, but it is equally clear that they have an area of overlap.

A third methodological principle is what Woods calls the Fundamental Law of Collective Bargaining. This is the "suggestion that disputes should always be settled by taking a middle position—and this we might note is the Fundamental Law on collective bargaining" (Woods 1992, 101). In other words, this principle stipulates that one split the difference that separates the two positions in a standoff.

Fourth, Woods considers what he calls the Pascalian Minimax Strategy. This principle enjoins us to "so constrain public policy that, of the options in question, it is best to back the one of least morally harmful consequences should it prove to have been the mistaken

---

[17] Zarefsky (2012, 82-83) is another writer on deep disagreements who has also shown an appreciation of Ramsey's Maxim, although under the labels of the rhetorical strategies of "incorporation" and "subsumption" of the two sides into a larger framework.

option" (Woods 1992, 103). In other words, we should "settle the issue in such a way as minimizes the realization of the greatest possible cost. The higher the cost the more it is mandatory to minimize the possibility of its exaction" (Woods 1996, 654). As the name suggests, the basic idea is to minimize the maximum possible harm.

Finally, there is the Last Gasp Dialectical Response in which the government and the dominant elites appeal to the citizens to keep the critical discussion going and not resort to violence. In Woods's words: "**LGDR**: *Last Gasp Dialectical Response*, 'Citizens ... What is at risk is that minimal social harmony which is the very condition of a good and peaceful life [and, more to the point, of the general practice of dialectical felicity]. The Government urges that the collective self-interest requires that tensions be lessened and that the most serious consideration be given by all to settling this matter [say, by a free vote in the House of Commons]" (Woods 1992, 105). Woods comments that in such an appeal to rationality, "the Government is openly proposing the voluntary collective downgrading of the dispute from a force five standoff to a force three standoff which fulfills the conditions on being a (merely) political disagreement. In this proposal, it is not suggested that people change their values but that, out of consideration for the commonwealth, they not act on them intractably or unlawfully" (Woods 1992, 105).

The last comment underscores both the power and the limitations of such methodological principles. Their power stems from the fact that such principles are explicitly formulated to deal with standoffs or disagreements that have reached an impasse when the participants attempt to revolve them by means of normal argumentation, which appeals to shared facts, preferences, and procedures. Their limitations are obvious from the fact that the nature of some standoffs is such that by definitions the participants rejects some of these principles. Obviously we cannot resolve a disagreement that involves the rejection of the means to resolve it; we cannot convince someone who refuses to be convinced. This enables us to see the plausibility of Woods's "*Limitation Theorem*: Standoffs of force five are logically irresolvable" (1996, 655), taking logic here to refer to both normal argumentation and appeal to methodological principles. And this also allows us to appreciate what I shall call Woods's Rhetoric Corollary, namely the thesis that quite properly "logic now defers to rhetoric" (Woods 1996, 655).

7.3.4 An Appreciation of Rhetorical Persuasion

About rhetoric, Woods has many insightful and wise things to say, and they constitute a third great improvement which he introduces into the discussion of deep disagreements and intractable quarrels. He is, of course, aware that philosophers in general, and logicians in particular, tend to dismiss rhetoric, under the influence and legacy of Socrates and Plato's hostility to sophists and sophistry. However, Woods's argument is difficult to fault:

If concerning any issue there should happen to be a fact of the matter, then it would be oddly illogical not to want to present the truth persuasively, that is in ways that maximize the chance of getting others to see that it is true. More interesting are those cases concerning which no consensus exists with regard to the question of truth. For wide ranges of such issues, giving up on critical discussion and going fishing is not a realistic option, and we are left with the hard question of what to do in the wake of intractable disagreement. Some issues require the fixation of belief, never mind that the matter at hand is underdetermined by the agreed-upon evidence ... So a central task of rhetoric is to specify measures for the fixation of belief about matters underdetermined by the agree-upon evidence and to establish that (and in what sense) such measures are acceptable. [Woods 1992, 107 n. 25; 1996, 655, n. 6]

Woods is also aware of the voluminous literature on the "social technology of persuasion" (1992, 106; 1996, 655), and of the fact that that it tends to be negatively critical of rhetoric. And he acknowledges that such literature in part reflects a modern social reality characterized by the phenomenon of the manipulation of beliefs and the "manufacture of consent," to use Noam Chomsky's expression (Herman and Chomsky 1988). And Woods admits that "there is evidence enough of thought-control to warrant our concern and chagrin" (1992, 107; 1996, 656). However, he is perceptive and judicious enough to realize that "much of the mass media bashing of recent years is half-baked, paranoid and politically self-serving" (Woods 1992, 107; 1996, 656). Or as I would put it, Chomsky's "manufacture of consent" is itself a good example of the problem against which he himself inveighs.

For Woods, the real challenge is to describe (and perhaps analyze and systematize) the rhetorical techniques and strategies that become common and necessary when a disagreement is so deep that it resists logical or rational resolution.[18] And such an account should do so without adding to, or exacerbating, the problem. Nor should it make facile and gratuitous assumptions about conspiracies by government officials, business leaders, or media moguls. In regard to the abortion controversy in Canada, Woods mentions the following rhetorical strategies used by dominant elites to defuse its force or depth:

1. Discredit the leaders as extremists ...
2. Demoralize the population ...
3. Marginalize the visible protagonists ...
4. Trivialize the contending values ... Invoke the non-cognitivism of moral principles. Emphasize their relativity.
5. Adjust the taxonomy. These extreme positions are also religious positions, fine as long as they do not intrude and, in any event, subject by implicit prior consent to the sanctity of the separation of Church and State.
6. Saturate communications with euphemisms, the more vapid the better. (Thus 'pro-choice', 'pro-life'). Keep disclosure of clinical details to a minimum ... Be tasteful.
7. Guilt by association. [Woods 1992, 107-8; 1996, 656-57]

This list is not meant to be exhaustive or definitive. Moreover, the various items would deserve more discussion not only than that quoted here (where I have omitted some parts), but also than that provided by Woods in his account. However, they do give us a sufficient idea of what he has in mind by rhetorical persuasion, as distinct from appeals to methodological principles and from logical argumentation. This is sufficient for the purpose of us glimpsing at his main line of argument. In fact, we are now in a position of being able to state a reconstruction of his main argument.

7.3.5 Reconstruction of Woods's Argument

Woods's argument may be reconstructed as follows: [W1] There are some disagreements that are not subject to rational resolution, but only to rhetorical persuasion. For [W11] there are disagreements that are standoffs of force five, [W111] such as the abortion controversy in Canada. Now, [W12] standoffs of force five are controversies where the two contending parties disagree about (i) particular facts, values, and procedures to which they might appeal to settle the issue; (ii) overlooking the disagreement; (iii) the propriety of third-party determination; (iv) the propriety of a political solution legislated by the government; and (v) whether the opposite position is a real possibility. And [W13] such

---

[18] Such a challenge would be later appreciated and in part met by Godden and Brenner (2010) and by Zarefsky (2012); in particular, the latter elaborated eight kinds of strategic moves.

standoffs are not subject to rational resolution because [W131] they are not resolvable by dialectical, or logical, argumentation, and [W132] they are not resolvable by appealing to methodological principles. However, [W14] such standoffs are amenable to rhetorical persuasion, partly because rhetorical techniques are [W141] unavoidable and [W142] desirable when rational methods (of argumentation and methodology) fail.

Now, it is a mere mechanical procedure to construct the structure diagram for this argument. It is the one shown in Figure 7.2.

Figure 7.2

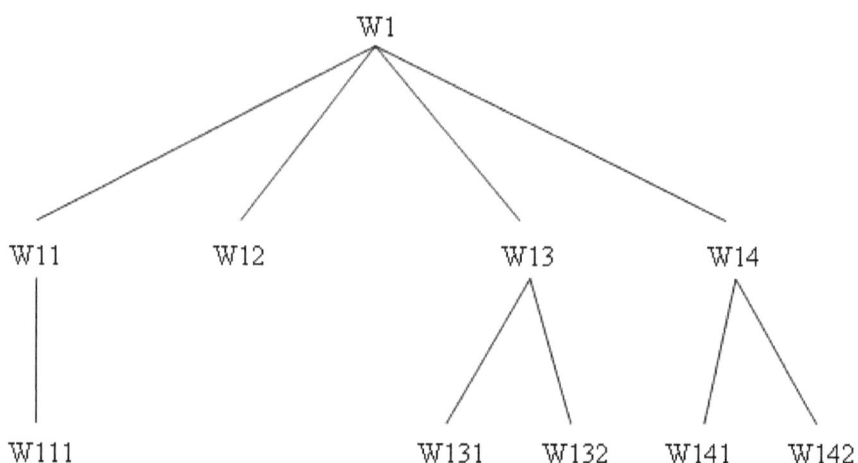

Woods's argument is stronger than Fogelin's argument in support of the same conclusion. Its superior strength derives partly from the greater strength of Woods's standoffs of force five, as compared to Fogelin's deep disagreements. For as mentioned earlier, Fogelin's deep disagreements basically correspond to standoffs of force one, and Woods's standoffs of force five are four levels deeper or stronger; that is, they are characterized by four additional conditions, each of which blocks a potential route to rational resolution. The greater strength of Woods's argument also derives from the fact that he explicitly considers appealing to methodological principles as a possible means of rational resolution, and explicitly argues for their ineffectiveness in the case of force-five standoffs. Finally, a third merit of Woods's argument is that he makes an explicit attempt to support the rhetoric part of the main conclusion, and his attempt involves a serious and nuanced view of the pitfalls as well as inescapability of rhetoric.

7.4 Criticism of Woods's Argument

7.4.1 Logical and Methodological Rationality of Rhetorical Persuasion

Despite these strengths, there are difficulties with Woods's argument. One is that, as I shall argue next, rhetorical techniques can be interpreted in large measure as appeals to methodological principles, or as special cases of normal argumentation, and hence their use does not fall outside the scope of rationality. Another difficulty is that Woods underestimates the power of appeals to methodological principles, especially Ramsey's Maxim, and to show this I will soon discuss an important and substantial example of the utilization of this maxim. Thirdly, appeals to methodological principles can be seen as special cases of argumentation, namely meta-argumentation, and hence such appeals do not really fall outside the domain of the logic or dialectic of dispute resolution either. Fourth, what really follows from Woods's argument is not that some standoffs are incapable of rational resolution, but rather that they are not de facto resolved, because their participants are incapable of rational behavior. Finally, Woods too neglects the existence of *ad hominem* argumentation (in Johnstone's sense) and its power in the rational resolution of disputes by dialectical, logical, or argumentative means;[19] I shall have more to say about this when I discuss explicitly Johnstone's account of philosophical controversies.

Let us begin such criticism by focusing on Woods's Rhetoric Corollary. This is the claim that in standoffs of force five "logic ... defers to rhetoric" (Woods 1996, 655), which in turn corresponds to proposition W14 in the above reconstruction of his argument. Given his nondismissive, nuanced, and appreciative conception of rhetoric, it seems to me that this claim cannot be construed as a skeptical or anti-rationalist position. And this, in turn, should be apparent from the way I have reconstructed the supporting subargument (W141, W142, W14). Even so, Woods may not be doing full justice to the nature and role of rhetorical persuasion. For if we look at his own examples of rhetorical techniques, taken from the abortion controversy in Canada, it is not obvious that they really fall outside the scope of rationality in the sense of methodology and argumentation. Let us examine those techniques.

The first one is to "discredit the leaders as extremists" (Woods 1992, 107). It seems obvious that this can be interpreted as a straightforward application of Woods's own Maxim of Moderation.

The second rhetorical technique is to "demoralize the population" (Woods 1992, 107). Here we must first elucidate the meaning of this notion, which is relatively obscure in his earlier discussion (Woods 1992). However, later Woods (1996, 658) explains that he means demoralization not in the ordinary sense of a lowering of the spirits, but rather in the etymological sense, stemming from the French Revolution, which is tantamount to "de-moralization." That is, to demoralize the population with regard to a given issue is to manage to have the population abandon the moral or ethical standing or problematicity of the issue, together with the concomitant tendency to be absolutistic and not to compromise. Thus, it seems to me that this technique is in accordance with Woods's own Last Gasp Dialectical Response, which attempts to downgrade a moral standoff (of force four or five) to a political or lower standoff.

The third technique is to "marginalize the visible protagonists" (Woods 1992, 107). Here, Woods himself elaborates it as an application of Ramsey's Maxim, when he says: "Take care to identify Ms. X as a member of the Feminist Caucus and Mr. Y as a Mormon bishop. For all its dialectical limitations, endorse and promote Ramsey's Maxim: the most sensible solution is a compromise. It sounds *so* right" (Woods 1992, 107). Although Woods mentions explicitly only Ramsey's Maxim, his talk of compromise suggest that perhaps what is in operation is also the Fundamental Law of Collective Bargaining.

Fourth, to "trivialize the contending values" (Woods 1992, 107) can also be seen as an application of the Maxim of Moderation and/or the Last Gasp Dialectical Response. This emerges from Woods's own elaboration: "Trivialize the contending values, not only as extreme but as *moral*. Thus moral choices are private, a matter of personal opinion only, are not to be taken up invasively ('forcing your personal values upon me'). Invoke the non-cognitivism of moral principles. Emphasize their relativity" (Woods 1992, 107). It should

---

[19] This critical remark applies only to Woods's articles on standoffs of force five, for in his more ambitious and articulated work on conflict resolution in the abstract sciences, he (Woods 2003, 17-30, 36) elaborates and exploits what he calls a method of costs and benefits whose essential content is precisely *ad hominem* argumentation in the sense of Johnstone and Locke.

be noted that here I am not claiming that moral absolutism is incorrect and moral relativism correct; rather I am saying that the rhetorical technique of "trivializing" values is an example of the methodological strategy of appealing to the Maxim of Moderation and/or the Last Gasp Dialectical Response.

Fifth, there is Woods's (1992, 107-8) technique of adjusting the taxonomy. For example, in the abortion controversy, one might argue that the pro-life position is a religious doctrine, insofar as it is based on Holy Scripture; that the pro-choice advocacy of a legislated solution that protects individual freedom of choice is a political position; and that the principle of separation of Church and State dictates the rejection of pro-life prohibitions and the adoption of freedom of choice. I am not claiming that this argument is conclusive or correct, but only that it is an argument, and that it involves a re-adjustment of the issues, insofar as it is based on the separation of Church and State—a principle which is perhaps not the most obvious one to appeal to in this particular controversy. It follows that such "re-adjustments" are legitimate argumentative strategies, and so this rhetorical technique can be seen as falling within the scope of rationality by way of logic or dialectic, that is, even more directly than if it were by way of methodology.

The sixth rhetorical technique is a cluster that involves such practices as using euphemisms, minimizing the discussion of repulsive or sickening clinical details, and complying with standards of tastefulness. These practices can be interpreted as special cases of the avoidance of extremes of various kinds, and hence as applications of Woods's Maxim of Moderation. Of course, sometimes such devices may be carried too far, and thus become potentially self-referentially inconsistent violations of this maxim; or they may be problematic in other ways. However, once again, the question here is not whether or not the Maxim of Moderation is being used in a justifiable or correct manner, but whether or not it is being appealed to or used. And while the answer to the former question is unclear or debatable, the answer to the latter question is obviously positive.

Finally, Woods (1992, 188) mentions "guilt by association." The example he gives is very revealing. In Romania in the 1980's, "President Ceaucescu ran a monstrous tyranny. He imposed absurd and burdensome reproduction targets on Romanian women, in an effort to increase the population substantially. Abortions were also forbidden" (Woods 1992, 108). My hunch is that here we have an incorrect or illegitimate attempt to criticize the pro-life position on the grounds that Ceaucescu's Romania subscribed to it. However, the attempt is an argument. Even if this is a weak, worthless, or incorrect argument, it is still an example of argumentation. The argument could be reconstructed in various ways, as an argument from analogy, an inductive generalization, and an argument against the person (in the sense of the inverse of an argument from authority[20]). Once again, this makes guilt by association an argumentative strategy, and hence subject to rational adjudication.

7.4.2 The Power of Ramsey's Principle

My next criticism is that Woods does not do full justice to the power of appealing to methodological principles in general, and to Ramsey's Maxim in particular. This maxim is the principle that in deep disagreements one should explore the possibility that the truth lies in the denial of some assumption presupposed by both sides. Let us recall that Woods quotes a version of this maxim advanced by Ramsey in a work on foundations of mathematics and the philosophy of logic. Ramsey was discussing the dispute between Bertrand Russell and W.E. Johnson about whether general terms can be subjects as well as

---

[20] For this notion of argument against the person, see M. Salmon 2002, 122-26, and W.C. Salmon 1984, 101-4.

predicates of atomic propositions. This context is no accident. In fact, the foundations of a discipline typically involve fundamental issues, so fundamental as to generate intractable disputes and apparently irresolvable disagreements.

However, as mentioned above, Woods illustrates this maxim with an example from the abortion controversy, in which both sides tend to advocate theses having the form of universal generalizations that are unqualified or exceptionless. Thus the solution proposed by Ramsey's Maxim, namely to formulate one's theses as nuanced existential generalizations, turns out to be an illustration of the Maxim of Moderation as well. Although it is clear that Woods realizes that these two maxims are distinct and merely overlap rather than being equivalent, it is equally clear that it would be desirable to discuss some examples of Ramsey's Maxim that are independent of the Maxim of Moderation.

Now, there happens to be a brilliant discussion of such examples in the context of the philosophy of the social sciences. It is found in Piotr Sztompka's (1979) *Sociological Dilemmas*.[21] Sztompka does not even mention Ramsey's name, but rather follows an approach which he labels "dialectical," in the Marxian or Hegelian sense of this term. However, terminology aside, the similarity or identity of what Woods calls Ramsey's Maxim and what Sztompka calls a dialectical approach is clear and unmistakable, as can be seen from the following summary.[22]

The sociological dilemmas examined by Sztompka are the following six controversies: science versus humanities (or the naturalism question); science of man versus science of society (or the reductionism issue); knowledge versus action (or the activism controversy); detachment versus bias (or the axiologism problem); man as object versus man as subject (or the determinism dispute); and society as a whole versus society as an aggregate (or the collectivism controversy). Sztompka formulates these dilemmas as follows. The naturalism dilemma is not about the true character of the natural sciences, but about the general character of the relationship between the social and the natural sciences. The reductionism problem is not about the ontological status of individuals and collectives, which is a separate issue, but about the relationship between the sciences of psychology and of sociology. The activism dispute is not about whether a social scientist may get involved in extrascientific activities as an ordinary citizen, or whether he has an ultimate practical motivation, or whether effective social action needs to be grounded on sound knowledge, or whether applied sociology is a legitimate branch, or whether sociological research is itself a type of practice, which are all uncontroversial matters; what is not uncontroversial is whether pure sociological theories should contain practical elements. The axiologism dilemma is easily confused with the previous one; the difference is that the present issue concerns whether or not sociological research can avoid making extrascientific value judgments. The question of determinism is neither whether mankind has a universal nature or merely a history, nor whether human action is determined internally or externally, nor whether it is determined consciously or subconsciously, nor whether it is impulsive or deliberate, but rather whether human action is or is not controlled by human beings. The collectivism dilemma is neither the moral-political question of the value of individual rights vis-à-vis society, nor the issue of liberalism and totalitarianism, nor the psychological problem of two opposite personality traits, but rather the question of the ontological status of supra-individual entities.

Having defined the problems, Sztompka proceeds toward their solution as follows. He first identifies what he calls "common meta-assumptions" which both contending parties make, that is the presuppositions they both share. Both naturalists and anti-naturalists

---

[21] Another analogous example is found in Agassi 1977.
[22] For more details, see Finocchiaro 1980c.

presuppose (Sztompka 1979, 69) that the two classes, natural sciences and social disciplines, have pronounced internal similarities and pronounced external divergences; that the potential methodological models are only within the class of natural sciences; that methods have an inherent value independent of the problems that need solving; and that the methods of the natural sciences constitute a monolithic unit to be accepted or rejected in toto. Similarly, both reductionists and antireductionists assume (Sztompka 1979, 114-15) that psychology and sociology are internally homogeneous and externally heterogeneous; that there is only one relevant direction of reduction, sociology reduced to psychology; that reduction would have to be complete; and that it must be a matter of principle rather than being contingent; moreover, it is also presupposed that definitional reduction of concepts and explanatory reduction of laws go hand in hand. The meta-assumptions common to both contending parties in the activism controversy (Sztompka 1979, 165-66) are that the social world exists independently of the thinker or agent, as nature does; that it exists as a constant, static, and unchanging entity; that it exists "out there" external to those who want to understand it and to those who want to change it; that understanding and practical action are separate dichotomous activities; and that there is a dichotomy between descriptive and normative statements. In the axiologism dilemma, both parties presuppose (Sztompka 1979, 219) that objectivity is an abstract property of scientific research and results determinable by an analysis of their internal structure; that it consists of lack of bias; and that bias is essentially related to value-judgments. In the determinism dispute, both parties assume (Sztompka 1979, 272) that mankind and the environment are separate entities, and that human nature is definable in terms of abstract rather than relational properties. Finally, both collectivists and individualists (Sztompka 1979, 305) conceive of social entities as substances rather than structures.

The next crucial step in Sztompka's approach is to reject these presuppositions by arguing in favor of principles contrary to them. The new principles dissolve the dilemmas, make possible a distinct model of sociological inquiry, and yet contain in each case elements from both sides of the dilemma. His principle of "integralism" states that the social sciences can and should apply the methods of empirical science, but not those of natural science (Sztompka 1979, 76). His principle of "separatism" claims that the *meaning* of sociological concepts can be reduced to that of terms pertaining to individuals and their activities, but that sociological *laws* are not derivable from psychological ones (Sztompka 1979, 119). What he calls "constructivism" holds that a sociological theory must contain information about the regularities existing in the world and contribute directly to its transformation (Sztompka 1979, 172); moreover, it must contain both "categorical" and "normative" propositions (Sztompka 1979, 17 3- 74) . His principle of "commitment" asserts that sociology cannot and should not avoid value-judgments, but that both its methods and its results can be unbiased (Sztompka 1979, 222). According to the principle of "creativism" man enters into creative, productive relationships with his natural and social environments, but these relationships are shaped under the impact of specific and historically varied contingencies over which he has no control (Sztompka 1979, 273-74). Finally, "structuralism" holds that there is no specific social substance, only individual human beings, and special social structures, representing the network of human relations (Sztompka 1979, 308-9).

Sztompka claims that this type of social inquiry is correct in the sense that it represents the most promising and fruitful paradigm. He also claims that such sociological inquiry resolves the dilemmas, and does so through a "dialectical overcoming" (Sztompka 1979, 34) of them.[23] Here my point is not to say that Sztompka's approach to sociology is

---

[23] For further elaboration, see Sztompka 1979, 32-36.

flawless, nor to claim that his conception of Hegelian dialectic is the most tenable one.[24] Rather I want to say that his analysis of sociological dilemmas is sophisticated, intelligent, and well-argued. Further, I want to stress that Sztompka's method of dialectical overcoming consists of (a) the identification of meta-assumptions common to the two sides of a controversy and (b) their rejection through the adoption of principles contrary to them. Finally, I want to point out that Sztompka's principle of dialectical overcoming corresponds to what Woods calls Ramsey's Maxim.

7.4.3 The Meta-argumentative Rationality of Methodological Appeals

My third criticism of Woods's argument is similar to my first. That earlier objection was that rhetorical appeals, such as the ones mentioned by Woods, if analyzed properly, can often be seen to be either methodological appeals or appeals to reasons and evidence; and so they fall within the scope of rationality. Similarly, now I want to stress that although appeals to methodological principles represent an important level of rationality, they are often instances of argumentation, specifically meta-argumentation.

Let us see this by reference to some of what Woods himself says regarding the Maxim of Moderation and the abortion controversy. In appealing to this maxim, one step is to note that the pro-life or anti-abortion position is often stated in terms of the universal generalization that abortion is *always* wrong. Another step notes that the pro-choice or pro-abortion position is typically stated as the universal generalization that the freedom to choose whether to have an abortion is inalienable, i.e., *always* legitimate. The advocate of moderation then says that the arguments on both sides are weak or flawed insofar as their conclusions are extreme theses of this sort. Next, the moderate claims that if one's conclusion is formulated in a qualified or nuanced manner, then it becomes self-consistent to claim that abortion is *sometimes* right and *sometimes* wrong. Additionally, such a moderate conclusion can be justified by means of arguments that are revisions of the previous extremist arguments when the latter have been stripped of their untenable extremist pretensions. Now, such considerations are an argument for the moderate position based in large measure on an interpretation and criticism of the typical pro-choice and pro-life arguments; and clearly such an argument is a meta-argument.

The same point can also be made by reference to the principles of open-mindedness and fair-mindedness. These normative principles prescribe that in controversies one should have these attitudes. Recall that to be open-minded means that one acknowledges that there are contrary arguments. To be fair-minded means that one acknowledges not only that there are contrary arguments, but also that some of them have some worth (although less than the good arguments favoring one's own position or against the opposite one). Practicing such principles involves, respectively, requiring of oneself that one learn and master the content and structure of the contrary arguments, and that one make an effort to appreciate their strength before attempting to refute them. Appealing to these principles may involve pointing out to one's opponents that they do not know the arguments favoring one's own side, or that they know those contrary arguments but have misunderstood them or failed to appreciate them is some particular way. All this clearly is argumentation consisting of identification, interpretation, and evaluation of ground-level arguments; and that's what meta-argumentation is.

---

[24] For some criticism (together with an appreciation) of Sztompka's account, see Finocchiaro 1980c. For an account of what I regard as the most adequate concept of (Hegelian) dialectic, see Finocchiaro (1988; 2005a, 207-30).

7.4.4 De Facto vs. Rational Resolution, Again

My fourth difficulty with Woods's argument may be elaborated as follows. Recall that force-five standoffs are disagreements in which one or both of the contending parties are closed-minded. Recall also that closed-mindedness is the unwillingness or inability to acknowledge that the opposite position is a real possibility, i.e., to take seriously the opposite position, i.e., to acknowledge the existence of contrary arguments (namely, arguments in favor of the opposite position or against one's own). I believe that what really follows from this is not that such standoffs are incapable of rational resolution, but rather that they are definitionally incapable of *actual* resolution. For insofar as we can plausibly claim that closed-mindedness is a questionable attitude from the point of view of rationality and that open-mindedness and fair-mindedness are desirable requirements of rationality, we can also say that if the contending parties of a force-five standoff were more or fully rational, their disagreement could probably be rationally resolved. In short, force-five standoffs are not subject to resolution if, insofar as, and because the contending parties behave irrationally; they are a reminder of the empirical existence of human irrationality, not a limitation on the normative ideal of rationality.

This criticism can be analyzed as follows. Let us focus on the following subargument of Woods's over-all argument: [W13] such standoffs are not subject to rational resolution because [W131] they are not resolvable by dialectical, or logical, argumentation, and [W132] they are not resolvable by appealing to methodological principles. I am saying that [W133] in a sense it is true that force-five standoffs are not resolvable by dialectical or logical argumentation, in the sense of argumentation guided by the attitude of closed-mindedness but not by the attitude of open-mindedness or fair-mindedness. Similarly, [W134] it is true that such standoffs are not resolvable by appealing to methodological principles, as long as we do not include open-mindedness and fair-mindedness among our methodological principles. However, it is also true (or at least it could be argued) that [W135] the attitude of closed-mindedness is irrational; that [W136] open-mindedness and fair-mindedness are requirement of rationality; and that [W137] dialectical or logical argumentation guided by the principles of open-mindedness and fair-mindedness would be likely to resolve the disagreement. What follows is that [W13'] force-five standoffs (as defined by Woods so as to include closed-mindedness) are by definition not susceptible to de facto resolution, but the disagreements underlying such standoffs are such that it is rational to treat them with open-mindedness or fair-mindedness and thereby attempt to resolve them where possible. It follows further that although proposition W13 is true in a literal, but inconsequential and insignificant, sense, in its substance it is false and its opposite true.

This analysis is meant to elaborate the *ad hominem* character of this criticism, *ad hominem* in Johnstone's sense. That is, this criticism is an argument that attempts to derive a conclusion critical of Woods's proposition W13 (and so not acceptable to him in that sense) from premises that are relatively uncontroversial (and so are acceptable to him in that sense) but go beyond the premises in the original argument. And this characterization brings us to my fifth and last criticism of Woods's argument, but also to the long anticipated topic of Johnstone's account of philosophical controversies, which also deserves examination for its own sake.

7.5 Johnstone on Philosophical Controversies

7.5.1 Definition, Examples, Theses

Let us begin by showing that Johnstone is addressing the same problem as those scholars who have dealt with deep disagreements, intractable quarrels, and force-five standoffs, albeit with slightly different but partially overlapping terminology.

For Johnstone, "fundamental philosophical" disagreements are disagreements such that each of the two contending parties thinks that (a) it is inconceivable "what it would be like for his opponent's statement to be true" (Johnstone 1959, 1); (b) "his [own] position includes all the relevant evidence" (1);[25] (c) "no statement adducing evidence against it [one's own position] is possible" (1); (d) "to attack [his] view by appealing to [counterinstances] is to beg the question" (1-2); (e) one "can reduce any allegedly [disconfirming] factor to a [confirming] factor" (2); (f) one cannot "interpret the disagreement as consisting of the fact that the other has made a statement incompatible with the one he has made" (1); (g) one cannot without great difficulty understand the position of the other party (1); and (h) one's own view "is thus not logically commensurate with views that oppose it" (2).

Even without much analysis of these defining conditions, it seems obvious that fundamental philosophical disagreements so defined are at least one level deeper than Fogelin's deep disagreements, one category stronger than Woods's force-five standoff, and one way more unmanageable than Friemann's intractable quarrels. For Johnstone's talk of inconceivability, unfalsifiability, incomprehensibility, and incommensurability suggests that the contending parties cannot even communicate meaningfully and effectively with each other, let alone resolve their differences rationally. But perhaps such talk is an exaggeration (or an inflated conceptualization), even from his own point of view, that is, from the point of view of the disputes he has in mind.[26]

In fact, examples of fundamental philosophical disagreements given by Johnstone are the disagreements over whether "the individual is the product of his heredity and environment" (1); over whether existentialists are right that "authentic existence is more important than technological progress" (2); over the viability of naturalistic epistemology, namely the view that knowledge is merely a natural process in which the human organism copes with the environment (69-70); over whether empiricists are right as against idealists, that all ideas derive from sense impressions (Johnstone 1978, 1); and over realism versus functionalism in logical theory and the philosophy of logic (Johnstone 1978, 46-51).

Regarding such fundamental philosophical disagreements, Johnstone holds three main theses of increasing specificity. First, [J1] "the only proper response to disagreement as radical as that found with respect to philosophical positions is participation in genuine controversy" (3). Second, [J2] "not only is it impossible to cross the abyss that separates opposing philosophical positions unless the partisans of these positions are willing to argue with each other, but also argument is the sole medium through which a position can communicate its content" (3). Third, [J3] "the abyss that separates conflicting philosophical systems precludes any use of *argumentum ad rem* [i.e., argumentation other than *ad hominem*] ... Thus every valid philosophical argument is *ad hominem*" (3-4).

7.5.2 Philosophical Controversies as Bilateral and Argumentative

To justify the first thesis, Johnstone undertakes a series of critiques of alternative

---

[25] In this section, unproblematic or unambiguous references to Johnstone 1959 will be given by means of simple numerals in parenthesis.
[26] On the other hand, the notion of incommensurability has also been a natural one to use for other scholars trying to come to grips with some radical disagreements occurring during scientific revolutions. See, for example, Kuhn 1962; Feyerabend 1962.

views. First, "the simplest sort of theory is that on which the apparent disagreement is purely verbal ... But ... it fails in not being able to account for the disagreement ... between those who maintain that philosophical disagreement results from [verbal preferences] and those who maintain that it does not" (10).

The second theory claims that philosophical disagreement "is really a kind of game; the motive is to win the game" (10). But this theory is inadequate because "a philosophical dispute, rather than being governed by fixed rules [as is the case for games], represents the effort of each disputant to enforce his own rules" (12).

Third, there is the view that philosophical "disagreement between two persons arises from the fact that each speaks a different language ... [and so] the disagreement is only apparent" (12). The merit of this theory "lies in the fact that it begins to do justice to the systematic nature of philosophical commitment" (12). However, "disagreeing philosophers do not literally speak different languages. For to the extent that different languages are genuinely different, two persons who cannot speak each other's language cannot disagree" (12).

Next, some have argued that philosophical disagreements "are occasioned not by the situation in which one disputant misunderstands another, but rather by that in which at least one disputant misunderstands the grammar of the philosophical terms he employs" (13). This account contains the insight that "the criticism of disagreements has come to be seen as a normative inquiry" (13). Unfortunately, the account has the serious drawback that "the aspect of controversy has now disappeared" (13) from philosophical disagreement, for "in identifying criticism with correction, the theory in question does not correctly construe the grammar of 'correct'; it is inadequate by its own criterion of adequacy" (14).

According to the fifth theory, in philosophical disagreement "criticism assumes the form of dialectic, the co-operative attempt to attain an articulate grasp of the ideal by finding through argumentative discussion a universe of discourse common to the philosophical statements whose apparent opposition initiated the discussion" (15). The difficulty with this is that "dialectical discussion as such involves no guarantee that the ideal universe of discourse to which it conducts its practitioners will in every case be the same" (15), and so it can lead to "radical pluralism ... an expressive or emotive view of metaphysics" (16). Hence, this theory "fails to do justice to the factor of negation occurring in all genuine disagreement" (16).

Now, "the effort to restore negation, and to criticize philosophical commitment while maintaining its idiosyncratic nature, results in the transformation of intuition into pure reason [in the Kantian sense of] the extension of reason beyond the bounds of possible experience" (16). This account too has merits: "one undeniable virtue of this critical theory of philosophical disagreement is its articulation of the insight that every philosophical thesis is disputable" (17). However, it too is inadequate, for it betrays "a systematic failure to afford any basis for distinguishing between controversy and contentiousness. When disagreement poses a threat [as it does in genuine controversy], that is a serious business which contrasts with the sportive or splenetic activity of disagreeing for the sake of disagreeing" (17).

For Johnstone, this difficulty is inescapable "so long as the object of philosophical commitment is taken to be merely a thesis" (17). What is needed instead is a step in the direction of argumentation. That is, "if commitment to a philosophical thesis is understood as resulting from the attempt to solve a problem, then, even though the thesis as such may bear no internal relationship to the personality of the individual committed to it, the fact of commitment is essential to the person" (18). But, "one does not express one's commitment at all except in communicating it to others capable of taking issue with it" (19). It follows that "the committed individual must therefore act in such a way as to maintain both the integrity of his own expression and his respect for his interlocutors" (19). In conclusion:

> Here, then, is a confirmation of the *bilateral* aspect of philosophical disagreement. This is a defining characteristic of controversy, as distinct from instruction, persuasion, correction, contentiousness, or the mere disparity of points of view. And once philosophical controversy has been so characterized, it becomes clear that it can originate only from the collision of beliefs that are *systematically* structured. For the mutual respect on which it depends is an appreciation on the part of any one interlocutor of the commitment of each of the others as a coherent whole, rather than as a set of independent tenets. [Johnstone 1959, 19]

It is important to stress a number of features of this argument for Johnstone's first thesis, J1. It consists of a series of criticisms of alternative views. Although these views are rejected, they are also typically appreciated as containing some merit, virtue, or insight. Indeed, he constructs his own theory in large measure based on the insights extracted from the rejected views. This is an elegant way of incorporating both open-mindedness and fair-mindedness into one's own argumentation. The argument has another neat feature: the alternative views are typically criticized on their own terms, for failing to satisfy some of their own requirements. This means that Johnstone's argument has a key aspect of *ad hominem* argumentation, in the sense of Galileo, Locke, and Whately. From another point of view, the argument is also an attempt to practice the (bilateral, argumentative, and genuinely controversial) theory of philosophical disagreement which Johnstone is elaborating and justifying in this very context. That is, the argument is self-referentially consistent in a way that suggests the general desirability of such a feature.

If there is a weakness in Johnstone's argument, it is that here his open-mindedness does not go far enough, for it does not encompass the reasons that have been given or might be given for the alternative views. However, this criticism is itself in accordance with Johnstone's principles, and so to the extent that it is cogent, it reinforces an essential element of his overall position. In short, this is an *ad hominem* argument, in Johnstone's sense, against his first argument.

7.5.3 Communicating One's Position

Johnstone's second thesis is that [J2] argumentation is the only means of communicating the content of one's own position in a fundamental philosophical disagreement. His argument is the following.

He begins by defining and clarifying what it means for the truth of a statement to be *relative to argument*. "The truth of a statement is *relative to argument* when it is impossible to think of the statement as true without at the same time thinking of an argument in its favor, and it is impossible to think of it as false without at the same time thinking of an argument against it" (23). Scientific statements are *not* in this category. For example, consider the statement that the Earth revolves around the Sun in the period of one year; we can think of its truth and falsity independently of any confirming and disconfirming arguments. Mathematical statements are not in this category either. In mathematics, this claim can be supported in part by Gödel's theorem. However, Johnstone clarifies, the truth of mathematical statements is relative to *assumptions*, e.g., the truth of the Pythagorean theorem is relative to Euclid's parallel postulate.

Next, Johnstone examines a number of philosophical statements which are, have been, and can be the subject of disagreement. They are statements such as the following: "the good is the object of desire, all men are created equal, every event has a cause, the real is the rational, and the universe exhibits design" (25). Then he argues that the meaning of such statements cannot be understood "merely by analyzing the words that occur in them" (29), for this analysis will lead to either tautology, inconsistency, or infinite regress. On the other hand, he also explains how such statements can be understood on the basis of how they originated, of the problems they were meant to solve, and of the arguments justifying them. He concludes that "the argument for a philosophical statement is always a part of its meaning ... [and] the argument against a philosophical statement is always a part of its meaning" (32).

At this point one could object that Johnstone's account implies that "philosophical statements in reality belong to formal science" (33). His answer relies on his previous thesis about the genuinely controversial nature of philosophical disagreements, which implies that "philosophical arguments must have a controversial aspect" (37), which the formal sciences presumably do not have.

A crucial aspect of this argument is an inductive generalization from the examples of philosophical statements which Johnstone examines. Therefore, this aspect of his argument depends on how typical or representative they are of the class of philosophical statements. My judgment is that they are indeed typical of a subclass of philosophical statements, which belong to such branches of philosophy as first philosophy, speculative philosophy, theoretical philosophy, and metaphysics. However, they are not typical of such branches as philosophy of science and of art, applied ethics, and the historiography of philosophy.

Moreover, Johnstone's argument depends on equating the notions of the origin of a philosophical statement and the problem which it was intended to solve with the notion of an argument justifying the statement. I do not think he could be correctly charged with committing the genetic fallacy. However, this aspect of his argument could stand more elaboration.

7.5.4 The Effectiveness of Ad Hominem Argument

Now, given that fundamental philosophical disagreements must be argument-centered (in order to be bilateral and genuinely controversial), and that they must be communicated through arguments (in order to be over the truth or falsity of philosophical statements), so far it is an open question whether and how the abyss that separates the contending parties can be bridged, i.e., whether and how they can be resolved. And here lies the function of Johnstone's third thesis that [J3] the resolution lies with *ad hominem* argumentation, although admittedly it is beyond the power of argumentation *ad rem*.

Johnstone's argument (57-80) for this thesis consists of an analysis of three cases: the disagreement between Aristotle and Eudoxus over whether pleasure is the chief good; the disagreement between Berkeley and the materialists over whether external bodies are the causes of human ideas; and the disagreement between naturalists and anti-naturalists over the naturalistic basis of all knowledge. Johnstone reconstructs Aristotle's key critical argument against Eudoxus, Berkeley's main critical argument against the materialists, and the chief anti-naturalist critical argument against naturalism. In each case he shows (argues) that the critical argument is successful and that it is *ad hominem*. It follows that *ad hominem* arguments of this sort are a rational means of resolving fundamental philosophical disagreements. And more generally, *ad hominem* arguments can rationally resolve (or at least go a long way toward resolving) deep disagreements, force-five standoffs, and intractable quarrels.

Here it must suffice to examine only the disagreement between Aristotle and Eudoxus. Eudoxus had held that [E1] pleasure is the chief good, on the grounds that [E11] "any good thing—e.g., just or temperate conduct—is made more desirable by the addition of pleasure" (64). Aristotle criticized Eudoxus by pointing out that a similar argument had been advanced by Plato to show that [P1] *wisdom* is the chief good; for [P111] "the pleasant life is more desirable with wisdom than without" (64), and indeed [P11] "if wisdom be added to *any* good thing—not just to the pleasant life—the result is more desirable" (64). So far, Johnstone has just given an interpretation of a passage in Aristotle's *Nicomachean Ethics* (1172b, 9-35). Next, he advances an evaluation.

Johnstone's evaluation is that "Aristotle's criticism is devastating" (64) insofar as the effect on Eudoxus's position is that "there does not even seem to be any way in which he could *revise* it to meet the criticism" (64). In other words, Aristotle's criticism is "maximally forceful" (65), insofar as "he shows that Eudoxus has defeated his own purpose" (65).

To justify this judgment, Johnstone's main argument consists of a reconstruction of Aristotle's criticism as an argument having the form of a reductio ad absurdum. Johnstone's reconstruction can in turn be reconstructed as follows: [A111] if Eudoxus's argument is correct, then pleasure is the chief good, because [A1111] the premise of his argument is indeed correct. But [A1121] if Eudoxus's argument is correct, then so is Plato's (i.e., the argument from P11 to P1), because [A11211] they have the same form. Now, [A1122] if Plato's argument is correct, then wisdom is the chief good, because [A11221] the premise of his argument (proposition P11) is correct. And [A1123] if wisdom is the chief good, then pleasure is not. Therefore, [A112] if Eudoxus's argument is correct, then pleasure is not the chief good. Hence, [A11] if Eudoxus's argument is correct, then pleasure both is and is not the chief good. Therefore, [A1] Eudoxus argument is incorrect.

This seems to be a plausible reconstruction of Aristotle's criticism, and so indeed it seems to be devastating and maximally forceful. In short, Aristotle's criticism is indeed successful.

But what is the structure of Aristotle's criticism? Obviously it is an argument, and a reductio ad absurdum. It is also a good example of a type which Johnstone labels *ad hominem*. This terminological choice has a long and important historical pedigree, which includes such thinkers as Galileo, Locke, Thomas Reid, and Richard Whately.[27] Whately's definition, often quoted by Johnstone is that "in the *argumentum ad hominem*, the conclusion which actually is established, is not the absolute and general one in question, but relative and particular, viz. not that 'such and such is the fact', but that '*this* man is bound to admit it in conformity to his principles of reasoning, or consistency with his own conduct, situation', &c."[28]

Johnstone (1978, 134) rephrases Whately's definition by saying that "*argumentum ad hominem* ... is precisely the criticism of a position in terms of its own presuppositions," in which he subsumes both propositions and arguments under the label of "position"; this formulation makes explicit the fact that *ad hominem* argument corresponds to a type of criticism that is usually called "internal." He also states that in philosophy an *ad hominem* argument is "an argument against a philosophical thesis [attempting to] exhibit that thesis as inconsistent with its own assertion or defense, or with principles that must necessarily be accepted by anyone who maintains the thesis" (Johnstone 1978, 45); because of its stress on inconsistency, this formulation makes explicit the logical aspect of *ad hominem* arguments. Finally, these formulations are meant to be equivalent to a still different one using the notion of a "self-defeating" position, as can be seen from this claim: "an argument that [purportedly] shows that a statement or argument defeats its own purpose is, to my way of thinking, precisely an *argumentum ad hominem*" (Johnstone 1959, 82).

Obviously, an *ad hominem* argument in this sense should not be confused with the

---

[27] Cf. Johnstone 1959, 73 n. 12; Finocchiaro 2005a, 329-39; Eemeren and Grootendorst 1993.
[28] Whately 1838, 196; quoted by Johnstone 1959, 73.

other type of argument denoted by the same term, that is the criticism of a claim by criticizing the circumstances or character of the arguer rather than the argument, reasons, or evidence advanced to justify that claim. Johnstone takes this to be so obvious that he hardly mentions this fact. On the other hand, there is another distinction which he is constantly making, namely that between *ad hominem* arguments and arguments *ad rem*. Johnstone is clear that an arguments *ad rem* is not a particular type of argument, but rather that "*argumentum ad rem* is a purely negative phrase. It denotes the entire spectrum of arguments other than *argumentum ad hominem*, or at least all those that are not obviously fallacious" (Johnstone 1978, 53). In positive terms, an argument *ad rem* is an "appeal to evidence" (Johnstone 1959, 3) or to allegedly "objective facts" (Johnstone 1959, 76).

Aristotle's criticism of Eudoxus is also a meta-argument. This is overwhelmingly obvious in my reconstruction of Johnstone's reconstruction, although less so in his own reconstruction per se.[29] However, not all *ad hominem* arguments need be meta-arguments. Almost all versions of his definition explicitly mention that the criticism may be directed at a particular thesis or assertion, as well as at its defense or supporting argument. Nevertheless, even in the former case, the *ad hominem* argument advances claims about the presuppositions of the thesis or its consistency with the assertion of the thesis, and such claims would also involve arguments, albeit less directly.

7.6 Conclusion: The Role of Meta-argumentation

I began this chapter by proposing to study meta-argumentation in the context of the cluster of cognitive phenomena that have been variously called deep disagreements, intractable quarrels, standoffs of force five, and fundamental philosophical controversies. The first step was to reconstruct Fogelin's argument that deep disagreements are not subject to rational resolution but only to rhetorical persuasion. His argument could thus be seen to be itself a meta-argument and to have some complexity.

Then I undertook a critical analysis of the many critiques that have been made of Fogelin's argument. Lugg (1986) argued that Fogelin's argument presupposes a static conception of rationality (which is incorrect) and fails to appreciate a dynamic conception (which is correct); Lugg's argument was interpreted as a meta-argument and evaluated as cogent. Davson-Galle's (1992) most telling point was that Fogelin overlooks the existence and power of internal criticism, in which one derives a conclusion not acceptable to an opponent from claims acceptable to him; this was regarded as an important insight, and indeed an intuitive formulation of a thesis elaborated at great length by Johnstone, who uses the label of *ad hominem* argument for such internal criticism. Feldman (2005) argued plausibly that Fogelin commits a fallacy of equivocation by trading on the ambiguity of the notion of rational resolution (which can mean resolution of the disagreement showing that one side is right and the other wrong, and also suspension of judgment in light of the demonstration that the arguments on both sides are equally good or equally bad); and hence resolution in the latter weak sense is available when the former strong resolution is not feasible. Turner and Wright (2005) argued that Fogelin presupposes an untenable conflation of rationality with argumentation; I argued that this criticism is essentially correct insofar as by argumentation they (Fogelin as well as Turner and Wright) mean simple argumentation, but that the other elements of rationality stressed by Turner and

---

[29] Johnstone (1959, 67) even goes so far as to question the meta-argumentative construal of Aristotle's criticism by saying that "Aristotle's argument is not *about* Eudoxus' argument, but rather *in terms of* it." But I have de-emphasized this aspect of his analysis because I find little plausibility in it.

Wright (experience, training, and education) are in large measure construable as complex argumentation. Campolo's (2005) argument was similar in content to that of Turner and Wright, but opposite in intent insofar as he tried to reinforce Fogelin's argument; that is, Campolo also stressed the distinction between argumentation and rationality, in order to strengthen Fogelin's thesis on the limitations of argumentation, rather than undermine his thesis on the limitations of rationality; but the distinction between simple and complex argumentation can also be applied to Campolo's argument with similar consequences as for the case of Turner and Wright. Friemann (2005) criticized Fogelin's argument based on an analogy between deep disagreements and intractable quarrels, arguing that deep disagreements can be rationally resolved (or at least become more tractable) by the help of a third party and the adoption of attitudes such as empathy and openness; I judged Friemann's argument cogent and important insofar as it stresses the existence, significance, and power of norms of rationality such as open-mindedness and fair-mindedness. Adams (2005) argued that argumentation is essential and desirable in deep disagreements, in order to determine that a given disagreement is indeed deep; this argument was deemed correct and elegant, and it was interpreted as showing that meta-argumentation is essential and desirable for this purpose. Phillips (2008) argued primarily that deep disagreements are very difficult to resolve in practice, but resolvable in principle, and so complex argumentation guided by proper principles is needed to overcome the practical difficulties; I found this distinction between practice and principle important and fruitful. Godden and Brenner (2010) argued that many deep disagreements have an element of conceptual disagreement, and that conceptual disagreements can be resolved by rational persuasion and persuasive argumentation. Finally, I objected that Fogelin has not shown the real existence of any disagreements that are deep in his sense, and that his supporting subargument is fundamentally flawed; for he merely asserts that some actual controversies (such as the one over affirmative action) reduce to disagreements over irreconcilable principles (such as the existence of group rights), without doing the hard meta-argumentative work of showing that all other arguments in the controversy reduce to that one.

Next, I examined Woods's reflections on what he calls standoffs of force five, and I call fierce standoffs. I reconstructed them as a complex meta-argument for a thesis that is essentially identical to Fogelin's, i.e., that there are some disagreements that are not subject to rational resolution but only rhetorical persuasion. However, Woods's argument is novel insofar as he argues that such disagreements are standoffs of force five, which are four levels deeper that Fogelin's deep disagreements. In fact, standoffs of force five are controversies in which the disputants disagree over (i) whether they share particular facts, values, and procedures to which they might appeal to settle the issue; (ii) whether to agree to disagree; (iii) whether to let a third party adjudicate; (iv) whether a political solution legislated by the government is appropriate; and (v) whether the opposite position is a real possibility. On the other hand, Fogelin's deep disagreements basically correspond to standoffs of force one, for which only condition (i) applies. Thus, Woods's argument is stronger than Fogelin's insofar as standoffs of force five contain several additional obstacles to the rational resolution of the disagreement. Woods's argument also has the additional merit that it explicitly considers a possibility overlooked by Fogelin, namely that of resolving disagreements by means of suitable methodological principles, and Woods explicitly argues that standoffs of force five cannot be resolved by appeal to methodological principles. Finally, Wood's argument is valuable insofar as, unlike Fogelin's, Woods explicitly argues for the rhetorical-persuasion clause of his main conclusion, and in doing so he displays a nuanced and sophisticated appreciation of rhetorical persuasion.

However, despite these strengths, Woods's argument is open to several criticisms, all pointing in the direction of weakening his main conclusion about the rational irresolvability of really deep disagreements, i.e., his limitation theorem on rationality. First, his own examples of rhetorical persuasion from the abortion controversy in Canada can be viewed as appeals to methodological principles or argumentative appeals to reasons, subject to various degrees of correctness or incorrect; it follows that when logic defers to rhetoric (to use Woods's own eloquent expression), there is less deference than it may appear and the rational practices of appealing to reasons and to methodological principles may be continuing unabated. Second, the methodological principle introduced by Woods himself and credited to Ramsey, is much more powerful as a means of resolving disagreements than Woods realizes; this can be seen by examining a significant utilization of it by Piotr Sztompka in the context of the philosophy of the social sciences; it follows that, again, rationality is less limited than Woods's limitation theorem suggests. In my third criticism, I argued that although methodological principles constitute an important level of rationality having some autonomy of its own, methodological appeals can be interpreted as instances of argumentation, specifically meta-argumentation. Finally, I objected that if we focus on the closed-mindedness that (by Woods's own definition) characterizes the participants in a force-five standoff, then what really follows from his argument is not that such standoffs are *rationally* irresolvable, but rather that they are *de facto* irresolvable and that they could be rationally revolved if the ideal of open-mindedness were followed.

Finally, I undertook an examination of Johnstone's account of fundamental philosophical controversies. The motivation for this undertaking was rooted in the previous investigations in several ways: fundamental philosophical controversies have some obvious similarities with deep disagreements, intractable quarrels, and force-five standoffs; Davson-Galle's most telling objection to Fogelin's argument amounted to pointing out that he neglects the power of *ad hominem* argumentation in Johnstone's sense of this notion; Adams's main objection to Fogelin (about knowing when disagreements are deep) was itself an elegant and cogent instance of *ad hominem* argumentation in Johnstone's sense; and another instance of Johnstonian *ad hominem* argument was my fourth criticism of Woods's argument, to the effect that what really follows from his argument is that force-five standoffs can be rationally resolved if but only if the participants adopt the rational ideals of open-mindedness and fair-mindedness.

The analysis of Johnstone's work revealed the following results. His definition of a fundamental philosophical disagreement is intended to apply to controversies common in the more speculative and abstract branches of philosophy, such as metaphysics. There are three main theses and corresponding arguments. First, he argues that such philosophical disagreements are, or ought to be, characterized by a bilateral or dialectical interaction of the disputants and by a systematic or argumentative elaboration of their views; his argument consists of criticism and appreciation of several alterative theories of philosophical disagreement. Second, he argues that in philosophical disagreements arguments are required in order to communicate one's own position in the first place; this argument consists of an elaboration of the consequences of the claims that philosophical disagreements are disagreements over the truth of philosophical statements and the meaning of philosophical statements is argument dependent. Third, he argues that *ad hominem* argumentation (in Whately's sense) can rationally resolve philosophical disagreements; his argument has the form of an inductive generalization from historical examples of disagreements in which one philosopher was able to devise a successful *ad hominem* argument against another. So reconstructed, Johnstone's account and arguments have considerable plausibility and promise, although my exposition makes clear that some qualifications and improvements are needed.

Deep Disagreements, Fierce Standoffs, Etc. 119

The upshot of this chapter is unified regarding the end, and manifold regarding the means. The end result is that radical disagreements are less intractable than commonly believed; that is, they are rationally resolvable to a greater degree than usually thought. To deal with them more effectively one should learn and master principles and practices such as the following: Ramsey's Maxim and the art of moderation and compromise it involves; the attitude or disposition of open-mindedness, namely learning and understanding contrary arguments; the attitude or disposition of fair-mindedness, namely showing some appreciation for contrary arguments before refuting them; internal criticism of the opposite position or *ad hominem* argumentation against it, *ad hominem* in Johnstone's sense; complex argumentation, consisting of multiple and long chains of supporting reasons and answers to objections; rational persuasion, which amounts to persuasive argumentation in which the key aim is to convince the other party; and meta-argumentation, namely to learn and master the art of arguing about arguments with as much care as that which many people advocate and display when arguing at the ground level about concrete or lower-level topics.

This last point of the end result substantiates the meta-argumentation approach by revealing the meta-argumentative aspects of deep disagreements. However, it is obvious that that approach is also substantiated by the fact than my analysis has proceeded by reconstructing as meta-arguments the views of Fogelin, Woods, Johnstone, and their critics. Nevertheless, despite this substantive and methodological role for meta-argumentation in deep disagreements, it is obvious that it is only one of several instruments useful for their rational resolution. Similarly, it should be clear that none of these seven principles or practices mentioned here are individually necessary conditions for the rational resolution of deep disagreements. Nor are they collectively sufficient. My claim is rather that these principles and practices are valuable instruments whose utilization increases the likelihood of rationally resolving a deep disagreement, while still not guaranteeing nor rendering certain its rational resolution.

7.7 Empirical Testing

It should be noted that in this chapter I have not carried out an empirical testing of the conclusions I have reached, but that it would be desirable and fruitful to do so. That is, it would be desirable to examine one or more actual deep disagreements to determine whether, and if so to what extent, they were rationally resolved by means of the principles of Ramsey, open-mindedness, and fair-mindedness, and the practices of *ad hominem*, complex, persuasive, and meta-argumentation. In light of the approach advocated and practiced in this book, stemming from the Toulminian applied logic, this requirement is extremely important.

My reasons for not doing such empirical testing *here* are in part practical considerations relating to the avoidance of prolixity and the requirements of manageability. However, some further comments are in order.

First, it should be pointed out that I *have* been following an empirical approach (as I advocate generally) at the meta-level of the study of the meta-arguments advanced by logicians and argumentation theorists in support of their own theoretical claims about deep disagreements, although admittedly I have not followed this approach at the ground level of argumentation involving deep disagreements about concrete issues. More importantly and crucially, I have elsewhere carried out several historical studies that may be seen as corresponding to and confirming the conclusions reached here.

Some of my studies relate to one of the most important deep disagreements in human history—the Copernican Revolution and Galileo's contributions to it. Now, it turns out that, in my account, the Copernican controversy was indeed resolved rationally, and that the seven principles and practices unearthed in this chapter played an instrumental and effective role. Indeed, I can add that I first came across these principles and practices in my studies of Galileo's contributions to the Copernican Revolution, and that this chapter's extraction of them from the literature on deep disagreements is a subsequent development made possible by that earlier work.[30]

Nor should we be satisfied with a single empirical-historical case study, however significant and instructive it may be. For the very features that make the Copernican Revolution so epoch-making, full of ramifications, and pregnant with meaning perhaps also make it special and unique, in the sense that its lessons are difficult to generalize or extrapolate to other deep disagreements. In this regard, I am encouraged by the fact that it turns out that many of those features and lessons recur and are applicable in another cause célèbre, which is admittedly related to but distinct from the Copernican Revolution.

I am referring to the subsequent Galileo affair, as distinct from the original affair. By the original Galileo affair I mean the controversy over the truth of the earth's motion that began around 1610 with his telescopic discoveries in astronomy and climaxed in 1633 with his trial and condemnation by the Inquisition. By the subsequent Galileo affair I mean the controversy over the rightness of his condemnation that began then and continues to our own day. Since the latter disagreement remains unresolved, one cannot strengthen my conclusions about deep disagreement by arguing that its rational resolution was due to the seven principles and practices mentioned here, as one can do for the case of the Copernican Revolution. However, one can try to argue that the so far unresolved subsequent Galileo affair could and would be resolved by greater adherence to and utilization of these practices and principles. And that is precisely the argument I have made in previous studies that focus on the comparison and contrast of the two affairs. In particular, the principles of open-mindedness and fair-mindedness turn out to be especially crucial. The key thesis is that "today in the context of the Galileo affair and the controversies over the relationship between science and religion and between institutional authority and individual freedom, the proper defense of Galileo should have the reasoned, critical, open-minded, and fair-minded character which his own defense of Copernicus had."[31]

Another similarly important deep disagreement was the twentieth-century controversy over the relative merits of communism and capitalism. Of course, this controversy was even more complex than the Copernican Revolution and the Galileo affair, having significant political, economic, and even military dimensions, besides its intellectual aspects. In this case, one issue could be whether the end of the Cold War, which put an end to the geopolitical and military aspects of the disagreement, also resolved the intellectual controversy; or whether the intellectual controversy had already been resolved, or remains yet to be resolved. In any case, the intellectual disagreement clearly had philosophical or conceptual aspects, as well as dimensions of political theory and economic theory.

Now, on this topic too, I have previously carried out some studies that happen to have

---

[30] The following specific references may be useful: for *ad hominem* argumentation, see Finocchiaro (1980a, 131-31, 231-31, 368-70, 4-2-3, 430; 2005a, 329-39); for complex argumentation, see Finocchiaro 1980a, 27-45, 311-31, 413-31; for persuasive argumentation, see Finocchiaro (1980a, 3-26, 46-66; 1997, 356-72); for meta-argumentation, see Finocchiaro (1980a, 103-41, 343-412; 1997, 335-56); for Ramsey's maxim, see Finocchiaro (1980a, 145-66; 1997, 335-56; 2005a, 405-8, 426-28); for open-mindedness and fair-mindedness, see Finocchiaro (1980a, 134-35, 177; 1997, 147-49, 339-41; 2010b, xxxvii-xliii, 129-34).
[31] Finocchiaro 2010b, x. See also Finocchiaro (2010b, 137-314) and the more factual and historical works on which this argument is based (Finocchiaro 1989; 2005b).

relevant and confirmatory implications, although some terminological transformations are needed to see the connection. My approach was to explore the nature of the capitalism-communism disagreement, the extent of its past resolution, and the prospects for its future resolution by means of a critical examination of the thought of Antonio Gramsci. He has traditionally been viewed as a Marxist, indeed as a patron saint of the Left. However, it is generally acknowledged that Gramsci somehow managed to understand and internalize both sides of the dispute. Thus, his thinking promised to offer a privileged point of view for understanding and resolving this deep disagreement.

Then, with regard to philosophical conceptual framework, I focused on the concept of dialectic and elaborated a Gramscian conception of dialectic that has elements drawn out of the thought of Hegel, Marx, and Benedetto Croce. So conceived the dialectic is a manner of thinking that emphasizes such practices as the avoidance of one-sidedness, the balance of extremes, and the search for diversity within unity and unity amidst diversity.[32] This manner of thinking corresponds basically, I believe, to Ramsey's Maxim as introduced above in the footsteps of Woods, as well as to Sztompka's approach to the resolution of sociological dilemmas summarized above as a significant, if implicit, utilization of that maxim, under the explicit label of dialectic.

Furthermore, with regard to political theory, it turns out that Gramsci's political theory can be reconstructed as being in the tradition of democratic elitism. To see the presence of such a democratic elitism in Gramsci, one needs a critical understanding of his intellectual relationship to Gaetano Mosca.[33] Now, Mosca is traditionally viewed a founder of the elitist school in political sociology, and indeed as a patron saint of the Right, although I argue that the more correct interpretation is as a leading exponent of democratic elitism. This tradition attempts to combine the concepts and requirements of democracy and elitism, rather than considering them to be opposites, as commonly done. Such a synthesis represents, in the domain of political phenomena, an instantiation or utilization of the dialectical manner of thinking, alias, again, Ramsey's Maxim.

Further empirical testing could come from the phenomenon of "scientific revolutions," in the sense of Kuhn (1962; 1970). It is obvious that including such a discussion here would have made this chapter too prolix or unmanageable, given that for half a century the reception and examination of Kuhn's work have been a phenomenon of unprecedented proportions in the annals of scholarship. Still, the similarity of issues should be clear from the following Kuhnian passages, which are as emblematic as they are revealing: "because it is a transition between incommensurables, the transition between competing paradigms cannot be made a step at a time, forced by logic and neutral experience. Like the gestalt switch [and the conversion experience], it must occur all at once (though not necessarily in an instant) or not at all ... Still to say that resistance in inevitable and legitimate, that paradigm change cannot be justified by proof, is not to say that no arguments are relevant or that scientists cannot be persuaded to change their mind ... We must therefore ask how conversion is induced and how resisted ... When asked about persuasion rather than proof, the question of the nature of scientific argument has no single or uniform answer ... Ultimately, therefore, we must learn to ask this question differently. Our concern will not then be with the arguments that in fact convert one or another individual but rather with the sort of community that always sooner or later forms as a single group" (Kuhn 1970, 150-53). Now, of course, Kuhn's views need to be properly understood and evaluated, but when this is done the result would also be along the lines of my conclusions about deep

---

[32] See the full elaboration in Finocchiaro 1988 (especially 179-80) and the summary in Finocchiaro 2005a, 207-30.

[33] See Finocchiaro 1999, especially 22-61, 84-143.

disagreements.[34]

Finally, one could explore the topic of political debates, that is disagreement in practical politics. There is evidence that it too can be studied in ways that are suggestive from the point of view of the limitations of ground-level argumentation and the power of meta-argumentation. This may be gathered from an essay by Dryzek and Niemeyer (2006).

7.8 Summary

This chapter has examined the views advanced by such scholars as Robert Fogelin, John Woods, and Henry Johnstone, Jr., about such topics as deep disagreements, intractable quarrels, standoffs of force five (dubbed "fierce" standoffs), and fundamental philosophical controversies. As much as possible, their views, and the critiques of them, were reconstructed as meta-arguments. From my analysis, it emerges that deep disagreements are rationally resolvable to a greater degree than usually believed, but only by using special principles and practices such as: the art of moderation and compromise (codified as Ramsey's Maxim); open-mindedness; fair-mindedness; complex argumentation; internal criticism, or *ad hominem* argument in the sense elaborated by Johnstone; rational persuasion, or persuasive argumentation; and meta-argumentation, in the sense elaborated in this book. This multi-faceted conclusion could and should also be tested by empirical studies of deep disagreements; some of these studies were summarized, sketched, and proposed here.

---

[34] Cf. Finocchiaro 1973, 188-98; 1980a, 202-23; 2005a, 386-429; 2010b, 37-64.

Chapter 8
Conductive Arguments, Pro-and-Con Reasoning, Etc.

This chapter will examine a cluster of cognitive phenomena that have been studied under various labels, such as conductive arguments, pro-and-con arguments, and balance-of-considerations arguments. These studies have been carried out by such scholars as Carl Wellman, David Hitchcock, Trudy Govier, and their many critics. It will emerge that this topic is especially pertinent to meta-argumentation for a substantive as well as for a methodological reason: not only because the scholarly views lend themselves to being reconstructed as meta-arguments, but also because conductive arguments have themselves a crucial meta-argumentative element.

8.1 Introduction

In 1971, Wellman coined the terms conduction, conductive argument, and conductive reasoning to characterize a type of argument or reasoning intended to be a distinct alternative to deduction, induction, and even abduction. He thought this new concept of conduction was needed in order to make sense of an important type of argument that is especially prevalent in ethics, but also common in many other fields. The problems discussed under this label involved issues such as the accumulation of evidence, the convergence of multiple reasons, the balance of pros and cons, and the possibility of good but nonconclusive inference. Wellman's account generated a considerable body of literature dealing with such problems and using his terminology.[1] This literature and these problems are sufficiently important to deserve continued attention, and the primary aim of this chapter is to undertake a critical examination of it.

I believe this is worth doing despite the fact that the topics and theoretical problems just mentioned have been fruitfully studied also by many other scholars without referring to Wellman or using the conductive terminology. An example of such studies is John Stuart Mill's essay *On Liberty*, but he is only a classic case.[2] This fact might lead some to regard the notion of conduction as simply a different way of talking about such problems, and thus to dismiss the views of Wellman and his followers and critics. However, the fact of the existence of this other literature leads me instead to want to explore the relationship between the two. Thus, after my critical examination of the literature on Wellman and conduction, I also plan to provide some perspective by briefly discussing the views of other scholars who have dealt with the same problems under different terminology. Such a perspective is a secondary aim of this chapter.

A third aim is to analyze some important actual or real cases of conductive argumentation, and relate them to the views of both Wellman and the other explicit conductive theorists on the one hand, and Mill and the other implicit conductive theorists

---

[1] It is useful to have a complete list of these works, dealing *explicitly* with conduction: Allen 1990, 1993, 2011; Battersby and Bailin 2011; Bickenbach and Davies 1997; Blair 2011; Blair and Johnson 2011; Ennis 2004; Fischer 2011, 2012; Freeman 1983, 2011; Govier 1980a, 1980b, 1980c, 1987, 1999, 2001 (pp. 392-412), 2010 (pp. 352-77), 2011; Hansen 2011; Hitchcock 1981, 1983, 1994; Jin 2011; Johnson 2000a (pp. 92-95), 2011; Johnson and Blair 2000; Kauffeld 2011; Kock 2011; Pinto 2011; Walton 2011; Wohlrapp 1995, 1998, 2008, 2011; Zenker 2009b, 2011.
[2] Other works cited or summarized later in this chapter are: Eisend 2006; Finocchiaro 2010b; Jacquette 2007d; Johnson 2000a, 84-88, 92-95; Kock 2003, 2007a, 2007b, 2007c; O'Keefe 1999; Scriven 1981. Regarding Mill, see chapter 10 below.

on the other. One of these cases is an opinion published in 2009 in the *New York Times* by columnist David Brooks about the U.S. heath-care reform bill, which at that time was being discussed in the U.S. Congress, and which soon thereafter was passed into law. The other case is the argument in favor of the Copernican theory of the earth's motion advanced by Galileo Galilei in his book *Dialogue on the Two Chief World Systems, Ptolemaic and Copernican* (1632).

The approach I am going to follow is, of course, a further instantiation of the Toulminian applied logic elaborated earlier (chapter 1), which in turn corresponds to the historical-textual approach I have used and reflected upon in other writings (cf. Finocchiaro 1980a, 293-431; 2005a, 14, 21-91). In this approach, one studies reasoning and argumentation found in actual, real, or realistic texts that have some historical importance, with the aim of deriving, testing, or refining general concepts and principles that may be normative or evaluative, as well as descriptive or explanatory. In the present investigation, the real arguments on which I am going to focus are those found in the theoretical literature dealing explicitly with conductive arguments, as well as in the two just-mentioned important actual examples of conductive argumentation dealing with concrete issues. The former focus represents, of course, the leap onto the meta-level and meta-analysis which is the main theme of this book.

However, in the context of the present chapter, that theoretical literature has an additional relevance. For besides containing important actual arguments and thus being amenable to the Toulminian applied-logic or my historical-textual approach, and besides containing meta-arguments and thus being amenable to the meta-argumentation focus, that literature contains many arguments that are themselves conductive, as some of the relevant authors have pointed out. In other words, the material we are dealing with consists not only of actual arguments (at least if that material is properly so analyzed), and not only of meta-arguments (given its theoretical nature), but also of conductive meta-arguments (at least in large measure). The self-reflective dimension of some meta-argumentation will be discussed more explicitly in the next chapter, but it should be noted that it is implicit in the present one, just as it was also implicit in some previous chapters (especially in my critical analysis of Johnson's arguments for his concept of argument in chapter 4).

Another preliminary point is worth making. It turns out that in the present case meta-argumentation is important also because conductive arguments themselves happen to have a significant meta-argumentative aspect; or at least, this is a *working hypothesis* which I would like to explore or test in this chapter. My theoretical assumption is that an essential characteristic of conductive arguments is a *balance-of-considerations claim* that must be implicitly made, explicitly formulated, or critically justified for the construction, interpretation, or evaluation of conductive arguments; and such a balance-of-considerations claim is an irreducible meta-argumentative aspect of conductive arguments. I stress that this assumption is a working hypothesis in the sense that it provides a substantive theoretical guide in this investigation of this chapter. Whether it is true, or to what extent it is true, is something to be determined at the end, or as a result, of this investigation. The presence of this working hypothesis should also clarify that although the applied-logic or historical-textual approach is a type of empirical approach, it is not empiricist in the sense of pretending to study the empirical material with a *tabula rasa*.

Finally, the topic of conductive arguments is the perfect occasion to make a clarification and simplification regarding the numbering or labeling of the various propositions whose interrelationship make up the propositional macrostructure of arguments, explained earlier (chapter 2) and used throughout this book. These qualifications relate to the difference between independent and linked support provided by two or more premises for a conclusion. Recall that when a proposition n is supported *independently* by two or more sets of premises, then letters should are used to distinguish one set from another, e.g., *na1, na2, na3, ..., nb1, nb2, nb3, ..., nc1, nc2, nc3*, etc. Such labeling of independent support is crucial for expressing and understanding the logical structure of an argument when there are sets of premises that provide independent support and sets that provide linked support. For example, suppose some conclusion C is supported by proposition P and by propositions Q and R in such a way that P supports C independently of Q and R, but Q and R are support C in a mutually linked manner, rather than independently of each other; then to reflect this structure properly, the systematic labeling would be: P = Ca1; Q = Cb1; and R = Cb2.

However, sometimes some simplifications are possible. For example, suppose that a conclusion C is supported by three propositions, P, Q, and R, and that all three provide independent support, or all three provide linked support. Then the strict systematic labeling would be Ca1, Cb1, and Cc1 if they all independently supportive; and it would be C1, C2, and C3 if they are all linked, or interdependently supportive. But there is no loss of clarity if in the former case we drop the 'a', 'b', and 'c' and use only sequential numerals; that is, if the three premises are labeled C1, C2, and C3 even when they are all independently supportive. In short, the different labeling of independently and interdependently supportive premises will be implemented only when necessary, which happens when a given step or subargument contains *both* independent and linked support.

Similarly, suppose conclusion C is independently supported by premises Ca1 and Cb1. Then, in a context where there are other propositions linked with these, say Ca2 and Cb2, it is important and necessary to keep both the letters and the numerals modifying the 'C'. However, in a context where there are no other propositions linked with Ca1 and Cb1, their independent support for C can be represented equally well by dropping the numerals and labeling them more simply Ca and Cb. In short, when a conclusion C is supported independently by a sequence of single premises, Ca1, Cb1, Cc1, etc., we can drop either each occurrence of the numeral 1 (resulting in the labels Ca, Cb, Cc, etc.), or we can drop the lower-case letters and use different numerals in sequence (resulting in the labels C1, C2, C3, etc.).

8.2 Wellman's Invention of Conduction

Wellman's main argument can be interpreted as an attempt to justify the claim that [W1] there is an important special type of argument called conductive argument. Its details can be reconstructed as follows.

[W11] Conductive arguments are arguments "in which 1) a conclusion about some individual case 2) is drawn nonconclusively 3) from one or more premises about the same case 4) without any appeal to other cases" (52).[3] [W12] Such arguments are common in ethics, but also in law, politics, and philosophy, and indeed in all fields where evaluations, recommendations, classifications, or interpretations are common.

However, [W13] conductive arguments are not deductive, since [W131] the nonconclusive clause (no. 2) of the definition implies that they neither are nor are claimed to be deductively valid, and [W132] deductive arguments are those that claim to be deductively valid (4).

[W14] Nor are conductive arguments reducible to deductive form (83), for several reasons. One is that [W14a1] conductive arguments usually involve the weighing of pros and cons, [W14a2] which has no place in deduction (25-28). To see why [W14a2]

---

[3] In this section, references to Wellman 1971 will be given by just citing the page number(s) in parenthesis, as done here.

"deductivism ... leaves no room for the weighing of pros and cons" (25), [W14a21] consider a typical conductive argument of what Wellman calls "the third pattern ... that form of argument in which some conclusion is drawn from both positive and negative considerations" (57). [W14a22a1] We could reconstruct it as a deductively valid argument whose premises would include all the pros and all the cons, as well as an additional premise that the pros outweigh the cons (or vice versa); [W1422a2] but such an additional premise would have to be justified by actually weighing the pros and cons, and not by deductive reasoning; "therefore, [W14a22] the difficulty has simply been pushed back one stage in the process of justification" (26). [W14a22b1] Or we could reconstruct the conductive argument as a deductively valid argument whose premises would again include all the pros and cons, but with the additional premise being a generalization covering and connecting all the premises and the conclusion; but [W14a22b2] such a generalization would have to be justified by means of an induction by enumeration based on particular cases; and [W14a22b3] to justify each of these cases we would need to engage in weighing the pros and cons for that particular case; therefore, again [W14a22] the nondeductive weighing of pros and cons has to be done at another stage of the process of justification.

Another reason why [W14] conductive arguments are not reducible to deductive form is that [W14b1] they are characterized by convergence of evidence, [W14b2] which deductive arguments do not exhibit (28). [W14b3] Convergence is the phenomenon of the accumulation of weight, i.e., the increase of logical force, i.e., the strengthening of the "implicative link" (29) resulting from different premises. [W14b21] There is no accumulation of evidence within a single deductively valid argument with multiple premises, because [W14b211] no individual premise alone lends any support to the conclusion, whereas [W14b212] all together they provide perfect support. [W14b22] For the case of a conclusion supported by several deductively valid arguments, there is still no accumulation of evidence or convergence, because [W14b2211] each argument provides conclusive support for the conclusion, and so [W14b221] the sum total does not exceed the support of any one argument. [W14b23] The point of multiple deductively valid arguments is to increase the "probative force" of the whole set. On the other hand, [W14b24] "where there is genuine convergence of evidence it is the logical force, not the probative force, of the argument that is increased with the addition of each new premise" (29).

[W15] Conductive arguments are not inductive either, since [W151] induction means "that sort of reasoning by which a hypothesis is confirmed or disconfirmed by establishing the truth or falsity of its implications" (32), and [W152] conductive arguments are not instances of the testing of hypotheses based on the testing of their consequences.

However, [W16] conductive arguments are capable of being "valid" or invalid, in a general sense of these terms meaning good or bad. For [W16a1] "to say that an argument is valid is to claim that, when subjected to an indefinite amount of criticism, it is persuasive ..." (110);[4] and [W16a2] "to say that an argument is persuasive is to say that it usually persuades one who accepts ... its premises, who rejected or doubted its conclusion just before being subjected to the argument, and who thinks through the argument" (91); and [W16a3] "by criticism I mean a process of thinking about and discussion of the argument" (92), such that "the process ... does not so much discover which arguments are antecedently persuasive or unpersuasive as make them persuasive or unpersuasive" (95).

---

[4] Here Wellman's sentence continues by saying "... for everyone who thinks in the normal way" (110). However, I am overlooking this qualification partly for simplicity's sake (but hopefully without oversimplification), and partly because he himself seems to have dropped it in a later work. There we find him saying simply that "the logical force of an argument is its psychological force after criticism" (Wellman 1975, 309).

Moreover, [W16b1] there is a practical guideline for checking the "validity" of conductive arguments, at least those of the "third pattern." [W16b11] The basic principle is that "one decides whether an argument is valid [i.e., good] by weighing the pros and the cons" (57). But [W16b12] this principle is not a quantitative method: "the weighing should not be thought of as putting each reason on a scale, noting the amount of weight, and then calculating the difference between the weight of the reasons for and the reasons against" (57). And [W16b13] the principle does not amount to a mechanical process: "nor should one think of the weighing as being done on a balance scale in which one pan is filled with the pros and the other with the cons" (57). Instead [W16b14] the principle requires an exercise in judgment: "rather one should think of the weighing in terms of the model of determining the weight of objects by hefting them in one's hands" (58). But we need one more refinement, i.e., "thinking through the arguments" (80): [W16b15] "suppose that I must estimate the relative weight of two piles of stones. In this case I am only strong enough to take one or two stones in a hand at a time. Hence I must lift the stones in each pile one after the other in order to estimate their total weight. Similarly ... it is usually necessary to turn over the pros and cons successively in one's mind" (58).

Any elaborate analysis and evaluation of Wellman's account is best postponed until we have examined the relevant views of other scholars who have already examined it. However, a few comments are relatively obvious and can be stated immediately. Wellman's definition of conductive argument (proposition W11) is largely stipulative since the term conduction is relatively, if not absolutely, new. The common occurrence of conductive arguments (proposition W12) is an empirical claim, but it is an extremely important one and grounds the importance of this class of arguments. The account is assuming a definition of deductive argument (proposition W132) that is relatively common but highly controversial. Similarly, Wellman is assuming a definition of inductive argument (proposition W151) that is highly idiosyncratic and highly questionable. Furthermore, he is using the term *valid* argument in the ordinary-language sense meaning *good* argument, but this contravention is best avoided since it contradicts standard technical logical terminology, and since several other ordinary terms are easily available and avoid confusion (such as good, cogent, strong, inferentially adequate, etc.). On the other hand, the content of his conception of good ("valid") argument, proposition W16a, is extremely interesting and suggestive insofar as it is involves the notion of ability to withstand criticism.[5] Finally, his normative notion of "thinking through an argument" to weigh the pros and cons (proposition W16b14) is not completely opaque or empty, since it is elaborated in terms of an enlightening analogy with estimating the weight of physical objects by lifting them with one's hands.

Before proceeding, it will be useful to draw the structure diagram for Wellman's argument, in order to provide a visual illustration of its propositional macrostructure corresponding to the numbering system I am using in these reconstructions. Drawing this particular diagram will also suggest that for the other arguments here such diagrams will not be really necessary and can be dispensed with.

This diagram will be drawn in two parts because subargument W14 is by itself so large and complex that the whole diagram for the whole argument would require more space than a normal printed page or computer screen. Thus the first part of the diagram represents the reconstructed argument minus the W14 subargument, although it contains proposition W14 as unsupported. The second part of the diagram represents the structure of the subargument supporting W14, and contains proposition W14 as the final conclusion. The two parts are

---

[5] This has also been noted by Walton (2011, 204-6), and labeled by him Wellman's "method of challenge and response," although Walton does not really elaborate.

represented, respectively, by the diagrams in Figures 8.1 and 8.2:

Figure 8.1

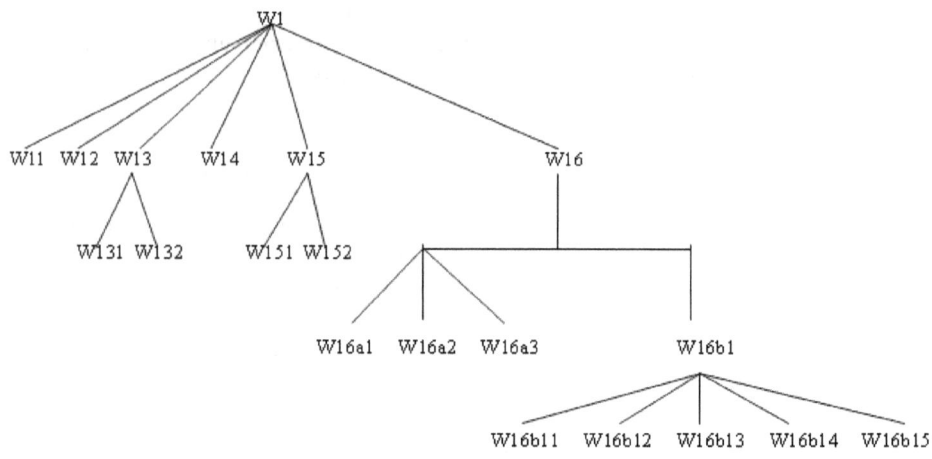

Figure 8.2

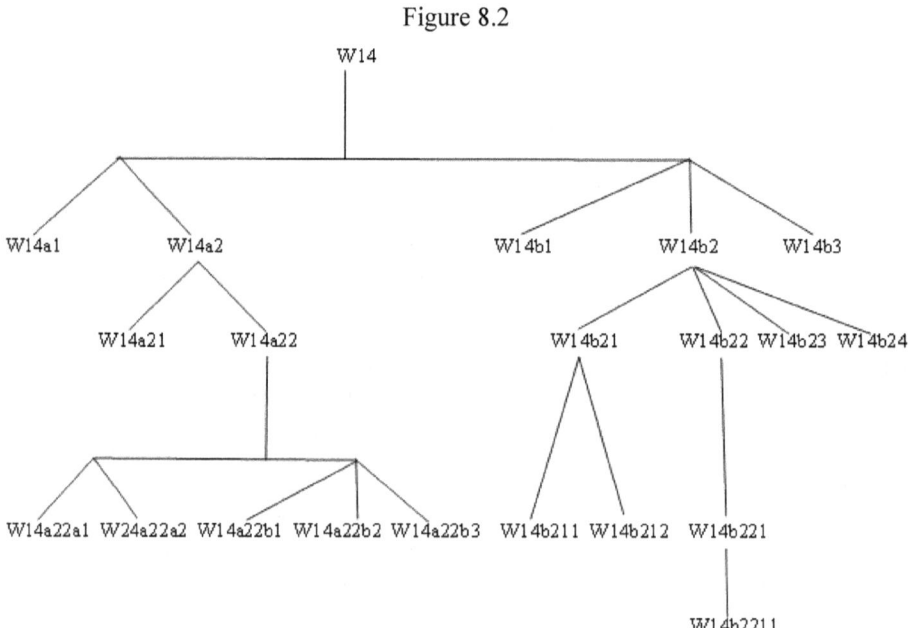

8.3 Hitchcock on Conductive Adequacy

In a number of works, David Hitchcock (1980, 1981, 1983, 1994) has elaborated a constructive interpretation of Wellman's account which I would reconstruct as follows.

Hitchcock's main conclusion may be taken to be the claim that [H1] there is an important and distinctive standard of appraisal of arguments, which is called conductive adequacy.

[H11] An argument is *conductively adequate* (1983, 105) if and only if the premises when true provide reasons for accepting the conclusion that are [H11a] nonconclusive, jointly as well as separately (Hitchcock 1983, 106), [H11b] separately relevant to the conclusion, and [H11c] mutually enhancing (1983, 51), i.e., cumulatively weighty (1983, 52), i.e., jointly supportive, i.e., jointly cumulative; and that [H11d] outweigh the considerations opposing it (1983, 105, 132; 1994).

[H12] Conductive adequacy is a very useful standard of appraisal because [H121] it is the most relevant one for arguments supporting evaluations, recommendations, interpretations, and classifications, and [H122] such arguments are ubiquitous (1983, 105-6, 130-34).

[H13] Conductive adequacy cannot be reduced to deductive validity because [H131] the needed extra premises would not be true or independently justifiable (1983, 131), or the conclusion could not be usefully qualified (1981).

[H14] Conductive adequacy is partly similar to, and partly different from, inductive probability. [H141] It is partly similar because [H1411] they are both special cases of deductive invalidity, and because [H1412] both conductively adequate and inductively probable arguments are non-monotonic in the sense that "new information independent of the truth value of the premises but relevant to the conclusion should lead us to re-examine our acceptance of the conclusion" (1981, 15). And [H142] they are partly different because [H1421] inductive probability is often quantifiable but conductive adequacy (by definition) never is, and because [H1422] the tacit premises added to make an argument more explicit often involve *ceteris paribus* clauses for conductive adequacy, but not for inductive probability (1981).

Note that in the way I have reconstructed Hitchcock's account of conduction, it has striking structural parallelisms with Wellman's account. Both accounts are meta-arguments whose main conclusions are existential generalizations; and in each case, the key premises are, in turn, a definition, an empirical claim, an anti-deductivist claim, and an anti-inductivist claim. And to Wellman's nuanced and complex anti-deductivist subargument, there corresponds Hitchcock's nuanced and complex anti-inductivist subargument. Moreover, it should be noted that the respective main conclusions, besides being existential claims, are from another point of view importance claims. This feature is itself important because at this point students of conduction will sense that arguments justifying claims of importance are typical cases of conductive arguments and/or arguments for which questions of conductive adequacy are especially relevant. This feature may be worth elaborating later, as our discussion proceeds.

The main substantive difference is that the focus for Wellman is conductive arguments and for Hitchcock conductive adequacy. That is, Wellman's conductive argument is an interpretive category that defines a particular type of arguments in terms of a number of properties specifying features of the argument's premises and conclusion, and such that some conductive arguments are good and some bad. However, Hitchcock's conductive adequacy is an evaluative category that defines a particular standard of appraisal, which may be compared and contrasted with other standards, and which may be applied in principle to all arguments, with the result that some arguments are conductively adequate and some conductively inadequate.

On the other hand, Wellman does explicitly articulate an evaluative guideline for conductive arguments, namely the weighing of pros and cons (proposition W16b11). And this corresponds to the main element of Hitchcock's standard, namely that the favorable reasons outweigh the opposing considerations (proposition H11d). Conversely, Hitchcock does not define a particular type of argument called conductive argument, but he does define (1983, 130-34) at least three particular classes of arguments for which the standard of conductive adequacy is especially relevant and appropriate, namely arguments that justify recommendations, evaluations, or classifications. And these three classes correspond to the prototypical cases of Wellman's conductive arguments. Moreover, in Hitchcock's definition of conductive adequacy, the first three clauses are essentially equivalent to a notion of conductive argument as a type of argument. In fact, we can preview immediately that they corresponds to Trudy Govier's definition of the conductive type of argument.

In any case, this (H) account is a good example of Hitchcock's usually insightful work. Those familiar with it, in particular his views on the distinction between deduction and induction, will recognize the familiar theme. That is, Hitchcock has argued plausibly and convincingly that deduction and induction are not interpretive but evaluative categories. They do not represent two distinct types of arguments defined in terms of discernible properties, but rather two distinct standards of appraisal that can be applied to all arguments, although the conditions and relevance of applicability depend on the context, and on the type of argument we are dealing with, where distinct argument types are defined independently of deduction and induction. Hitchcock recognizes that this approach is not totally original with him, but rather that it can be traced to philosophers such as Brian Skyrms (1966, 6-16; 1975), and I might add Robert Ennis (2001, 98). However, no one has stressed and exploited the evaluative approach to deduction and induction as Hitchcock has. In fact, his account of conduction may be regarded as an extension of that approach.

However, I think that the evaluative approach is much more feasible and viable for deduction and induction than it is for conduction. For in the former two cases, one can take as one's starting point the relatively clear and uncontroversial definitions of deductive validity and inductive probability; then the application of these definitions to any argument whatever is a feasible and highly valuable enterprise; furthermore, one can give relatively clear and uncontroversial definitions of various particular subtypes of arguments in ways that are completely independent of the notions of deductive validity and inductive probability; and finally, one can formulate rough guidelines for deciding which one of the two general standards to apply for which particular subtypes of arguments. However, in the case of conduction, this is not equally viable because the definition of conductive adequacy is not completely independent of the definition of the conductive type of argument. This difficulty I propose as an open problem for further investigation.

Moreover, even for the cases of deduction and induction, the evaluative approach may be deemed too self-limiting insofar as it fails to take seriously and exploit the possibility of formulating what might be called a conductive definition. A conductive definition of deductive argument and of inductive argument as types of argument would be a set of conditions whose application would yield an interpretive claim that a given argument is a deductive type of argument (or inductive, as the case may be) based on a *conductive* meta-argument. On the other hand, Hitchcock's criticism of the distinction between a deductive type and an inductive type of argument, and the usual critiques by other scholars (e.g., Ennis 2001), seem to presuppose that the interpretation or classification of an argument as deductive or inductive should be based on a deductive or deductively valid meta-argument.[6]

Such a conductive definition of deductive and inductive arguments has been provided by James Freeman, whose insight I am here adapting.[7] His conductive definition is this: "An argument is to be judged deductive (inductive) as the balance of deductive indicators

---

[6] For the case of induction, I have elsewhere elaborated a criticism of the deductivist definition of inductive argument and proposed what might be called an inductive definition of inductive argument; cf. Finocchiaro 1980a, 295-96.

[7] Freeman himself did not follow up on this insight of his, even when the occasion arose for him to study conductive arguments again; instead, on this more recent occasion (Freeman 2011) he tried to apply Toulmin's substantive model of the layout of arguments.

outweighs the balance of inductive indicators (the balance of inductive indicators outweighs deductive indicators). In particular, all things being equal, when an argument specifically claims that its premises guarantee the truth of its conclusion or when it belongs to a deductive family, it should be judged deductive. Similarly, when it claims that its premises only give evidence for its conclusion, or when it belongs to an inductive family, it should be judged inductive" (Freeman 1983, 9; cf. 1988, 225-29). Deductive indicators are, of course, words such as necessarily, entails, and proves; whereas inductive indicators are words such as probably, likely, and supports. And the deductive family includes argument types such as *modus ponens*, *modus tollens*, and categorical syllogism; whereas the inductive family includes induction by enumeration, statistical syllogism, and inference to the best explanation. Then Freeman insightfully asks: "what is the status of an argument, A, pray tell, which argues that a certain argument, B, is either deductive or inductive? Is A inductive or deductive?" (1983, 10). And he answers: "by taking account of various factors each of which is a relevant mark for the argument's being deductive or inductive ... such an argument, or much of the reasoning in it, is conductive" (1983, 10). I believe this is correct, at least in the usual case. But if and when we can apply the clauses "when it belongs to a deductive family" and "when it belongs to an inductive family," then the meta-argument will be deductive.

Now, if conductive definitions have some viability for defining the interpretive notions of deductive argument and inductive argument, then surely the possibility of a conductive definition of conductive argument deserves serious consideration. This would be a definition of a type of argument called conductive such that its application to determine whether a given argument is or is not conductive would involve a conductive meta-argument. Wellman's own definition of conductive argument (proposition W11) appears to be not conductive but rather deductive, because it consists of four individually necessary and jointly sufficient conditions, and so its application to a given argument would consist of a deductive meta-argument. As regards Hitchcock's account, he does not really advance a definition of conductive argument, but the one buried within his definition of conductive adequacy (namely propositions H11a, H11b, H11c, which as stated earlier correspond to Govier's definition) also looks like a deductive definition. In constructing a conductive definition of conductive argument, one would have to exploit notions such as conductive family and conductive indicators: the conductive family would include argument types such as balance-of-considerations arguments and pro-and-con arguments; and conductive indicators would be phrases such as *ceteris paribus*, other things being equal, and notwithstanding.

Finally, it should be noted that in this reconstruction of Hitchcock's account of conductive adequacy, I have reluctantly but deliberatively disregarded the more formal notion of conductive "validity" which he elaborates elsewhere (Hitchcock 1994). For it is difficult to disagree with Govier's (1995, 411) judgment that one remains "not convinced that the 'formalization' contributes significantly to our understanding. The need for intuitive notions such as negative relevance and positive relevance has not disappeared. Nor has the need to make judgments."[8] Indeed, Hitchcock himself (1994, 62) points out that "on this conception it is difficult to show that a conductive argument is invalid ... conductive arguments will turn out to be valid even when the reason given for the conclusion provides very weak support for it. Even worse, they will turn out to be valid even when there are unstated overriding reasons why the conclusion is false." Hitchcock does attempt to defuse this devastating implication, but the attempt only reinforces Govier's judgment.

---

[8] A similar criticism of Hitchcock's account has been advanced by Freeman (2011, 136).

8.4 Govier's Synthesis

Trudy Govier's main conclusion can be stated as essentially identical to Wellman's: [G1] there is an important special class of arguments called conductive arguments. However, the details of her argument are different because she rejects some of his claims and makes a number of modifications, while retaining the substance, as well as his general tone and flavor. Thus, it will be useful to begin with some of her critical arguments.

First of all, she rejects the individual-case clauses of Wellman's definition, namely propositions W11(1) and W11(3). Her argument here could be stated by saying that [G2] the individual-case clauses of Wellman's definition of conductive argument should be dropped because [G21] some arguments have all the essential features of conductive arguments, but have conclusions and premises that are generalizations. And she gives some examples (1987, 69). I find this argument cogent and essentially correct.

Another one of Govier's cogent critical argument is the one that questions Wellman's definition of induction. This argument could be reconstructed as follows: [G3] Wellman's definition of induction is both too narrow and too broad. For [G31] by induction he means, as we have seen, confirmation or disconfirmation of hypotheses based on establishing the truth of falsity of their consequences. But [G32] this definition excludes many types of argument generally recognized as inductive, such as inductions by enumeration and causal inferences; and [G33] it includes such obviously deductive arguments as refutations of hypotheses by *modus tollens* (1987, 67).

Going back to Govier's main argument, the first important element is her modified definition of conductive argument: [G11] a conductive argument is best defined as an argument that has the following features. First, [G111] the premises support the conclusion *nonconclusively* (1999, 155); that is, the premises do not deductively entail the conclusion, nor support it with the kind of evidence provided in strong inductive arguments, such as induction by enumeration, inference to the best explanation, or causal inference (2010, 352); I would rephrase this condition by saying that a conductive argument is one that is *neither deductively valid nor inductively probable* (or *strong*). Second, [G112] the premises support the conclusion *convergently* (1999, 156-57; 2010, 352-54, 375); that is, [G112a1] the premises are *separately relevant* to the conclusion (1999, 155; 2010, 38, 352-54), but [G112a2] they support the conclusion *cumulatively* (2010, 55, 353). Next, [G113] there are *counterconsiderations*, namely "claims negatively relevant to the conclusion" (1999, 155); and they are explicitly acknowledged or implicitly presupposed by the arguer (2010, 375). Finally, on several occasions Govier stipulates another condition, namely that [G114] the relevance of the premises to the conclusion is *a priori* or non-empirical (1987, 66, 69), and presumably non-inductive; also conceptual, normative, or criterial (1987, 70; 1999, 157); and nondeductive (1999, 157).

Regarding this last condition, the motivation and import are relatively unclear. It seems[9] that they involve the desire to distinguish conductive arguments from inductive arguments and not just from inductively probable arguments, together with the assumption that inductive arguments have an irreducibly empirical element. However, this assumption could be questioned by adapting L. Jonathan Cohen's (1986) plausible argument that analyses of concepts in analytical philosophy involve essentially inductive arguments whose conclusions are generalizations and whose premises are particular statements embodying particular linguistic intuitions (cf. Finocchiaro 2005a, 193-206).

Having defined conductive arguments in this manner, Govier is keen to show that

---

[9] Private communication from Trudy Govier.

[G12] "there really are such things as conductive arguments" (1999, 166), that indeed [G12a] such arguments are common in everyday life, [G12b] as well as in such fields as ethics, politics, law, philosophy, and literary criticism. We have seen that this thesis was also advocated by Wellman and Hitchcock, although they did not stress it and they were not as explicit. In particular, Govier deserves credit for having stressed and disseminated one particular strand of this subargument, involving a subthesis which is both little known and extremely important: that [G12c] conductive arguments are common in science, including the physical and biological sciences, and not just the social sciences (1987, 77; 2010, 298-302, 354). The argument is this (cf. Finocchiaro 1981, 1986; Laudan 1983; Pera 1994).

On many occasions scientists want to argue and need to argue that one theory is better than, or preferable to, or more acceptable than, another. They typically do so by appealing to a number of methodological principles, which formulate various desirable features, such as: empirical accuracy, explanatory power, predictive power, research fruitfulness or fertility, problem-solving effectiveness, simplicity, systemic coherence, mathematical elegance, and conceptual intelligibility. The usual situation is one in which one theory ranks higher than another with respect to some of these criteria, and the other theory does so with respect to other criteria. Moreover, normally these criteria are not weighted equally, but there are frequently individual differences in the weight or importance attached to these criteria. Thus the choice of one theory over another requires balancing the relative merits of the two theories. It follows that the justification of a particular choice consists of an argument that is nonconclusive, convergent, and mindful of counterconsiderations.

The conductive nature of such arguments for theory choice should not be surprising if we reflect on the fact that theory choice in science is really a special case of evaluation or recommendation, and we have already seen that conductive arguments are the norm in the justification of evaluations and recommendations in general. In fact, a similar kind of evaluation occurs in another common type of scientific argument, namely inference to the best explanation. For to show that a particular hypothesis is the best explanation of some data, one has to show both that it explains the data and that it explains them better than the available alternative(s). The latter claim (or series of claims) would have to be justified by some subargument(s) comparing the relative merits of the hypotheses, and such subargument(s) would normally be conductive. It follows that inferences to the best explanation have a component that is conductive.[10]

Continuing to echo, but also to clarify and amplify Wellman (his subargument W14), next Govier argues that [G13] conductive arguments should not be interpreted or assessed as enthymematic deductive arguments. For [G13a11] "possible additional premises are either false, unverifiable independently of a judgment about the individual case, or impossible to formulate in advance" (1987, 73), and so [G13a1] "the enthymeme approach makes an inference watertight at the cost of introducing an unknowable premise" (1987, 73). Moreover, [G13b1] such premises "may distort the original argument, which is typically not put forward as being conclusive" (1987, 73). Finally, [G13c1] the deductivist

---

[10] Thus it is not surprising that in the fifth edition of her textbook, in a new chapter on causal inductive arguments, in the context of a discussion of inference to the best explanation, Govier (2010, 302, 317n20) explicitly notes this connection between explanation and conduction. Nor should we find surprising the example given a long time ago by Angell (1964, 377-79), in the context of discussing arguments with multiple independent reasons, but without so much as using the words *conductive* or *convergent*; his example was Albert Einstein's (2005, 158-70) argument that his relativity principle was better than Isaac Newton's gravitation. See also, M. Salmon 2002, 260-65; W.C. Salmon 1984, 127-39; Scriven 1976, 217.

reconstruction is such that "one argument is turned into several" (1987, 73); and there are three reasons why [G13c2] one should not break down a conductive argument into several smaller ones. They are: "First, [G13c2a11] the diverse considerations in conductive arguments are characteristically put forward together ... [and so] [G13c2a1] their collective bearing on the conclusion should be taken into account when we are deciding whether to accept the conclusion. The second reason is that [G13c2b1] were we to break such a conductive argument into separate arguments, we would have to consider the various premises together when we arrived at the point of deciding how well the premises support the conclusion ... A third reason for marking conductive arguments as a distinct type is that [G13c2c1] a number of credible authors on normative reasoning and critical thinking (including Michael Scriven, James Freeman, Kurt Baier, and Stephen Thomas) have acknowledged their existence" (2010, 353-54).

Here it is worth noting that Govier herself characterizes this last subargument, in support of proposition G13c2, as a conductive argument (2010, 353). This note may serve as a second reminder that a relevant research project would be to analyze all these philosophical meta-arguments being reconstructed here from the point of view of conduction; this project would not be aimed at showing that all these arguments are conductive, but rather at determining which are and which are not. But note also that, since this point applies to all the constituent subarguments, this research project is an indefinitely long one.

As far as I can tell, Govier does not incorporate into her account an anti-inductivist thesis, analogous to Wellman's proposition W15 (which she criticizes as misconceived), nor one analogous to Hitchcock's more nuanced and sophisticated claim that there are both differences and similarities, proposition H14. This should not be surprising since she does not need such an anti-inductivist thesis; in fact, she makes conductive arguments non-inductive by definition, when she elaborates the first definitional clause, stipulating nonconclusiveness, by saying that conductive arguments are not only not deductively valid, but not even inductively probable or instances of the standard inductive forms, such as induction by enumeration, inference to the best explanation, causal inference, etc. However, she is concerned with distinguishing conductive from convergent arguments.

In fact, she has a subargument designed to show that [G14] conductive arguments should not be *equated* to convergent ones. Of course, it is true, by definition, that [G141] all conductive arguments are convergent. However, the converse does not hold: [G142] not all convergent arguments are conductive. For "[G142a1] it is possible to offer an argument which exemplifies the convergent support pattern, but in which there are several different premises, each of which, taken alone, deductively entails the conclusion. [G142a11] This is rather uncommon, but may occur, either because the arguer expects that some of his premises will be contested or because he is not aware that the entailment relationships hold and make some premises logically redundant" (1987, 70).[11] Similarly, "[G142b1] some inductive arguments use the convergent support pattern, particularly [G142b11] if a number of distinct, and apparently unrelated cases, are cited to support a generalization" (1987, 70). That is, such arguments would be convergent, but some deductively valid and some inductively probable; hence, they would violate the first ("nonconclusiveness") clause of the definition, and so they would not be conductive.

It must be pointed out that, in her published work, Govier does not give or construct actual examples of such arguments, with the properties of being convergent, but deductively valid or inductively strong, and so not conductive. She does not do so, not only

---

[11] Govier repeats this argument in an instructors' footnote in the latest edition of her textbook (2010, 376n2).

when she first advances the argument (1987, 70), but also in later editions of her textbook, where the argument is repeated in a footnote for instructors (2001, 411n2; 2010, 376n2).[12] I believe what is needed here is to find real or realistic[13] examples of such arguments, as a way of testing the existential claims in the subargument. The failure to find them would considerably weaken this argument. In such an inquiry it might also happen that we find candidates for such examples, but that a careful analysis reveals that they are not, after all, convergent or deductively valid or inductively probable. And this in turn would undermine the key thesis (proposition G14) in this subargument, namely the distinction between convergence and conduction. However, in terms of the overall account, this would only mean that the first two conditions (G111 and G112) of the definition are not distinct and should be collapsed into one. I am not sure the damage would be any greater than that.

Finally, like Wellman, but elaborating and amplifying, Govier argues that [G15] conductive arguments can be appraised. After all, they are arguments, and like all arguments [G151] we can evaluate the acceptability of each premise. Similarly, [G152] we can asses the relevance of each premise. And, as usual, [G153] we should try to determine the strength of each relevant premise as a reason for the conclusion. Then, moving in the direction of the more distinctive and specifically relevant manner of evaluation, [G154] we should try to determine the collective or cumulative strength of all premises taken together as support for the conclusion. Next, assess [G155] "whether any counterconsiderations acknowledged by the arguer are negatively relevant to the conclusion" (2010, 365); [G156] "what additional counterconsiderations, not acknowledged by the arguer, are negatively relevant" (2010, 365); and [G157] "how strong is each of the counterconsiderations as a reason against the conclusion" (1999, 170). Finally, [G158] "reflect on whether the premises, taken together, outweigh the counterconsiderations, taken together, and make a judgment. [G159] Try to articulate good reasons for that judgment" (2010, 364).

In elaborating these principles, Govier is clear that her goal is to steer a judicious middle course between two opposite and unsatisfactory extremes. On the one hand, we have the claim that there is nothing one can do to appraise a conductive argument than to think the argument through again and again. Sometimes Wellman is taken to be making this unhelpful claim, but we have seen above that his account does include more: he is explicit in formulating the guideline that the pros outweigh the cons (proposition W16b11); and he elaborates a hefting model of weighing (W16b15), which is very helpful. On the other hand, there is the other extreme of thinking that the strength of each premises and each counterconsideration can be measured; that the cumulative strength of each set can be added up; and that the two can be subtracted from one another to arrive at the net positive support or negative disconfirmation. Some of Govier's own language may give this impression, for example the talk of strength, of positive and negative relevance, and the distinction of the individual and the cumulative assessment of strength. However, her commentary on these rules makes it clear that she is rejecting this quantitative and

---

[12] In private communication, Govier has proposed the following: "an example of an argument with convergent support that I would not be inclined to call conductive would be one that was inductive, but had separately relevant premises ... Consider—James went to Western Canada high, the same school as Mary; James went to the University of Calgary, as did Mary; James was a high achiever at the U. of C., as was Mary; these are reasons to think that James is acquainted with Mary." However, the difficulty here is that although we can agree that this is a convergent inductive argument, to deny it conductive status begs the question. To me, this example seems to be as conductive as Govier's (1999, 168-69) much discussed example of the good-manager argument.

[13] With these words ("real or realistic") I want to echo Govier's (2000, 289-90) own approach, which I have independently advocated (Finocchiaro 1980a, 293-431; 2005a, 21-91), as have others (e.g., Fisher 1988, 2004).

mechanistic model (1999, 170).

Nevertheless, some questions arise. Some of these will be taken up when I discuss the critiques advanced by other philosophers. For now I want to raise the following point, which I do not think has been previously raised. Govier's principles of appraisal seem to contain an asymmetry between the treatment of positively relevant reasons and the treatment of negatively relevant counterconsiderations: whereas for the case of counterconsiderations one is supposed to think of other considerations besides those acknowledged by the arguer (proposition G155), before performing the balancing act of weighing the pros and the cons, for the case of favorable reasons there is no requirement to think of and take into account additional ones not stated by the arguer.

Her justification for this asymmetry is this: "to reflect on whether there are further considerations—not stated in the argument—that would count in favor of the conclusion and would outweigh any counterconsiderations ... takes you beyond appraising the stated argument. It moves you to a new stage where you are amending or reconstructing that argument ... when your real interest is whether you should accept the conclusion and not merely whether the conclusion is well supported by the particular argument you are evaluating" (2010, 366). In short, finding and taking into account additional evidence is not relevant to the evaluation of the original argument, but to the evaluation of the acceptability of the conclusion.

However, this justification presupposes that finding and taking into account counterevidence is relevant to the evaluation of the original argument, and not merely to the eventual evaluation of the conclusion. And then the issue become why this assumption should be accepted: why taking into account additional counterevidence is relevant, but taking into account additional evidence is not relevant, to the assessment of the original argument as given; why taking into account additional evidence is relevant only to the evaluation of the acceptability of the conclusion, but taking into account additional counterevidence is not limited in the same way. I don't think this issue has been properly addressed in the literature. I believe further work is needed to determine whether this asymmetry holds, and if so why.

8.5 Critiques of Govier's Argument

8.5.1 Allen's Objections and the Justification of Weighing

Some important critiques of Govier's argument have been advanced by Derek Allen.[14] In his critical study of her *Problems in Argument Analysis and Evaluation*, he elaborates a powerful criticism of Govier's anti-deductivist claim, proposition G13, and the supporting subargument. And since this subargument has three parts, so does the criticism. Here, Allen's key claim is that [A1] Govier's proposition G13 has not been justified, i.e., that the supporting subargument is not successful, or not cogent: "a reconstructive deductivist will take the enthymeme approach to conductive arguments ... Govier has not demonstrated that this approach is always mistaken. Nor has she demonstrated that we will always be in a better position to assess conductive arguments if we reject the enthymeme approach in favor of her own" (1990, 58). Allen's argument is the following.

For the purpose of this discussion, Allen asks to [A111] focus on the same example that had been considered by Govier and Wellman: (a) you should return the book because (b) you promised to do so. [A112] The deductivist could reconstruct this argument by adding two other premises: (c) other things being equal, you should keep your promises,

---

[14] Here I examine Allen 1990 and Allen 1993; but for some of his further reflections, see Allen 2011.

and (d) other things are equal. [A113] Note that premise (c) is not a categorical universal generalization, which would have been false, but that it is qualified by the *ceteris paribus* clause; however, premise (d) asserts that this clause holds in this case. [A114] The expanded, and qualified, argument is now deductively valid, but admittedly we may not always be in the position of knowing that premise (d) is true or acceptable. However, [A115] if we do not know whether in this case the other things are equal, then we are in no position to claim that the original argument, interpreted as a conductive argument, is conductively adequate, for, in Allen's clear and succinct reformulation, [A1151] conductive adequacy means, that "the premise outweighs ... any stated or unstated counterconsiderations ... that are negatively relevant to its conclusion" (1990, 56). On the other hand, [A116a1] if we have enough information to be able to say that the other things are equal, that would make the reconstructed argument one with acceptable premises, and thus sound or cogent (on account of the deductive validity); but [A116a2] the same information would also enable us to asses the original argument as conductively adequate. Allen concludes that [A11] "it matters not" (1990, 56), "it ... makes no difference" (1990, 57) whether we interpret and assess the argument conductively or deductively. That is, [A1] Govier has not shown that her approach is better than the deductivist one.

What I would add is that, as long as we do not claim Allen to have shown that the deductivist approach is better, his counterargument is cogent. In other words, Allen has shown that the inferential adequacy of a conductive argument corresponds to the acceptability of some of the premises in a properly nuanced deductivist approach. Thus, from the point of soundness or cogency, which involves both strength of the inference and acceptability of the premises, there is no advantage in either approach, unless or until we consider other factors. The situation is analogous to the deductivist reconstruction of traditional inductive arguments, which has long been known to be possible, but whose desirability or preferability has to be judged on other grounds (see, e.g., W.C. Salmon 1984, 18). And this leads to the other grounds on which Govier bases her anti-deductivist claim.

As we saw above, another reason she gives is that the deductivist reconstruction of a conductive argument turns one argument into several (proposition G13c1). Allen (1990, 57-58) argues cogently that [A2] this is not true. [A211] Consider a conductive argument with two premises: Ra1; Rb1; so C. Here [A212] Ra1 does not entail C, and Rb1 does not entail C. [A213] Suppose we give a deductivist reconstruction, which adds two premises: Ra2 such that Ra1 and Ra2 entail C; and Bb2 such that Rb1 and Rb2 entail C. The reconstructed deductive argument is: "Ra1 and Ra2; Rb1 and Rb2; so C." [A21] The reconstruction is no more two arguments than the original was. [A214] It is true, of course, that in the original argument, Ra1 and Rb1 are separately relevant and cumulative in weight, whereas in the reconstruction some of the premises are not separately relevant (e.g., Ra1 and Ra2, Rb1 and Rb2), and none are cumulative or mutually reinforcing. But [A215] this means that the original argument is, and the reconstruction is not, a conductive argument. [A216] It does not mean that the reconstruction is two arguments; to so conclude "is a non sequitur" (1990, 58).

With another powerful argument, Allen (1993, 116-18) has objected that [A3] Govier's definition of conductive argument should be amended to stipulate that each premise should not be deductively relevant to the conclusion, where deductive relevance means: a premise P is deductively relevant to the conclusion C if and only if P's relevance depends on another proposition Q such that P&Q entail C. This amendment may be included in the fourth main clause of Govier's definition, proposition G114. In fact, I have already included it in my reconstruction above, where that fourth clause includes a miscellaneous cluster of notions, such as non-empirical, *a priori*, conceptual, criterial, and normative. I was led to so include it by the fact that Govier (1999, 157, 178n6) has explicitly accepted Allen's amendment. For this reason, and also because of the difficulty and complexity of Allen's argument, here I shall omit presenting a reconstruction.

Finally, Allen (1993, 109-11) raises another insightful objection which I would interpret as another amendment to Govier's definition, regarding the counterconsiderations clause, proposition G113, but which she does not even mention, let alone accept. That clause stipulates, as we have seen, that there may be counterconsiderations; that is, explicit counterconsiderations may be present, but if they are not, implicit counterconsiderations are presupposed. Indeed, the distinctive conductive adequacy of such arguments depends on whether the favorable reasons outweigh the counterconsiderations; and normally the evaluation of such arguments involves explicit argumentation about what outweighs what. Now, when the counterconsiderations are explicit in the original argument, sometimes the argument may contain also an explicit justification of such a balance-of-considerations claim. This is especially true when there is just one explicit favorable reason and one explicit counterconsideration. Allen gives examples of this situation, taken from a decision of the Canadian Supreme Court declaring unconstitutional a rape-shield law (barring the admissibility of evidence about the previous sexual activity of a plaintiff in a rape trial); and the examples involve opinions in which different judges balance and weigh the value of protecting the victim and the value of ascertaining the truth. Now, Allen's key point is that [A4] such arguments need not, and, I would add, normally would not, be conductive; for [A41] these are arguments concluding that one value or factor outweighs another in importance, not arguments concluding the presence of a value or factor after assuming that it outweighs the other. In Allen's words: "According to Govier, conductive reasoning frequently involves the weighing of pros and cons ... reasoning of that sort is *based on* the judgment that one factor, or set of factors, outweighs another, which is to say that some such judgment *underlies* the inference to the reasoning's conclusion—rather than being itself the conclusion ... But it is worth adding that arguments that involve [or to be more precise: arguments that conclude with] the weighing of one factor, or set of factors, against another need not be conductive" (1993, 110, my italics).[15]

This issue has already arisen above, in one of Wellman's subarguments justifying his anti-deductivist claim (W14a). There he argued that the deductivist reconstruction of conductive arguments includes premises whose justification cannot be purely deductive but must include some conductive reasoning. In a sense, there Wellman is making a point which is the reverse of the one in Allen's last considered objection (A4). Allen is arguing that some conductive reasoning has parts that are not conductive; Wellman is arguing that some deductive reasoning has parts that are nondeductive but conductive.

At the moment, I am not ready to decide this issue. The question seems to be what type of arguments, if any, can be used to justify claims that one reason outweighs a counterconsideration. These are claims that are normally presupposed in conductive arguments; often explicitly stated in such arguments; sometimes explicitly justified in these arguments; and always to be explicitly assessed in the evaluation of them. Although this is an open question, if we recall my working hypothesis formulated in this chapter's introduction above, and if we note the nuanced character of that formulation, we can see that Allen's fourth criticism seems to confirm the first part of that hypothesis; that is, the claim that an essential characteristic of conductive arguments is a *balance-of-considerations claim* that must be implicitly made, explicitly formulated, or critically justified for the construction, interpretation, or evaluation of conductive arguments.

---

[15] See also Allen (2011, 188) for a reiteration and an additional illustration of this conclusion.

8.5.2 Ennis's Objections and the Detection of Convergence

Another philosopher who has advanced some noteworthy critiques of Govier's account is Robert Ennis (2001, 2004). One criticism is that [E1] "the definition of 'conduction' is needlessly narrow" (2004, 38), because [E11] "the convergence requirement seems unduly narrow" (2004, 35). For [E111] this requirement "eliminates a large number of arguments, including all that I have so far considered in this paper" (2004, 33); that is, cases of qualified reasoning such as [E111a] Ennis's raccoon argument, [E111b] Toulmin's Petersen argument, [E111c] Weddle's rain-prediction argument, and [E111d] Plantinga's Frisian-lifeguard argument. Ennis's raccoon argument is this: "raccoons rarely attack a human when they do not feel threatened and do not feel that their young are threatened. The raccoon that is ambling across the yard does not feel threatened by us, and it does not feel that its young are threatened—its young are not around. So the raccoon will probably not bother you, even though you are within fifteen feet of it" (2004, 25).

It might seem that this objection is analogous to Govier's own objection (G2) to the individual-case requirement of Wellman's definition, and that Ennis is in effect suggesting that the convergence requirement be dropped from the definition of a conductive argument. Or perhaps he is suggesting that we drop the language of "conductive" arguments, and instead talk of "qualified" reasoning. In either case, Ennis argument is committed to claiming that [E1121] qualified-reasoning arguments have all the essential characteristics of what Govier calls conductive arguments, and hence [E112] should be treated or categorized in the same way. This subargument can and should be added to or grafted onto the one reconstructed in the last paragraph, and may be regarded as a tacit part of that argument.

These are "formal" similarities between this objection by Ennis to Govier (E1) and the earlier objection by Govier to Wellman (G2). However, evaluatively speaking, one can raise the question whether Ennis's objection is relevant: whether it is true that qualified-reasoning arguments have all the essential characteristics of conductive argument, and whether we can conclude that the two classes of arguments should be categorized or treated in the same way. A key point I would make is that the examples of qualified-reasoning arguments mentioned here are cases or variations of "statistical syllogism," that is, arguments that apply an inexact, qualified, or statistical generalization to an individual case to reach a conclusion about that individual case. This is an extremely common and important type of argument because it is connected with questions of stereotyping, profiling, and prejudice (cf. Schauer 2003; Scriven 1976, 205-10). And it even shares a significant similarity with conductive arguments, insofar as the inferential adequacy of statistical syllogisms depends on the so-called requirement of total evidence,[16] and this requirement is analogous to the one requiring that counterconsiderations be taken into account, which is the key factor in conductive adequacy. However, I regard this problem as an open question, deserving of further study.

Ennis's second criticism argues that [E2] Govier's definition of conduction is difficult to apply because [E21] the convergence requirement is difficult to apply (2004, 35). For [E211] when an argument is such that more than one reason support the conclusion, often there is no indication "whether each reason is an INUS condition or an independent condition" (2004, 34); here, [E212] an INUS condition is an acronym Ennis adopts from John Mackie (1993) to mean a condition that is an *i*nsufficient but *n*ecessary member of an *u*nnecessary but *s*ufficient set.

---

[16] As acknowledge by Govier herself (2010, 269, 371); cf. W.C. Salmon 1984, 96-97.

That is, consider the argument: C because P, Q, and R. An instance might be Govier's own example: "she would be a good manager because she has considerable experience, she is very good at dealing with people, and she knows the business well" (1999, 156, 168). Govier starts by interpreting P, Q, and R as separately or independently relevant to C. And then, since they are mutually reinforcing, she concludes that the argument is convergent; and given the nonconclusiveness, she takes the argument to be conductive. Ennis is questioning the initial interpretation of the reasons as independently supportive: "each could be an INUS condition ... all three together constituting the set, making these conditions dependent on each other" (2004, 34).

I believe Ennis's question is well taken because the argument as stated does not convey the separation of the reasons as explicitly as would be done by another possible formulation, namely "C because P, because Q, and because R." Repeating the connective "because" would make it clear that one intended to treat each reason separately and independently. But as it stands, the meaning is ambiguous. Thus, Ennis's second criticism seems cogent.

Lastly, Ennis has another criticism that is longer and more complex to justify, but easy to state. The objection is that [E3] Govier's attitude toward deductive standards is inconsistent. That is, "although she apparently eschews the use of deductive standards in conduction, she needs them to show the relevance" (2004, 35) of the reasons. In particular, this criticism can be taken as directed at Govier's key anti-deductivist claim, proposition G13.

The argument is this. In evaluating the strength of reasons (proposition G153), and the same would also apply to counterconsiderations (G157), Govier claims that "what helps us to evaluate the strength of reasons is that reasons must have a degree of generality" (1999, 171); such generality is qualified by means of the clause "other things being equal"; "by spelling out qualified universals ... we are able to move beyond the apparently irreducible claim that 'P1 is relevant to C'" (1999, 171); and "a strong reason is one where the range of exceptions is narrow. A weak reason is one where the range of exceptions is large" (1999, 171). Ennis is pointing out that Govier explains the relevance of a reason in terms of whether it together with a qualified universal would deductively entail the conclusion when the qualifier is removed; and she explains the strength of a reason in terms of the number of exceptions that are covered by the qualification (the more exceptions, the weaker the reason; the fewer, the stronger). Schematizing further, Govier seems to conceive strength as follows:

Original conductive argument: P, so C.
Reconstruction for evaluation purposes:
    P
    Other things being equal, if P then C
    Probably, other things are equal.
    So, probably C.

The strength of P is being correlated to the degree of probability of the third premise in this reconstruction. Now, Ennis notes (2004, 34) that such a reconstruction is to be contrasted with one where the qualified universal would read, "Other things being equal, if C then P." This alternative reconstruction might seem strange and idiosyncratic, but it can be made more familiar and plausible by noting that it would conform to the form of reasoning called abductive, or inference to the best explanation. Thus, Ennis seems to be making a forceful point, and this criticism seems to have some cogency.

8.5.3 Zenker's Objections and the Role of Ceteris Paribus

A more radical criticism of Govier's method for evaluating the strength of reasons has been advanced by Frank Zenker.[17] His own critical conclusion is that [Z1] "*ceteris paribus*, c-p generalizations are irrelevant to evaluating conductive argumentation" (Zenker 2009b, 11), where a *c-p generalization* is an obvious abbreviation meaning a generalization qualified by means of the clause *ceteris paribus* or *other things being equal*. Zenker advances four reasons against Govier's method of evaluating the strength of reasons in terms of the number of exceptions to the c-p generalization.

First, [Z11] this method says nothing about the comparative evaluation of the strength of the pro reasons as a whole and the con reasons as a whole; it only pertains to the comparative strength of reasons within each group (Zenker 2009b, 3-4).

Second, [Z12] Govier's method presupposes an incorrect analysis of c-p generalizations. [Z121] Govier's presupposition is that c-p generalizations sometimes warrant and sometimes do not warrant transitions from the particular reason to the particular conclusion. But [Z122] the correct analysis is that they do not allow such transitions unless pertinent objections are answered (Zenker 2009b, 4-6).

Thirdly, [Z13] the introduction of c-p generalizations is not necessary. Rather, [Z131] it is better to introduce the individualized associated conditional. [Z1311] Then, working at this individual level, a strong reason will be one embedded in an argument that meets the objections advanced against it (Zenker 2009b, 6-8).

Finally, [Z141] working at the same individual level, it can happen that a reason is stronger than another even if the former has a larger class of exceptions. Thus [Z14] the strength of a reason in a conductive argument cannot be a function of only the exception class (Zenker 2009b, 8-9).[18]

This criticism does show, in my opinion, that the concept of a c-p generalization and its connection to conductive arguments deserve more and deeper analysis than that provided by Govier. Furthermore, it seems likely that the analysis is to be carried out along a different direction than that pursued by Govier. Indeed it is not surprising that someone who has carried such a comprehensive study of *ceteris paribus* (Zenker 2009a) would discover such limitations. However, the other side of this coin is that such criticism of Govier's account seems somewhat unfair insofar as it focuses a very small part. This is the part pertaining to proposition G153, the elaboration of which I did not include in my reconstruction above, precisely because it did not seem to be sufficiently important or central to Govier's account.

On the other hand, two other aspects of Zenker's criticism reinforce some aspects of Wellman's original account that were modified or not elaborated by Govier. For example, when Zenker says that c-p generalizations allow the inference of the conclusion from the reason if and only if the pertinent objections are answered (proposition Z122), this is reminiscent of (although not identical to) Wellman's definition of "validity," proposition W16a1: "to say that an argument is valid is to claim that, when subjected to an indefinite amount of criticism, it is persuasive" (Wellman 1971, 110). Similarly, when Zenker emphasizes the individualized associated conditional and case-by-case reasoning, this corresponds to Wellman's (1971, 73-82) stress on reasoning without rules or criteria,

---

[17] Besides Zenker 2009a and Zenker 2009b, to which I refer here, see also Zenker 2011. Zenker's criticism, in turn, has been criticized by Fischer (2011, 95-98), although I am not convinced by the latter.

[18] A similar criticism of this thesis of Govier's has also been advanced by Freeman (2011, 136-38).

reflected even in the fourth clause of his definition of conduction, proposition W11 above.

The upshot of my considerations is that Zenker's criticism is not as negative and destructive as it may seem at first, but rather reflects an approach to conductive arguments different from Govier's.

8.5.4 Wohlrapp's Dynamical Approach to Conduction

There is another one of Govier's critics whose criticism reflects an alternative approach to conductive arguments: Harald Wohlrapp (1995, 1998, 2008, 2011).[19] Indeed, the alternative nature of Wohlrapp's criticism is more marked and striking than was the case for Zenker's.

Referring to Govier's account, Wohlrapp claims that "there are at least two reasons to [WO1] reject this view. First, [WO1a1] it contains a misunderstanding of the relationship between deductive schemes and other possibilities for passing from sentences to other sentences. [WO1a11] Deduction is not just one possibility among others, but is fundamental in the sense that it is presupposed by all other inference schemes like inductions, etc., whereas these are not presupposed by deduction" (Wohlrapp 1998, 341-42). There seems to be no explicit elaboration of this subargument in Wohlrapp's English-language publications, although the deductivist character of his alternative account (to be summarized below) is relatively obvious.

Wohlrapp's second reason to reject Govier's account is that "[WO1b1] this view fixes argumentation theory in a quasilogical and nondynamic perspective, [WO1b2] viz. in an unfruitful dichotomy of structural and procedural perspectives. [WO1b11] The process of argumentation seems here to be no more than a sequence of inference steps, where each step can be isolated and analyzed by itself. [WO1b3] I want to plea for a more differentiated and realistic view in which procedural and structural elements of argumentative speech are integrated and where premises and conclusion of an argumentation form a 'retroflexive' system of mutual support" (Wohlrapp 1998, 342).

Wohlrapp then goes on to sketch his dynamic account, thus providing an articulation and justification of the claim just made (proposition WO1b3). However, before reconstructing this dynamic account, let us mention a third criticism, pertaining to Govier's method of evaluating the strength of a reason based on the size of the class of exceptions to the associated c-p generalization. He too finds it unsatisfactory, but it is revealing that he expresses his criticism in terms of it being unfruitful, at least relatively speaking, namely relative to his alternative dynamic approach: "[WO1c1] The question 'how many exceptions-to-be-respected are there for the argument's associated general principle?' is not productive ... [WO1c11] The exceptions are not countable and [WO1c12] if an intuitive estimate shall suffice, then it remains unclear what is the significance of such an estimate. [WO1c13] If we disregard all these puzzles for a moment: how would we proceed with these somehow determined numbers ? ... would we really calculate them arithmetically?" (Wohlrapp 2008, 11-12). For Wohlrapp, these are rhetorical questions.

Let us now reconstruct Wohlrapp's more "differentiated," and "realistic," and "fruitful" account that "dynamically" combines "structural" and "procedural" elements. The key thesis is that [WO2] a conductive argument (as usually defined, i.e., a pro-and-con argument) should be interpreted as a temporary stage in the process of argumentation that needs to be completed and modified in various ways to result in a good and more stable argument. Or in his own words, "as long as we have to deal with a construction featuring

---

[19] Wohlrapp's criticism has also been criticized by Fischer (2011, 91-103; 2012, 132-34, 140); but it focuses on other issues and does not affect my analysis here.

open counterarguments, the thesis [i.e., the conclusion] either is not a valid orientation, or we have not yet completed the argumentation ... A conductive argument is nothing but a state of argumentation assessed at some point in time" (Wohlrapp 2008, 21).

The completion and modification processes are distinct from each other and internally multi-faceted. [WO21] A conductive argument should be "completed" in three ways (Wohlrapp 2008, 16-17): one way is [WO211] to pair each pro reason with a con reason, searching for unstated ones if need be; another is [WO212] to check the acceptability of each pro and con reason, reconstructing the supporting subarguments if need be; the third way is [WO213] to elaborate the point of view (called the "frame") of each reason, by reconstructing the unstated assumptions and latent structure if need be. [WO22] Furthermore, a conductive argument should be "modified" in four ways aimed at integrating the points of view (the "frames") used in the pro and con reasons (Wohlrapp 2008, 17-18): one way is to [WO221] criticize and if necessary replace a given frame or point of view; another is to [WO222] rank the various frames for their importance, if possible; a third is to [WO223] harmonize, if possible, the various frames so that they are no longer incompatible, but still remain distinct; the fourth way is to [WO224] elaborate a combination or synthesis of the frames. Wohlrapp (2008, 17) is at pains to point out that [WO225] such a frame integration should not be merely a "reconciliation" of interests and differences, in the sense that "existing" interests and differences are "negotiated," but rather a critical reconciliation that involves evaluation and modification.

The rest of Wohlrapp's main constructive argument consists of an analysis of two of Govier's own examples, the nanny argument in the dialogue between husband and wife, and the euthanasia-legalization argument. The analysis is designed to show that [WO3] these two conductive arguments can be completed and modified in accordance with the various principles just stated, i.e., propositions WO211, WO212, WO213, WO221, WO222, WO223, WO224, and WO225.

In a sense, the analysis of those two arguments does illustrate and support these dynamical principles. The analysis is insightful and plausible, and so Wohlrapp's argument has some cogency, and his dynamical account receives some confirmation. However, does the argument show that his dynamical account is "more realistic" than Govier's, as claimed in proposition WO1b3? Taking realism to mean at least empirical adequacy, this claim is questionable. In fact, I believe Wohlrapp (1998, 347) is correct when he himself admits that "this is not the normal case. Usually, we hang around halfway and we comfort ourselves with the quantitative metaphor having produced a reasonable balance of so many arguments." That is, in cognitive practice conductive arguments as such have a reality which the dynamical approach perhaps fails to capture, or at least captures less faithfully than the static approach. Thus, my conclusion is that more empirical work is needed to test whether the dynamical account is sufficiently descriptive and not excessively prescriptive.

Finally, Wohlrapp advances a claim that may be regarded as a corollary of what I have called his key thesis above, proposition WO2. The corollary is that [WO4] in conductive argumentation, "the conclusion reached with the arguments [i.e., reasons] presented is not the result of a weighing, whatever that be. The result was reached via an evaluation that arranged the arguments [reasons] into a thetical construct and, by means of continuing their discussion, examined them with respect to their suitability for supporting the thesis [i.e., conclusion]" (Wohlrapp 2008, 21). The term "thetical construct" is, of course, both a neologism and obscure. The German phrase, *thetischen Konstruktion*, is slightly clearer insofar as "construction" (*Konstruktion*) seems more appropriate than "construct." The adjective "thetical" (*thetischen*) suggests a meaning of pertaining to "thesis" (*These*).[20]

---

[20] In private communication, Harald Wohlrapp has confirmed this particular suggestion.

Thus, I decipher this corollary claim, WO4, as follows. I believe "thetical construct" refers to a network or structure of logically or probatively interrelated propositions; the "discussion" consists of the processes of completion and integration that make up his alternative procedure for the evaluation of whether or to what extent the conclusion is thereby supported; and the procedure is the one mentioned in propositions WO21 and WO22 above. In fact, in another place Wohlrapp (1998, 349) gives this other formulation of what I take to be the same claim: [WO4] "the secret of a successful argumentative conduction is frame unification. The nature of the inference step is the formulation of ... a (complex) frame which integrates, on the one hand, our positions and, on the other, the different realms in which we place" the issue of the argument. This corollary thesis represents Wohlrapp's interpretation of the nature of conductive inference: the crucial element is not a "weighing" of the pro and con reasons, but an evaluation of them ideally carried out in accordance with the dynamical principles sketched. The only thing I would add here is that this thesis is also in need of further empirical support.

8.6 David Brooks on the Health-care Bill

It is now time to examine some relevant empirical material, namely some conductive ground-level argumentation, so to speak. On December 18, 2009, the *New York Times* published two op-ed columns about the health-care reform bill that had been widely discussed the whole year and would be voted upon in the U.S. Senate a few days later. One opinion was in favor of the bill and was authored by Paul Krugman (cf. Appendix 2). The other was against the bill and was authored by David Brooks (cf. Appendix 1). Besides the obvious difference in substantive content, the two arguments make an interesting and instructive contrast in argumentative style, structure, and power, so much so that they would reward close study on the part of logicians and argument theorists. However, given our present focus on conductive arguments, I shall limit myself to Brooks's argument.

Brooks's column is reproduced verbatim in Appendix 1. However, for convenience of analysis, and to facilitate reference, I have added in brackets a label for each paragraph, from the letter 'a' to 'q'. As is customary for newspapers, each paragraph is very short, consisting of either just one sentence or at most a few. Thus, it will also be convenient to assign a label to each sentence within a given paragraph, consisting of the paragraph letter plus a numeral corresponding to the place of the sentence in the sequence within the paragraph. For example, the fifth and last sentence of paragraph 'g', "but if you've got cancer, you want surgery, not nasal spray," will be given the label 'g5'. However, note that I have not actually inserted these numerals in Appendix 1, because I did not want to overburden the eye of the reader, and the work can be quickly done in one's head with the mind's eye.

It is worth saying at the outset that whatever one may think of the substantive content of Brooks's conclusion, and of the strength of his argument, his essay is a model of clarity. It is equally obvious, from the context, the style, and the structure, that here we have a conductive argument, indeed a paradigm example of conductive argument.

Before we get involved into the details, let us have a quick glimpse at the whole. Brooks is recommending rejection of the bill, on the basis of several reasons to oppose the bill, but with an awareness that there are also several reasons to favor the bill, and in light of a judgment that the bad aspects of the bill outweigh the good ones. However, note that since Brooks's conclusion is a negative recommendation, the premises *supporting* his conclusion are the reasons *against* the bill; the counterconsiderations or objections to his conclusion are the reasons *for* the bill; and the balance-of-considerations claim that the strength of the argument's premises outweighs the strength of the objections is equivalent

Conductive Arguments, Pro-and-Con Reasoning, Etc. 145

to the claim that the reasons *against* the bill outweigh the reasons *for* it. In short, there is a kind of reversal of sign that must be kept in mind depending on whether we take the point of view of the pros and cons of Brooks's conclusion or the point of view of the pros and cons of the bill.

With these preliminaries and clarifications in mind, Brooks's argument may be reconstructed as follows.

[B] The Senate health care bill should not be passed (cf. p1), for although [B1] there are at least four reasons for passing the bill, [B2] there are at least six reasons for not passing it, and [B3] the latter outweigh the former, i.e., the reasons for not passing outweigh the reasons for passing the bill (cf. p2).

The main reasons to favor the bill are: first, [B1a] "it would provide insurance to 30 million more Americans" (a1); second, [B1b] the bill addresses "the deficit issue seriously" (b1), relatively speaking; third, [B1c] the bill contains very many "little ideas in an effort to reduce health care inflation" (d1); and fourth, [B1d] "if this fails, it will take a long time to get back to health reform" (f1).

The first one of these reasons is left unsupported, being uncontroversial in the context of these discussions. But the other three are supported by subarguments that are themselves relatively complex, and some of which may themselves be conductive. For example, Brooks claims that [B1b] relatively speaking, the bill takes "the deficit issue seriously" because [B1ba] it is much more fiscally responsible than the prescription drug benefit passed during the Bush administration (b2); [B1bb] the extra costs to cover the uninsured are partly offset by Medicare cuts and tax increases (b3); and [B1bc] "the bill won't explode the deficit" (c3); but [B1bd] "the bill is not really deficit neutral" (b1). This seems to be an attempt to support the bill's fiscal seriousness nonconclusively by means of three separately relevant and jointly cumulative considerations, while acknowledging one objection.

The main reasons to oppose the bill are: first, [B2a1] "it does not fundamentally reform health care" (g1); second, [B2b1] "it will cause national health care spending to increase faster" (i1); third, [B2c1] "the bill sets up a politically unsustainable situation" (j1); fourth, [B2d1] the bill regulates about 17% of the economy, and [B2d2] "you can't regulate 17% of the U.S. economy without a raft of unintended consequences" (l); fifth, [B2e1] "it will slow innovation" (m1), since [B2e11] "government regulators don't do well with disruptive new technologies" (m2); sixth and finally, [B2f1] "if this passes, we will never get back to cost control" (n1).

These considerations are premises meant to support Brooks's negative recommendation not to pass the bill. Like the counterconsiderations, and even more so, these premises are individually part of relatively complex subarguments designed to justify them. Some of this structure I have already included, being relatively straightforward, as in B2d and B2e. Let us look at the other more complex subarguments.

To show why [B2a1] the bill does not really reform health care, Brooks argues as follows. [B2a11] The current system is financially opaque, insofar as patients are insulated from the cost of their medical decisions. [B2a12] It contains perverse financial incentives, insofar as providers are rewarded for providing more services and penalized for being more efficient. [B2a13] These aspects of the system are the root cause of the increase and acceleration of health care costs. [B2a14] The bill does not change these aspects of the system because [B2a141] it embodies a gradualist "Burkean" approach, [B2a142] which is effective for minor imperfections, but [B2a143] not for essential flaws.

In Brooks's second subargument against the bill, the reason he mentions, i.e., the acceleration of health care spending (B2b1), involves only an intermediate effect. The end result will be that [B2b2] health care spending will squeeze out all other government spending, especially at the state level (i3).

Brooks's third subargument tries to justify his third reason against the bill, that [B2c1] it will produce "a politically unsustainable situation" (j1). That is, "[B2c1111] the demand for health care will rise sharply. [B2c1112] The supply will not ... As a result, [B2c111] prices will skyrocket while efficiencies will not. [B2c11] There will be a bipartisan rush to gut reform" (j2, 3, 5, 6).

Finally, Brooks's sixth subargument is a two-step argument, one of which is an argument by analogy. [B2f111] Health care reform for American society is like developing proper eating habits for children; for example, expanding medical coverage is like eating dessert, while cost control is like eating spinach (n2). [B2f112] Children are motivated to first eat beneficial but tasteless food like spinach by being allowed to later eat less beneficial but more tasty food like dessert; they could not be motivated to eat spinach later by first eating dessert (n3). So, [B2f11] American society cannot be motivated to control costs later by expanding coverage first. But, [B2f12] the bill amounts to first expanding coverage and expecting to control costs later. So, [B2f1] "we will never get back to cost control" (n1).

In the reconstruction above, the third premise immediately supporting the main conclusion is the balance-of-considerations claim that [B3] the reasons for not passing outweigh the reasons for passing the bill. A striking feature of Brooks's argument is that this claim is explicitly made by the arguer when he states that if the bill is passed "the few good parts of the bill will get stripped out and the expensive and wasteful parts will be entrenched" (p2). Equally striking, is the fact that Brooks does not explicitly argue in support of this claim; instead he seems to suggest that the claim involves a pure or irreducible judgment call (o3). Moreover, he also seems to suggest that the difference of weight is small, when he says "I flip-flop week to week and day to day" (o2).

On the other hand, implicitly, he seems to have an argument in mind when in the context of this claim he calls attention to the fact that this is "a health care bill without systemic incentives reform" (p2). I believe this is a reference to the lack of fundamental reform, elaborated in the first subargument against the bill (B2a). So the argument is perhaps that the bill's lack of fundamental reform causes its drawbacks to overwhelm its advantages. That is, the bill's cons outweigh the bill's pros because the bill lacks fundamental reform and without it the cons will prevail over the pros. The bill's pros are: [B1a] expanded insurance coverage, [B1b] relative fiscal responsibility, [B1c] numerous promising piecemeal experiments, and [B1d] the attempt to break self-fulfilling cycle of failure. The bill's cons are: [B2b] tendency to monopolize government expenses, [B2c] political unsustainability, [B2d] unintended consequences, [B2e] slowing of innovation, and [B2f] reversed motivational sequence.

Then the main argument would perhaps have the following structure: [B'] the bill should not be passed because [B'1] it does not fundamentally reform health care and [B'2] without it the bad aspects of the bill will prevail over its good aspects. [B'21] The bad aspects are: [B'21a] tendency to monopolize government expenses, [B'21b] political unsustainability, [B'21c] unintended consequences, [B'21d] slowing of innovation, and [B'21e] reversed motivational sequence; [B'22] the bill's lack of fundamental reform will tend to strengthen these. [B'23] The good aspects are: [B'23a] expanded insurance coverage, [B'23b] relative fiscal responsibility, [B'23c] numerous promising piecemeal experiments, and [B'23d] the attempt to break self-fulfilling cycle of failure; [B'24] the bill's lack of fundamental reform will tend to weaken these.

There are problems with this reconstruction. First, this reconstruction (B') is not a conductive argument. The final subargument (B'1 and B'2 to B') has two premises that are not separately relevant but rather dependent on each other. The justification of the second premise (B'2) has four premises that are also linked and not separately relevant. This B'2 subargument seems to be a predictive argument that reaches a conclusion on the basis of an analysis of conflicting trends and consequences. It is the type of argument on which Allen's fourth critique (A4) focused.

However, this is not really a problem for this reconstruction, for there is no reason why Brooks's argument has to be a conductive argument. Instead it could be taken to be a problem for Wellman's and Govier's definitions of conductive argument. But even that is questionable since it is obvious that their definitions do not have to apply to all arguments; conductive arguments may be common, but they are certainly not the only type of argument. Brooks's argument may very well be one of these other types. But perhaps a trace of a problem remains, insofar as Brooks's argument seems to have the typical or essential characteristics of a conductive argument, and yet this reconstruction does not exhibit them.

A more serious problem with this reconstruction (B') emerges if we focus on the justification of the strengthening claim [B'22] and the weakening claim [B'24]. It is easy to see, in the context of this discussion, that the bill's lack of fundamental reform would strengthen its tendency to monopolize government expenses, its political unsustainability, and the reversed motivational sequence. On the other hand, the points about unintended consequences and the slowing of innovation seem to be independent of the lack of fundamental reform, and indeed Brooks's corresponding subarguments above exhibit this independence. Similarly, it is easy to see that the bill's lack of fundamental reform will tend to weaken its relative fiscal responsibility and the numerous promising piecemeal experiments, but the expanded insurance coverage and the breaking of the self-fulfilling cycle of failing to reform are effects or features of the bill that are independent of the lack of fundamental reform. I suppose the issue here could be formulated by saying that the B' reconstruction violates the principle of charity.

Thus, let us try another reconstruction that avoids this uncharitable flaw. Brooks's argument is best reconstructed as follows: [B''] The bill should not be passed. Admittedly, [B''a1a] it will expand insurance coverage and [B''a1b] would represent a break with the self-fulfilling cycle of failing to reform, and [B''a1c] these are good things. However, [B''a2a] the bill lacks fundamental reform, [B''a2b] will have massive unintended consequences, and [B''a2c] will slow innovation; [B''a2d] these are bad things, and [B''a3] they outweigh the good. Moreover, [B''b1] without fundamental reform the bill's *other* bad aspects will prevail over the bill's *other* good aspects. [B''b11] The other bad aspects are: [B''b11a] its tendency to monopolize government expenses, [B''b11b] its political unsustainability, and [B''b11c] the reversed motivational sequence; [B''b12] these will be strengthened by the lack of fundamental reform. [B''b13] The other good aspects are: [B''b13a] its relative fiscal responsibility and [B''b13b] the numerous promising piecemeal experiments; and [B''b14] these will be weakened by the lack of fundamental reform.

In this reconstruction, there are two independent subarguments, B''a and B''b, supporting the main conclusion B''. The second (B''b) subargument has the same structure as that of the previous main argument B', and so it is not a conductive argument. But the first (B''a) subargument is a conductive argument, essentially identical in structure to the final step of the original (B) reconstruction. This (B''a) subargument uses and explicitly states, but does not justify, the balance-of-considerations claim (B''a3). Notice that the aspects of the bill are not only being subdivided into pros and cons, but that each of these groups is being subdivided into those that are not affected by the lack of fundamental reform and those (the "others") that are. A conductive argument is being given involving the former, whereas the latter are involved in a predictive argument based on the analysis of conflicting trends and consequences.

8.7 Logical Structure and Visual Representation

Let us now study the logical structure of Brooks's argument by means of the visual representation provided by structure diagrams. I shall focus on the B version, reconstructed above (as distinct from the B' and B'' versions). Such a diagram is useful partly for the usual reason of helping us understand and evaluate the argument. But in this case the diagram is especially valuable because it suggests that the framework elaborated in this book is powerful enough to deal with the problem of how to logically structure and visually represent conductive arguments.[21]

The basic diagram for Brooks's (B) argument is given in Figure 8.3, where I have neglected premises deeper than the third level beneath the final conclusion (that is, propositions B2a141, B2a142, B2a143, B2c111, B2c1111, B2c1112, B2f111, B2f112). These are omitted from this diagram for simplicity's sake, but there is no loss of generality.

Figure 8.3

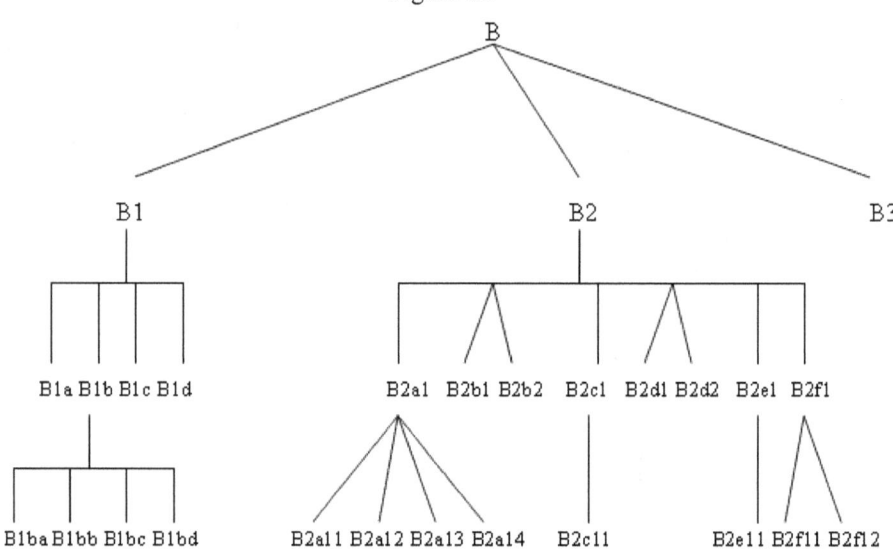

This diagram illustrates visually, vividly, and explicitly what is implicit in the standard labeling of the propositions in my reconstruction and in my commentary in the previous section. That is, a conductive argument (of Wellman's third kind) has typically three main strands that are linked or interdependent with one another: the pro reasons or supporting premises, the con reasons or counterconsiderations, and the balance of considerations.

The strand with the pro reasons consists of a *pro-reasons acknowledgment claim* (here,

---

[21] This is a problem that has received considerable discussion among writers on conductive arguments, such as Hansen (2011, 35-42), Jin (2011, 24-30), and Govier (2011, 270-75). I believe that my account overlaps in large measure (although not completely) with Jin's and Hansen's, as well as with Wohlrapp's emphasis on the "modification and completion" of conductive arguments (discussed in section 8.5.4 above); and furthermore, that my account essentially solves the problem, stressed by Govier, of representing both the convergence of some aspects of conductive argumentation and the linkage of other aspects; and finally, that my analysis in this section also corresponds to the way I analyzed, in chapter 3.4, the example of the simplicity argument for terrestrial rotation, at the point where it takes into account a theological objection to the principle of simplicity.

Conductive Arguments, Pro-and-Con Reasoning, Etc. 149

B2), supported by particular premises (here, B2a, B2b, B2c, B2d, B2e, B2f), which are positively relevant to the final conclusion (here, B), but which are independent or separate and convergent or cumulative with respect to the support provided for that pro-reasons acknowledgment claim. These premises, in turn, may or may not be themselves supported.

The strand with the con reasons consists of a *con-reasons acknowledgment claim* (here, B1), supported by particular premises (B1a, B1b, B1c, B1d), which are negatively relevant to the final conclusion, but which are independent or separate and convergent or cumulative with respect to the support provided for that con-reasons acknowledgment claim. And again, these premises may or may not be themselves supported.

Thirdly, the strand with the balance of considerations includes *a balance-of-considerations claim*[22] to the effect that one group of reasons outweighs the other. This claim may be a mere judgment call left unsupported, as is the case in this reconstruction (B3). Or it may be justified by means of subarguments that will normally not be conductive. Rather such justification will normally consists of reasoned analysis of the pro and con reasons, which is to say meta-argumentation. But there may be other ways of accomplishing such justification, and the topic deserves further research.[23]

According to the Wellman-Govier account, it's as if they were considering condensed, abbreviated, or simplified versions of real pro-and-con arguments, such as Brooks's. Their conductive arguments usually skip the first level of reasoning that involves the pro-reasons acknowledgment claim, the con-reasons acknowledgment claim, and the balance-of-considerations claim (here, B1, B2, B3). They have the final conclusion connected directly with the pro-reasons (here, B2a1, B2b1, B2c1, B2d1, B2e1, and B2f1) and with the con-reasons (here, B1a, B1b, B1c, B1d). This forces Govier (1985b, 259-61; 2010, 355-57) to invent a wavy line to represent the connection between the con reasons and the final conclusion, which connection is the opposite of the connection between the pro reasons and the final conclusion (represented by the usual solid line); for the usual solid line represents confirmation, whereas the wavy line represents disconfirmation. Moreover, the Wellman-Govier conductive arguments typically have to ignore the fact that some of the pro reasons and some of the con reasons are linked or interdependent with other propositions (here B2b2, B2d2), in order to provide their independent and separate support or disconfirmation to the conclusion.

Thus, it seems that the analysis of conductive arguments proposed here has distinct theoretical and pedagogical advantages.

8.8 Other Views: Conduction under Various Aliases

The discussion so far makes it obvious that Wellman's account of conductive arguments has been influential insofar as it has generated a considerable body of literature. Moreover, the discussion leaves no doubt of the importance of the concepts defined, principles formulated, and issues raised in this literature. The details of my discussion, however tedious at times, make it clear that this literature contains novelties, insights, and promises that merit further dissemination, elaboration, and research.

On the other hand, a number of caveats are in order. First, it is obvious that although Wellman may have invented the notion of conduction, he did not invent the practice of conductive argumentation, any more than the theorists of prose and other literary genres

---

[22] What I am here calling "balance-of-considerations claim" corresponds to what Jin (2011, 27) calls "on-balance consideration"; and to what Hansen (2011, 38-42) labels "on-balance premise"; and to what Zenker (2011, 75-81) labels "on balance principle."
[23] For some relevant studies, see Pinto 2011 and Fischer (2011, 2012).

were the first to actually speak prose. Second, it would be a mistake to think that nowadays, after the growth of such philosophical work on conductive arguments, good conductive argumentation can occur only if one explicitly uses the technical terminology and conceptual framework of conduction. The example of Brooks's health-care argument, and its elegance, sophistication, and brilliance (whatever one may think of the acceptability of his conclusion) provide an antidote against such a mistake.

Thirdly, and more importantly in the present context, we should not pretend that Wellman and his followers are the first or only ones to have theorized about or studied conductive arguments. To do so would lead us to neglect other scholarly work which, without so much as using the word conduction and its cognates, contains contributions to the same area of cognitive reality. Some of these contributions deserve mention here.[24]

One classical source would be John Stuart Mill, in particular the theory of argument in his essay *On Liberty*, especially chapter 2 entitled "Of the Liberty of Thought and Discussion." One strand of his argument there involves the view that "on every subject on which difference of opinion is possible, the truth depends on a balance to be struck between two sets of conflicting reasons" (Mill 1951, 128). This view is elaborated in the context of an attempt to show the importance of considering objections to one's own conclusions, besides the supporting reasons (cf. chapter 10 below).

Other sources are closer and more current. One is Scriven's (1981) account of the "weight and sum" methodology for evaluating such things as programs, products, proposals, and personnel. In the typical case, several distinct "dimensions" are involved and need to be weighted, and for each dimension the entry being evaluated is "scored" for its "performance." Both the dimension weights and performance scores are assigned numbers, such as 1-5 or 1-10. Then "the total score for each entry is obtained simply by multiplying the weights by the performance scores and summing them" (Scriven 1981, 86).

Scriven is concerned to point out that this methodology is extremely important and useful, but also extremely tricky. For example, care is needed regarding questions such as the following: the numbers used in the two scales and the relationship between them; the number of dimensions, the cumulative effect of small differences in performance scores, the possible existence of big differences within a dimension, and the relationship among these three things; the possible need to lump some dimensions together or subdivide others, before the evaluation process is completed; the possibility that some dimension embodies a necessary condition, whose non-satisfaction pre-empts and swamps all other weights; and the difference between pairwise comparison and general comparison of entries.

It seems obvious that the problems Scriven is dealing with overlap to a large extent with the ones examined by writers on conductive arguments. In other words, what Scriven calls evaluative arguments correspond to what Wellman and his followers call conductive arguments, and the methodology of former overlaps with the methodology of the latter.

Another nearby and current source is Ralph Johnson's account of the dialectical tier of argumentation and of the dialectical obligations of arguers. Although he (Johnson 2000a, 92-95) has explicitly attempted to elaborate his own critical appreciation of Govier's account of conductive arguments, it seems to me that the most crucial point has been left unsaid.[25] It is this. Suppose one follows Johnson is analyzing the nature of argumentation

---

[24] Other contributions which deserve further study from this point of view, but which here cannot be even summarized and must be merely cited, are: Perkins (1985a, 2002); Perkins, Allen, and Hafner 1983; Perkins, Farady, and Bushey 1991; Pollock (1974, 33-49, 300-340; 1995, 38-42, 85-140); Slob 2006; Verheij 2006, 194-98; and Voss, Perkins, and Segal 1991.

[25] More recently, however, Johnson has discussed more explicitly this issue of the relationship between Govier's "conductive" arguments and his own "dialectical" arguments; see Johnson 2011.

in such a way that, besides having an illative core consisting of reasons supporting the conclusion, arguments possess a dialectical tier consisting of replies to objections and criticism of alternative positions. (Of course, one need not follow Johnson completely in his conception of argument, but may do so partially, as I argued in chapter 4 above; but those differences are not relevant to the present issue.) And suppose one again follows Johnson in conceiving the appraisal of argumentation as consequently requiring the appraisal of the dialectical tier, involving such questions as: exactly what objections must an arguer reply to; exactly what alternative positions must one criticize; must an arguer, for example, anticipate objections? It seems to me that such a dialectical approach to the analysis and appraisal of argumentation is one designed precisely to deal with what Wellman and Govier call conductive arguments. In effect, given his dialectical definition of argument, Johnson is treating all arguments (in his sense) as conductive arguments (in Govier's sense). The overlap is so deep and pervasive that he does not need to examine conductive arguments in a particular section of his theory; they lurk everywhere throughout his account.

Actually, Johnson (2000a, 92-95) comes close to admitting as much, although not in such words. This occurs when he articulates his appreciation that "the truth behind conductivism" is that argumentation is a special case of reasoning (studied in informal logic), but distinct from entailment (studied in formal deductive logic) and from inference (studied in inductive logic). However, my main point here is that Johnson's theory of argument is a good example of a contribution that deals with the same problems examined by conductive theorists, but without using the conductive terminology; instead it uses dialectical terminology.

Another such significant and current example is the work of Christian Kock (2003, 2007a, 2007b, 2007c, 2011). He has studied what he calls "deliberative argumentation," which he defines as argumentation about "proposals." Unlike propositions, proposals are neither true nor false, but rather right or wrong, or more or less good. Such arguments are common in political debates and, indeed, in all practical argumentation. Kock claims that deliberative argumentation has properties such as the following: "There will always be several good but contradictory arguments. Contradictory arguments do not cancel each other. A good argument never entails a policy by necessity or inference. Contradictory arguments often rely on plural values which are not objectively commensurable. Contradictory arguments must nevertheless be compared for choices to be made. Choices rely on individuals' value commitments and are subjective. Debates between exponents of opposite policies cannot be expected to lead towards agreement, but must help other individuals consider and compare the pro and con arguments relating to a policy" (Kock 2007a, 238). These claims make it obvious that, as Kock himself has pointed out, his account is similar to that of Wellman and Govier, although of course there are differences. Clearly, Kock's work exemplifies well the possibility of an account of conductive arguments that uses different terminology but shares conceptual and substantive content.

Finally, there is the instructive case of the literature on two-sidedness, as distinct from one-sidedness. Such literature spans a very broad spectrum, ranging from pedestrian and formulaic empirical studies by marketing, consumer, and communication researchers to lofty and sublime[26] analyses by philosophers. Starting with the former, one might begin by

---

[26] Among the sublime, I have in mind the Hegelian concept of dialectic, which can be connected with the idea of avoiding one-sidedness (see Finocchiaro 1988, 143-230); but obviously even a summary of this connection would beyond the scope of the present work. On the other hand, note also the possible connection between Hegelian dialectic (at least as interpreted by Gramsci) and the resolution of deep disagreements, mentioned in chapter 7.7 above.

noting that if conductive arguments are as ubiquitous and effective as claimed by their proponents, then they should have been discovered and exploited also by advertisers. Indeed, this happens to be the case. The terminology used is "two-sided advertising": advertising in which some negative information, and not only positive information, about a product is given in the advertisement. The practice has also become the subject of an extensive social-science literature. Fortunately, there are at least two "meta-analyses" that draw their conclusions based in part on an interpretive summary of the literature. Eisend (2006) reviews twenty-five studies of two-sided vs. one-sided advertising, and also does a so-called regressive analysis of them. O'Keefe (1999) is a more ambitious review, examining eighty-nine studies of two-sided vs. one-sided messages in both advertising and non-advertising contexts; he focuses on a so-called random-effects analysis.[27]

One of Eisend's general conclusions is that two-sided advertising is more persuasive that one-sided advertising. Some quotations will give us an idea of nuances and qualifications to this general conclusion, as well as give us a flavor of the potential relevance to conductive argumentation: "Two-sided messages work, particularly they enhance source credibility, reduce negative cognitive responses, and have positive impact on brand attitude and purchase intention ... Two-sided messages help to mainly improve source credibility and to reduce negative cognitions, whereas the impact on attitudes and purchase intentions proved to be weaker" (Eisend 2006, 195-96). On the other hand, O'Keefe argues that this is true only in the context of advertising messages. However, in non-advertising contexts, two-sided messages are not always more persuasive. It depends on whether they are refutational or nonrefutational, refutational two-sided messages being those which not only mention the negative information or opposing arguments, but also refute or criticize them. In fact, it turns out that "for non-advertising messages, refutational two-sided messages are significantly more persuasive than one-sided messages, and nonrefutational two-sided messages are significantly less persuasive than one-sided messages" (O'Keefe 1999, 231). Although there are many more details and nuances, and although a philosopher would want to adopt a critical stance toward such empirical work, it is clearly suggestive and one can ignore it only at one's own risk.

Moving on to a philosophical example, Dale Jacquette has also studied conductive arguments without using Wellman's terminology of conduction, and has done so in the context of an examination of the tactics of the ancient Greek sophists. Jacquette discusses the problem of whether there are two sides to any issue, and what to make of the sophists' practice of switching sides. His resolution is, I believe, insightful and correct. It is this: "what is objectionable is that ... the sophist argues pro and then argues con, and leaves it at that, omitting any assessment of the strengths and weaknesses of the arguments ... This is the sophists' error, that they defend a proposition, and then criticize it and defend its negation, as though the two were always in an even standoff" (Jacquette 2007d, 125-26).

In the spirit of such investigations, which are relevant to the study of conductive arguments but do not use the conductive terminology, I would next like to summarize and adapt some of my own work relating to Galileo (Finocchiaro 1980a, esp. 27-45; 1997, 1-7, 309-56; 2010b, esp. xiii-xliii, 1-134). I hope to be able to show that it has obvious relevance to conductive arguments.

8.9 A Classic Example: Galileo's *Dialogue*

In 1543, Copernicus published a book elaborating a world system whose key thesis

---

[27] Although O'Keefe 1999 is the most directly relevant study for my purpose here, other details may be found in O'Keefe (2002, 2012).

was that the earth moves by rotating on its own axis daily and by revolving around the sun yearly. Copernicus's accomplishment was really to give a *new argument* in support of an *old idea* that had been considered and almost universally rejected since the ancient Greeks. He demonstrated that the known facts about heavenly motions could be explained in quantitative detail if the universe is structured geokinetically and heliocentrically; and further that this explanation was more coherent (and also simpler and more elegant) than the geostatic account.

Despite its novelty and significance, however, as a proof of the earth's motion Copernicus's argument was inconclusive. First of all, his argument was clearly an abductive one, i.e., an inference to the best explanation. Moreover, there were many powerful arguments against terrestrial motion that had accumulated for two millennia. In summary, the earth's motion seemed epistemologically absurd because it flatly contradicted direct sense experience. It seemed empirically false because it had astronomical consequences that were not observed to happen: for example, similarities between the earth and the planets; phases for the planet Venus; and annual parallax in the fixed stars. The earth's motion also seemed physically impossible because it supposedly contradicted the laws of motion of the available (Aristotelian) physics, and the most incontrovertible mechanical phenomena (such as vertical fall). And it seemed religiously heretical because it conflicted with the literal meaning of Scripture, such as the passage about the miracle of stopping the sun, in Joshua 10: 12-13.

Thus, the Copernican Revolution required much more than Copernicus's own argument. First, the geokinetic hypothesis had to be supported not only with new theoretical arguments, but also with new empirical evidence. Galileo's telescopic discoveries provided such novel evidence.

Second, the earth's motion had to be not only constructively supported with new arguments and evidence, but also critically defended from the host of powerful old and new objections. Galileo answered the observational astronomical objections by showing that almost all the empirical consequences implied by Copernicanism were visible with the telescope, although still invisible with the naked eye. He answered the scriptural objections by arguing that Scripture is not a scientific authority, and so scriptural passages should not be used to invalidate astronomical claims that are proved or provable. And he answered the physical mechanical objections by articulating a new physics centered on the principles of conservation and composition of motion.

Third, the defense of the geokinetic hypothesis required not only the destructive refutation of those objections, but also the appreciative understanding of their strength. Galileo was keen on this, and so in his writings we find the anti-Copernican arguments stated more clearly and incisively than in the works of Aristotelians advocating the geostatic system.

Fourth, Galileo also realized that his case in favor of Copernicanism was not absolutely conclusive or decisive because, for example, his telescope failed to reveal an annual parallax of the fixed stars, which was a consequence of the earth's annual revolution around the sun.

In short, Galileo's key contribution to the Copernican Revolution was to elaborate an argument for the geokinetic thesis that stressed reasoning and observation judiciously guided by the ideals of rational-mindedness, open-mindedness, and fair-mindedness. Galileo presented such an argument in his book *Dialogue on the Two Chief World Systems, Ptolemaic and Copernican* (1632). And he not only constructed, presented, and published such a piece of argumentation, but he showed a rare and keen awareness of what he was doing and of the desirability of doing so.

I want to illustrate this metacognitive awareness by quoting his explicit formulation of the three key principles just mentioned. What I label rational-mindedness is the principle that when one is considering a new doctrine "one [should] examine with the utmost severity what the followers of this doctrine know and can advance [in its favor], and that nothing be granted them unless the strength of their arguments greatly exceeds that of the reasons for the opposite side" (Galilei 2008, 165). What I label open-mindedness is the attitude exemplified by the Copernicans insofar "the followers of the new system [of the world] produce against themselves observations, experiments, and reasons much stronger than those produced by Aristotle, Ptolemy, and other opponents of the same conclusions" (Galilei 2008, 216-17). And what I call fair-mindedness is the principle prescribing that "when one presents arguments for the opposite side with the intention of confuting them, they must be explained in the fairest way and not be made out of straw to the disadvantage of the opponent" (Galilei 2008, 283).

Expressed in the terminology of the present subject, my claim can be expressed by saying that I have been arguing (however cryptically and succinctly) that the main argument in Galileo's *Dialogue* is a conductive argument. For Galileo justifies the conclusion that the earth moves (1) nonconclusively by means of (2) several distinct and separately relevant subarguments that (3) cumulatively reinforce each other and that are formulated in the context of a (4) strong acknowledgment of counterconsiderations and objections. Moreover, I would also claim that Galileo's argument is conductively adequate, although here I have not given reasons to justify this claim.

Furthermore, however convinced I am of my interpretation and however strong my meta-argument may be, it would be ironic if in the present context I forgot that the field in which I am presently operating is that of interpretive studies, and that arguments supporting interpretations are typically conductive arguments. Additionally, besides being in a context of interpretive studies, I am dealing here with a question of classification: how to classify Galileo's own main argument in the *Dialogue*. Finally, since the issue is whether his argument is a conductive one, I am interested in exploring a definition of conductive argument that would be analogous to Freeman's definition of deductive and inductive arguments, mentioned above. This would be a conductive definition of conductive argument, where the various requirements to be satisfied are not individually necessary and jointly sufficient conditions, but rather separate indices the accumulation of which can adequately justify our interpretive conclusion, without guaranteeing its truth or rendering it inductively probable. Thus, I feel it is only prudent to give some other reasons of a different sort, as well as to acknowledge some opposing reasons.

So far, I have given two separately relevant (sets of) reasons. One is that many facts about the Copernican Revolution and about Galileo's work can be explained or understood in terms of the interpretation that his key contribution to that episode was to have advanced a conductive argument for the geokinetic thesis. Another reason is that on several occasions, by explicitly formulating the principles of rational-mindedness, open-mindedness, and fair-mindedness, Galileo showed that he was reflectively aware that what was needed and all that was possible at the time was to give a conductive argument for the earth's motion.

A third reason is one which I appropriate from another scholar, named Filippo Soccorsi (1947), who did not even mention the word conduction and advanced his interpretation in the 1940's. This was long before Wellman and his followers, as well as long before the present writer, whose background in both argumentation studies and Galileo studies makes it unsurprising and predictable that he would explore the potential interpretation of Galilean arguments by means of informal-logic categories. Translated into conductive terminology, Soccorsi's interpretation amounts to the claim that many features about the Inquisition trial and condemnation of Galileo in 1633 (starting with the very fact the trial occurred at all) can be understood if we construe the argument in the *Dialogue* as a self-conscious conductive argument, because this type of argumentation was bound to be misunderstood by the Church authorities and manipulated by his enemies. Here we must recall that the *Dialogue* is the book published in 1632 that triggered the trial and condemnation by the Inquisition the following year. Keeping my terminological translation in mind, Soccorsi's words are worth quoting precisely for their lack of loaded logical and informal-logical terminology, which lack makes the substance of his insight even more striking. Referring to Galileo's arguments in the *Dialogue*, Soccorsi claims:

If no one of the arguments constituted a rigorous demonstration, still their combination ... was not devoid of persuasive force, especially for a mind that could comprehend the synthesis and penetrate it with the new point of view of the new mechanics ... This observation allows us to explain how it was possible that there was a profound misunderstanding between Galileo and those old-fashioned minds ... which ... at most were concerned with asking the experts whether any one of the proofs was conclusive: in this manner they missed the persuasive force of Galileo's arguments. [Translated and quoted in Finocchiaro 2005b, 291]

A fourth and final reason for my interpretation comes from the full title of Galileo's book. As was customary in the seventeenth century, books tended to have very long titles that were descriptive of the content and form of the work. The full title of Galileo's *Dialogue* is very revealing since it contains many words and phrases that may be taken as conductive indicators. It reads: *Dialogue by Galileo Galilei, Lincean Academician, Extraordinary Mathematician at the University of Pisa, and Philosopher and Chief Mathematician to the Most Serene Grand Duke of Tuscany; where in meetings over the course of four days one discusses the Two Chief World Systems, Ptolemaic and Copernican, proposing indeterminately the philosophical and natural reasons for the one as well as for the other side* (cf. Finocchiaro 1980a, 12-18; 1997, 359-60; 2005b, 133). Here, Galileo's "indeterminately" is equivalent to the "nonconclusively" of Wellman and his followers. And the Galilean talk of "reasons for the one as well as for the other side" is equivalent to the reasons and counterconsiderations, or pros and cons, of the conductive theorists.

Of course, such language was connected with the historical fact that Galileo had been prohibited by officials and institutions of the Catholic Church to hold or defend the geokinetic thesis. Moreover, such a title had fateful historical repercussions. That is, during the proceedings of the 1633 Inquisition trial, Galileo was charged and convicted in part for having tried to give the misleading impression that his argumentation was "indeterminate," but in reality favoring the Copernican side. In conductive terminology this means that he was blamed for not having limited himself to presenting the pros and cons of the geokinetic thesis, but for having dared to make the judgment that the strength of the pros outweighed the strength of the cons, and indeed for having argued convincingly to justify this judgment. Using Jacquette's terminology mentioned above, we can say that one alleged crime of the *Dialogue* for which Galileo was convicted was that he refused to be a sophist!

Finally, I am aware that various reasons have been given to try to show that Galileo's argument is not conductive, but rather deductive. For example, some scholars begin by stressing the fact that (1) the book's most prominent argument in favor of the earth's motion is one based on the existence and properties of the tides and on trying to show that they are caused by the earth's motion; (2) originally Galileo wanted to entitle the book *Dialogue on the Tides*, but (3) Church authorities objected to such a title, forcing him to adopt the title mentioned above. These scholars then go on to (4) explain this fact by saying that (5) he regarded this argument to be conclusive ([6] as one can gather from many passages in the book), and (7) originally he wanted to mention the tides in the title in order to give an indication that the book contained a demonstrative proof of the earth's motion. And so the book's original but ultimately discarded title is evidence that its main argument is deductive.

Now, the factual claims (1, 2, 3) made in this argument are indeed correct, but the explanation of them lies elsewhere; indeed the claims (5, 6, 7) making up the explanans are untenable. For Galileo did not regard the tidal argument as conclusive and deductive; the book's passages usually mentioned as examples of deductive indicators merely indicate that he regarded this argument as very strong, indeed his strongest. Instead he was keenly aware that the tidal argument was an inference to the best explanation; his original title was meant to stress precisely the hypothetical and explanatory nature of the book. In light of the restrictions under which he was operating, including a prohibition to defend the earth's motion, a stress on hypothetical or explanatory reasoning was much more prudent, reasonable, and innocuous. Moreover, although the tidal argument is prominently displayed and highly regarded by Galileo, the book makes it crystal clear that it is only one of several arguments in favor of the earth's motion, and the context of multiple argumentation is an indication that no one particular argument is being advanced as conclusive. I think such a connection between multiple and conductive argumentation is, *ceteris paribus*, a sound general principle of interpretation, and its soundness has been recognized by several scholars in other contexts, especially in purely philosophical argumentation such as the mind-body problem (Dauer 1974, 131; Govier 1980a, 14).

Another reason against the conductive status of Galileo's argument involves the fact that the earth's motion contradicted many biblical passages when literally interpreted (e.g., Joshua 10: 12-13). Then some scholars claim that in such cases of conflict, the operative hermeneutical principle was the one stipulating that biblical passages about natural phenomena must be interpreted literally unless there is a conclusive scientific proof to the contrary. Presumably the *Dialogue* was trying to provide the conclusive proof required by this principle in light of the contradiction between the earth's motion and the literal interpretation of Scripture. An inconclusive, conductive argument would not have been sufficient.

However, Galileo did not really accept this hermeneutical principle in the form just stated, which amounts to a biconditional. That is, it claims that if there is a conclusive proof of a natural phenomenon, then biblical passages to the contrary should be interpreted nonliterally; and if there is no such proof, then relevant biblical passages must be interpreted literally. He did accept the first conditional, making conclusive proof a sufficient condition for nonliteral interpretation; indeed this sufficiency principle was universally accepted, and had been applied centuries earlier to the case of the earth's shape, which was demonstrably spherical, rather than flat as stated or implied in many biblical passages. However, he rejected the other conditional, making conclusive proof a necessary condition for nonliteral interpretation. His critical argument (cf. Finocchiaro 2010b, 243-48) began with the universally accepted sufficiency principle. Then he went on to ask for an explanation why such a principle holds, why in cases of conflict between a conclusive physical proof and the letter of a biblical text, priority is given to the former. His explanation was basically that Scripture is not a scientific (or astronomical) authority, but only an authority on matters of faith and morals. But then Galileo went on to consider what follows from the denial of the scientific authority of Scripture. What follows is that biblical statements have no probative weight in scientific investigation, and hence scientists ought to be free to search for natural truth by engaging in argumentation, presenting all available evidence, and determining what is the conclusion supported by the better arguments. Thus, the *Dialogue* did not have to be a conclusive proof, a deductive argument, and if the state of inquiry enabled one to formulate an inconclusive, conductive argument, there was nothing wrong with that. In particular, there was nothing wrong with weighing the evidence, to see whether one side or theory was stronger than the other. And as discussed earlier, this is precisely what Galileo did.

8.10 Conclusions: Progress, Problems, Prospects

A main strand of this chapter has been a discussion of the views on conductive arguments by Wellman, Hitchcock, Govier, Allen, Ennis, Zenker, and Wohlrapp. By and large, I interpreted Hitchcock's and Govier's views as constructive elaborations of Wellman's; Allen's, Ennis's, and Zenker's views as a series of mostly negative criticisms of Govier's views; and Wohlrapp's views as consisting partly of negative criticism, but primarily of a constructive alternative position. In accordance with the Toulminian applied-logic and my historical-textual approaches, I interpreted and reconstructed all their views as a series of arguments. Moreover, since their arguments are theoretical, they obviously constitute meta-arguments, and hence my analysis was also a case study in meta-argumentation. Now, let us recall that a meta-argument was defined as an argument about one or more arguments, as distinct from a ground-level argument, which is about a subject matter other than argument. Then it is useful to explicitly coin the term conductive meta-argument and define it as a meta-argument that is about one or more conductive arguments, or as a meta-argument that is itself conductive. These notions enable us to conceive this strand of this chapter as an applied-logic and historical-textual study of conductive meta-argumentation.

My meta-argumentative reconstructions attempted to be quite comprehensive and well-documented, by including all the relevant views of these scholars as well as appropriate references and quotations. By contrast, my evaluation of these reconstructed meta-arguments was more selective, indirect, implicit, incomplete, and tentative. Nevertheless, the results of my evaluation are perhaps more interesting, important, and suggestive than the results of my interpretation.

For example, despite the impression conveyed by many commentators on Wellman, including his followers, he does have a guideline for the evaluation of conductive arguments, namely the model of comparing the weight of bodies by lifting them with one's hands; and this deserves further exploration. Moreover, Wellman has two arguments for his anti-deductivist claim (W14) that raise important and relatively open questions: one about the justification of balance-of-considerations claims (subargument W14a), the other about the non-convergence of deductive arguments (subargument W14b).

Out of Hitchcock's meta-arguments, there emerge two problems for further investigation. One is whether it is possible to give a definition of conductive adequacy as a type of appraisal that would define it independently of the definition of the conductive type of argument. The other issue is that more work needs to be done to explore the possibility of a conductive definition of conductive argument; this would be analogous, but of course substantively different, from Freeman's conductive definition of deductive and inductive arguments.

Out of Govier's meta-arguments, besides the critiques by other scholars to be highlighted presently, two points deserve to be stressed. One is her claim that some significant scientific arguments, those justifying theory choice, are conductive; this is important, insightful, and correct. But there is also the problem of the asymmetry between positively relevant reasons and negatively relevant counterconsiderations: her principles of evaluation seem to assume that counterconsiderations not mentioned in an argument are relevant for its evaluation, but unmentioned positive reasons are not relevant to the evaluation of the argument, but only to the evaluation of the acceptability of the conclusion; the problem is that it's not clear why there should be such an asymmetry.

Combining one of Allen's, of Ennis's, and of Wohlrapp's meta-arguments, we have a reaffirmation of the deductivist approach. Allen's first criticism is a cogent argument that Govier's justification of her anti-deductivist claim is not cogent. Ennis has a cogent argument that she uses deductive standards in her evaluation of the strength of reasons, which is inconsistent with her attempt to avoid deductivism. And I also reported Wohlrapp's claim, but not his supporting argument, that she fails to appreciate that deduction is presupposed by induction and by conduction, but not vice versa.

Next, we can combine one of Wellman's important subarguments (W14a), one of Allen's critiques (A4), one of the main strands of Wohlrapp's alternative account, and one main point emerging from my analysis of Brooks's health-care argument. I believe we get a confirmation of my working hypothesis, adumbrated in this chapter's introduction, to the effect that conductive arguments have a crucial and irreducible meta-argumentative aspect, namely a *balance-of-considerations claim* that must be implicitly made, explicitly formulated, or critically justified for the construction, interpretation, or evaluation of conductive arguments. Of course, the status, use, and justification of balance-of-considerations claims is problematic. It is not exactly clear how one justifies such claims; what type of arguments one would use in such a justification, in particular whether they would be conductive arguments; and consequently, whether in a conductive argument, if the argument is to remain of the conductive type, we must limit ourselves to merely using such claims, and not get involved in a non-conductive justification. Obviously more work needs to be done in this regard.

Another important open question is that of the relationship among conduction, convergence, and multiple argumentation. This emerges from Wellman's subargument (W14b) that deduction allows no place for the convergence of evidence, from Govier's subargument (G14) that not all convergent arguments are conductive, and from Allen's criticism (A2) of Govier's thesis that a deductivist reconstruction of a conductive argument turns one argument into several.

A common criticism of Govier's account has focused on her method for evaluating the strength of reasons in terms of the range of exceptions to the corresponding c-p generalization. Ennis objected that this method presupposes a deductivism which she seems generally to want to reject. Zenker objected that this method is misconceived because c-p generalizations have nothing to do with conductive arguments. And Wohlrapp finds the method unfruitful, at least as compared to his alterative account, which bypasses such specific estimates of strength. Although I do not think the method was intended to be taken seriously, after such criticisms, it is difficult to so take it.

Another strand of this chapter has been the analysis of some actual important examples of conductive ground-level arguments, in accordance with my historical textual approach. Brooks's argument against the U.S. Senate health-care bill is an obvious example of a conductive argument. Its conductive character is obvious both from its subject matter and from its form. Its elegance and sophistication make it deserving of further interpretation, evaluation, and analysis. This is especially true if we contrast his argument with that of Paul Krugman. On the other hand, Galileo's argument for the motion of the earth has been widely studied for about four centuries. It is of epoch-making importance both for its role in the Copernican Revolution and in the Galileo affair. Although it has never been interpreted as a conductive argument, I have argued that it can be so interpreted; and although I have no hesitation in characterizing this meta-argument of mine as a conductive one, I believe it is very strong. Brooks's and Galileo's arguments show that the practice of conductive argumentation is prior to, is relatively independent of, and perhaps has primacy vis-à-vis the theory of conductive argumentation.

Similarly, the theory of conductive arguments has much to learn from other studies which examine the same cognitive phenomena under different labels. This claim seems a well justified conclusion from the third strand of this chapter, which briefly examined Mill's plea for liberty of argument in *On Liberty*, Scriven's "weight and sum" methodology, Johnson's account of the dialectical tier and dialectical obligations, Kock's account of deliberative argumentation, Eisend's and O'Keefe's meta-analyses of the social-scientific literature on one-sided vs. two-sided messages and advertising, Jacquette's discussion of the problem of two sides to any issue, and my study of open-mindedness and fair-mindedness in Galileo's work. Admittedly, I had had this hunch all along, since the beginning of this investigation; that is, the hunch that conductive argumentation, besides being practiced before and independently of conductivist studies, had also been studied without the conductive terminology. This hunch was another working hypothesis, so to speak, and it too is now reinforced.

8.11 Summary

Conductive arguments are arguments whose conclusion is justified neither deductively nor inductively, but based on separately relevant pro reasons, while acknowledging reasons against. This chapter has reconstructed the conductive-argument literature as a series of meta-arguments advancing theoretical conclusions about conductive arguments. It has analyzed two ground-level conductive arguments, by David Brooks about the U.S. health-care reform, and by Galileo about the earth's motion. It has summarized some literature discussing the same theoretical issues without explicitly using conductive terminology. The meta-argumentation perspective also emerges insofar as an important feature of conductive arguments is a balance-of-considerations claim whose justification requires meta-argumentation.

Appendix 1

"The Hardest Call," by David Brooks, *The New York Times*, December 18, 2009

[a] The first reason to support the Senate health care bill is that it would provide insurance to 30 million more Americans.

[b] The second reason to support the bill is that its authors took the deficit issue seriously. Compared with, say, the prescription drug benefit from a few years ago, this bill is a model of fiscal rectitude. It spends a lot of money to cover the uninsured, but to help pay for it, it also includes serious Medicare cuts and whopping tax increases — the tax on high-cost insurance plans alone will raise $1.3 trillion in the second decade.

[c] The bill is not really deficit-neutral. It's politically inconceivable that Congress will really make all the spending cuts that are there on paper. But the bill won't explode the deficit, and that's an accomplishment.

[d] The third reason to support the bill is that the authors have thrown in a million little ideas in an effort to reduce health care inflation. The fact is, nobody knows how to reduce cost growth within the current system. The authors of this bill are willing to try anything. You might even call this a Burkean approach. They are not fundamentally disrupting the status quo, but they are experimenting with dozens of gradual programs that might bend the cost curve.

[e] If you've ever heard about it, it's in there — improved insurance exchanges, payment innovations, an independent commission to cap Medicare payment rates, an innovation center, comparative effectiveness research. There's at least a pilot program for every promising idea.

[f] The fourth reason to support the bill is that if this fails, it will take a long time to get back to health reform. Clinton failed. Obama will have failed. No one will touch this. Meanwhile, health costs will continue their inexorable march upward, strangling the nation.

[g] The first reason to oppose this bill is that it does not fundamentally reform health care. The current system is rotten to the bone with opaque pricing and insane incentives. Consumers are insulated from the costs of their decisions and providers are punished for efficiency. Burkean gradualism is fine if you've got a cold. But if you've got cancer, you want surgery, not nasal spray.

[h] If this bill passes, you'll have 500 experts in Washington trying to hold down costs and 300 million Americans with the same old incentives to get more and more care. The Congressional Budget Office and most of the experts I talk to (including many who support the bill) do not believe it will seriously bend the cost curve.

[i] The second reason to oppose this bill is that, according to the chief actuary for Medicare, it will cause national health care spending to increase faster. Health care spending is already zooming past 17 percent of G.D.P. to 22 percent and beyond. If these pressures mount even faster, health care will squeeze out everything else, especially on the state level. We'll shovel more money into insurance companies and you can kiss goodbye programs like expanded preschool that would have a bigger social impact.

[j] Third, if passed, the bill sets up a politically unsustainable situation. Over its first several years, the demand for health care will rise sharply. The supply will not. Providers will have the same perverse incentives. As a result, prices will skyrocket while efficiencies will not. There will be a bipartisan rush to gut reform.

[k] This country has reduced health inflation in short bursts, but it has not sustained cost control over the long term because the deep flaws in the system produce horrific political pressures that gut restraint.

[l] Fourth, you can't centrally regulate 17 percent of the U.S. economy without a raft of unintended consequences.

[m] Fifth, it will slow innovation. Government regulators don't do well with disruptive new technologies.

[n] Sixth, if this passes, we will never get back to cost control. The basic political deal was, we get to have dessert (expanding coverage) but we have to eat our spinach (cost control), too. If we eat dessert now, we'll never come back to the spinach.

[o] So what's my verdict? I have to confess, I flip-flop week to week and day to day. It's a guess. Does this put us on a path toward the real reform, or does it head us down a valley in which real reform will be less likely?

[p] If I were a senator forced to vote today, I'd vote no. If you pass a health care bill without systemic incentives reform, you set up a political vortex in which the few good parts of the bill will get stripped out and the expensive and wasteful parts will be entrenched.

[q] Defenders say we can't do real reform because the politics won't allow it. The truth is the reverse. Unless you get the fundamental incentives right, the politics will be terrible forever and ever.

Appendix 2

"Pass the Bill," by Paul Krugman, *The New York Times*, December 18, 2009

A message to progressives: By all means, hang Senator Joe Lieberman in effigy. Declare that you're disappointed in and/or disgusted with President Obama. Demand a change in Senate rules that, combined with the Republican strategy of total obstructionism, are in the process of making America ungovernable.

But meanwhile, pass the health care bill.

Yes, the filibuster-imposed need to get votes from "centrist" senators has led to a bill that falls a long way short of ideal. Worse, some of those senators seem motivated largely by a desire to protect the interests of insurance companies — with the possible exception of Mr. Lieberman, who seems motivated by sheer spite.

But let's all take a deep breath, and consider just how much good this bill would do, if passed — and how much better it would be than anything that seemed possible just a few years ago. With all its flaws, the Senate health bill would be the biggest expansion of the social safety net since Medicare, greatly improving the lives of millions. Getting this bill would be much, much better than watching health care reform fail.

At its core, the bill would do two things. First, it would prohibit discrimination by insurance companies on the basis of medical condition or history: Americans could no longer be denied health insurance because of a pre-existing condition, or have their insurance canceled when they get sick. Second, the bill would provide substantial financial aid to those who don't get insurance through their employers, as well as tax breaks for small employers that do provide insurance.

All of this would be paid for in large part with the first serious effort ever to rein in rising health care costs.

The result would be a huge increase in the availability and affordability of health insurance, with more than 30 million Americans gaining coverage, and premiums for lower-income and lower-middle-income Americans falling dramatically. That's an immense change from where we were just a few years ago: remember, not long ago the Bush administration and its allies in Congress successfully blocked even a modest expansion of health care for children.

Bear in mind also the lessons of history: social insurance programs tend to start out highly imperfect and incomplete, but get better and more comprehensive as the years go by. Thus Social Security originally had huge gaps in coverage — and a majority of African-Americans, in particular, fell through those gaps. But it was improved over time, and it's now the bedrock of retirement stability for the vast majority of Americans.

Look, I understand the anger here: supporting this weakened bill feels like giving in to blackmail — because it is. Or to use an even more accurate metaphor suggested by Ezra Klein of *The Washington Post*, we're paying a ransom to hostage-takers. Some of us, including a majority of senators, really, really want to cover the uninsured; but to make that happen we need the votes of a handful of senators who see failure of reform as an acceptable outcome, and demand a steep price for their support.

The question, then, is whether to pay the ransom by giving in to the demands of those senators, accepting a flawed bill, or hang tough and let the hostage — that is, health reform — die.

Again, history suggests the answer. Whereas flawed social insurance programs have tended to get better over time, the story of health reform suggests that rejecting an imperfect deal in the hope of eventually getting something better is a recipe for getting nothing at all. Not to put too fine a point on it, America would be in much better shape today if Democrats had cut a deal on health care with Richard Nixon, or if Bill Clinton had cut a deal with moderate Republicans back when they still existed.

But won't paying the ransom now encourage more hostage-taking in the future? Maybe. But the next big fight, over the future of the financial system, will be very different. If the usual suspects try to water down financial reform, I say call their bluff: there's not much to lose, since a merely cosmetic reform, by creating a false sense of security, could well end up being worse than nothing.

Beyond that, we need to take on the way the Senate works. The filibuster, and the need for 60 votes to end debate, aren't in the Constitution. They're a Senate tradition, and that same tradition said that the threat of filibusters should be used sparingly. Well, Republicans have already trashed the second part of the tradition: look at a list of cloture motions over time, and you'll see that since the G.O.P. lost control of Congress it has pursued obstructionism on a literally unprecedented scale. So it's time to revise the rules.

But that's for later. Right now, let's pass the bill that's on the table.

Chapter 9
Self-referential Arguments

We have already examined several theoretical contexts where meta-argumentation is involved in an important way: definitions of the concept of argument; articulation and classification of methods for evaluating and criticizing arguments; the role of argumentation in deep disagreements; and the need of balance-of-considerations claims in pro-and-con or conductive arguments. And we have also seen the fruitfulness of taking seriously the task of identifying, interpreting, reconstructing, evaluating, and analyzing such meta-arguments. Another theoretical context especially pregnant with meta-argumentation is the one where theorists advance arguments that are either intended to exemplify their own theoretical principles, or are *de facto* subject to interpretation or evaluation in light of their own principles. Such arguments may be called self-referential arguments. This chapter will examine three instructive examples of such self-referential arguments. The first is an argument by Blair (2002) about the notions of argument and logic in logic textbooks. The second is an attempt by Eemeren and Houtlosser (2003a) to analyze fallacies in general, and the *tu quoque* in particular, as derailments of strategic maneuvering. And the third is a discussion by Johnson (2007a) of the normative principle of anticipating objections as a requirement for good argumentation.

9.1 Blair on 'Argument' and 'Logic' in Textbooks: *Argumentum Ad Hominem*?

In an essay entitled " 'Argument' and 'Logic' in Logic Textbooks," J. Anthony Blair (2002) reports on his studies of the notions of logic, argument, and related concepts in logic textbooks of the second half of the twentieth century. And he advances some interesting (meta)arguments concerning important and perennial issues.

1. *Reconstruction*. Blair's key thesis may be formulated as the claim that (a) logic is a part, but only a small part, of the philosophical study of argument. From this he infers the corollary that (b) we ought not to equate logic and the philosophy of argument. To justify his key thesis Blair begins by (c) identifying logic with the conception of logic found in logic textbooks of the previous fifty years; discovering that (d) the textbook conception is that logic is the theory of the deductive validity of arguments (or the theory of deductively valid arguments); and thus concluding that (e) logic is essentially equivalent to the theory of deductive validity. And then he argues that (f) the philosophical study of argument must focus on "full-bodied arguments"; that (g) the interpretation, analysis, and evaluation of full-bodied arguments raise many issues beyond those of deductive validity; and hence that (h) the philosophical study of argument must study many aspects of arguments besides deductive validity.

In reconstructing Blair's argument in this manner, I am simplifying the discussion somewhat by not taking into account the distinction between reasoning and argument. I do not deny this distinction, but none of the issues I want to raise here hinge on this distinction.

My reconstruction is meant to make clear and explicit that Blair's final conclusion is proposition (b); that his key thesis (a) is an intermediate proposition immediately supporting (b) and in turn supported in a linked manner by propositions (e) and (h); that (c) and (d) are two linked propositions supporting (e); and that (f) and (g) are linked reasons for (h). This is to say that Blair's argument may be visually represented by the structure diagram in Figure 9.1.

Self-referential Arguments

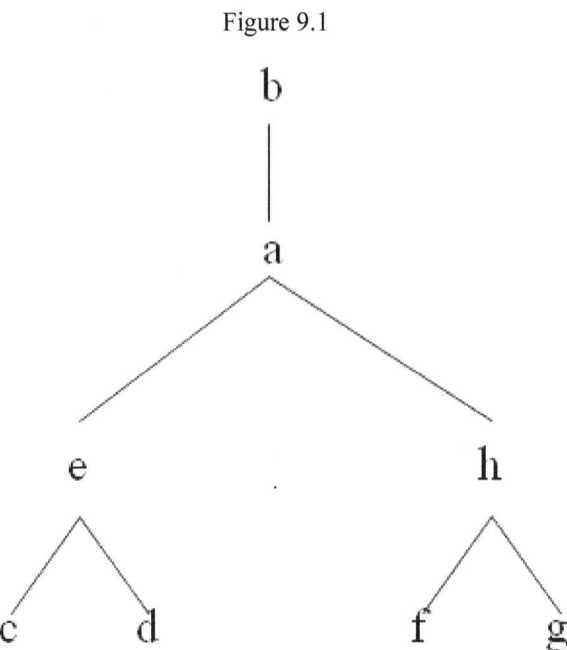

Figure 9.1

Moreover, as reconstructed, Blair's argument appears to be deductively valid, and so any disagreements will ultimately hinge on the truth or acceptability of propositions (c), (d), (f), and (g). Of these, although (c) and (f) are presented as unsupported, (d) and (g) are supported by means of subarguments and so their acceptability will depend on the correctness of their respective supporting subarguments.

2. *Scope of logic*. I begin with an issue involving a disagreement which may be mostly, if not exclusively, semantical or verbal. In fact, in a sense I agree with Blair's key thesis that logic is only a part of the philosophical study of argument. But my inclination is to formulate this claim by saying that formal deductive logic is only a part of logic, thus equating logic in general with what Blair calls the philosophy of argument, and equating his term logic with my phrase formal deductive logic. Blair seems to assume that the word logic means what formal deductive logicians say it means; he seems to attach a particular meaning to the word. This assumption is embodied in his proposition (c), which, while not arbitrary, clearly could be questioned.

The alternative, broader conception of logic could be defended by a historical argument consisting of elaborating the connotation the word has had throughout most of the history of philosophy. This is the conception one could elaborate on the basis of such works as Antoine Arnauld and Pierre Nicole's *Logic or the Art of Thinking* (1662), Benedetto Croce's *Logic as the Science of the Pure Concept* (1909), and John Dewey's *Logic: The Theory of Inquiry* (1938). However, in this context this issue needs no elaboration; it is a familiar point that has been discussed by such well-known authors as Stephen Toulmin (1958, 185-88), as we saw in chapter 1 above. To conclude my first comment, the question I should like to raise is whether logic should be equated with formal deductive logic or with the philosophy of argument.

3. *Full-bodied arguments*. On the other hand, even without accepting such a generalized conception of logic, there is an intermediate conception which affects the correctness of Blair's claim (d). His supporting argument is a quasi-empirical survey of

recent textbooks, which may be interpreted as an inductive generalization argument. And a further conclusion he draws from (d) is an explanation of why recent philosophy Ph.D. recipients or candidates end up teaching the theory of deductive validity when asked to teach a course on reasoning, argumentation, or critical thinking. The question I should like to raise here has two aspects: one is the fairness or representativeness of Blair's sample from the point of view of induction, and the second is the difficulty of his hypothesis explaining a particular aspect of the phenomenon that might be described as the relative neglect of induction. Let me elaborate briefly.

Suppose we were to add the following books to Blair's list: Morris Cohen and Ernest Nagel's *Introduction to Logic and Scientific Method* (1934), Irving Copi's *Introduction to Logic* (1st edn. 1953, 9th edn. 1994), Wesley Salmon's *Logic* (1st edn. 1963, 3rd edn. 1984), and Merrilee Salmon's *Introduction to Logic and Critical Thinking* (1st edn. 1984, 4th edn. 2002). A common feature of these four books is that they define logic as the study of the relationship between the truth of the premises and the truth of the conclusion of an argument, in such a way that two important special cases are immediately distinguished, namely deductive validity and inductive correctness. Here inductive correctness is taken to be the property that an argument has when, although it is deductively incorrect, it is such that if all the premises are true then the conclusion is probably (but not certainly) true. And this distinction is not only presented abstractly at the beginning of these books, but it is upheld throughout their subsequent presentation, in the sense that they devote about the same amount of space to induction as they do to deduction. If one were to take such books seriously, one would end up practicing and teaching a version of logic which, while less general than that of Dewey, would be broader that the theory of deductive validity. Hence the question arises, what else (what other assumptions) enable Blair's prospective teachers to equate the study of argument with the study of deductive validity.

Perhaps, their argument has been historically influenced, or might be otherwise strengthened, by the following one found in the preface to W.V. Quine's *Philosophy of Logic* (1970), which otherwise shows an unusual and surprising sensitivity to induction. As we saw in chapter 1, he begins by quoting the following ostensive definition of logic by Lewis Carroll: " 'Contrariwise', continued Tweedledee, 'if it was so, it might be; and if it were so, it would be; but as it isn't, it ain't. That's logic'." Then Quine goes one to clarify:

> We shall be occupied in this book with the philosophy of logic in substantially Tweedledee's sense of the word 'logic'. This is not the invariable sense of the word. Precedent could be cited for applying the word collectively to two dissimilar studies: deductive and inductive logic. The philosophy of inductive logic, however, would be in no way distinguishable from philosophy's main stem, the theory of knowledge. What arrogates a distinctive bit of philosophy to itself is deductive logic, the discipline that Tweedledee had in mind.
>
> If pressed to supplement Tweedledee's ostensive definition of logic with a discursive definition of the same subject, I would say that logic is the systematic study of logical truths. Pressed further, I would say that a sentence is logically true if all sentences with its grammatical structure are true. Pressed further still, I would say to read this book. [Quine 1970, xi]

This issue of the inclusion or exclusion of inductive correctness is important because in his "critique" (i.e., his subargument for [g]), Blair mentions types of arguments and of considerations some of which (although not all) could be covered under the heading of inductive correctness. For example, "arguing for or against the legal guilt of an accused in a criminal trial" (8)[1] consists primarily of arguments which can hope at best to be inductively correct; and the same would apply to what Blair calls "arguing for a fact or construction

---

[1] In this section, references to Blair 2002 are given by just indicating the page numbers in parenthesis.

upon the facts (for example, when one argues that global warming is, or is not, a pressing problem, or that pollution abatement programs are working, or that the bird singing by the lakeshore is a Yellowthroat warbler)" (8). Similarly, consider Blair's principle number 7, to the effect that "the primary goal of an argument might be to weaken or strengthen the audience's adherence to a point of view, rather than to convince the audience for all time of its falsity or its truth. In such contexts, arguments that fall short of a decisive demonstration might well succeed in achieving their purpose" (10). Now, inductive correctness is by definition a matter of degree and is designed explicitly to deal with such situations.

However, even if we add the study of the inductive correctness of arguments to the study of their deductive validity, there is still much more that would need to be done in the philosophical study of argument. Thus Blair is still importantly right. Or to be more precise, an important part of Blair's argument as reconstructed above is essentially correct, that is proposition (h) and the subargument going from (f) and (g) to (h). When we add a reference to induction, the last two propositions may be reformulated as (g′) and (h′), and we then get the following modified argument: (h′) the philosophical study of argument must include many other aspects of arguments besides deductive validity and inductive correctness, because (f) the philosophical study of argument must focus on "full-bodied" arguments, and (g′) the interpretation, analysis, and evaluation of full-bodied arguments raise many issues beyond those of deductive validity and inductive correctness.

4. *The arguer*. Next, I wish to make some remarks designed to strengthen this argument (but also to qualify other aspects of Blair's discussion). I agree with his emphasis on what he calls "full-bodied" arguments, and I also agree with the several types of examples he mentions. In particular, I agree that an important type is what he describes as "arguing for a thesis in a scholarly paper, such as this one" (8). We have already seen that, as Blair himself recognizes, the argument in his paper has an empirical component, which would raise issues of inductive correctness along the lines already alluded to. However, what I want to do now is to reinforce claim (g′) by using Blair's own argument as an example of a full-bodied argument and subjecting it to an analysis that uses some of Blair's own (plausible) additional principles (8-10).

One of these principles (number 1) is that in the context of (full-bodied) argumentation "who the arguer is makes a difference" (8). This principle allows me to present in the form of a potential criticism a puzzle I have about one aspect of Blair's paper. My puzzle stems from the things he says about the motivation underlying it. Blair tells us that his paper was in part motivated by his administrative experience of being involved in hiring recent philosophy Ph.D. recipients or candidates to teach courses on reasoning, argumentation, and critical thinking. He reports that in these courses such instructors tend to teach elementary symbolic logic, namely "virtually the same subject-matter as is taught" (1) in courses bearing the label "logic" at his university.

Now, I must confess that I would find such a situation intolerable. Such prospective instructors should be given appropriate instructions. They should be instructed to study the catalogue description of the introductory reasoning course, which is: "An explanation of, and practice in, the basic knowledge, skills and attitude which are essential components of reasoning well." Blair should have pointed out to them that this description is such that their focus should not be deductive validity, but rather full-bodied arguments and their analysis by means of a much wider set of principles, such as those discussed by him in this essay (Blair 2002). In short, he should tell them to study some version of this essay.

What is the upshot of these considerations? The point is that there is a tension between Blair's claim (h) that the philosophical study of argument must include much more besides deductive validity, and the behavior, action, or conduct exhibited in his administrative experience. I am not clear at the moment about exactly where and how Blair's argument is affected, but it is clear that this is a criticism of Blair's "position" and that such criticism is an instance of *ad hominem* argument. It is equally clear that my *ad hominem* argument is not of the "abusive" or "poisoning-the-well" type. Perhaps it could be regarded as being of the "circumstantial" type, in which case its correctness is an open question that would require further analysis.

Or more promisingly and interestingly, the considerations that underlie this criticism add further support to Blair's own thesis (g) that the study of full-bodied arguments must include more than questions of deductive validity (and [g'] questions of inductive correctness), such as considerations of "who the arguer is." Thus, my analysis here also suggests that the first in Blair's list of additional principles relates to the notion of *ad hominem* argument in the Whately-Johnstone sense.[2]

5. *The audience*. According to Blair's plausible account, a second nondeductive, noninductive principle required in the philosophical study of argument is that "who the audience or interlocutor is makes a difference: it can impose constraints on the argument" (9). Let us apply this principle to Blair's own argument.

The relevant audience is, narrowly conceived, the participants at the 2001 conference of the Ontario Society for the Study of Argumentation, where Blair's paper was originally presented; more broadly, the audience consists of informal logicians and argumentation theorists everywhere, whether or not present at that conference. Let us examine, therefore, what elements of Blair's argument would be accepted by such scholars prior to and independently of his argument. I would venture to guess that, *in the sense in which Blair intends* the propositions making up his argument, this audience already accepted his key thesis (a), his further conclusion (b), as well as all the other propositions except for proposition (c) and the intermediate conclusion (e) it helps to support. In other words, the main questionable premise is the claim that identifies logic with the conception of logic found in the logic textbooks of the past fifty years. This means that he is trying to argue for an already accepted (or relatively acceptable) conclusion on the basis of a premise that is not accepted or relatively unacceptable. This seems to be to reverse of what one usually does in an argument, which is to render a claim more acceptable to an audience on the basis of claim(s) that it already accepts.

This sounds paradoxical, but perhaps Blair's argument may be rescued as follows. If Blair's argument is otherwise sound; if he has shown to this audience that if (c) is true then (a) would be true; and if this audience already believed (a) to be true; then Blair has rendered (c) more acceptable to this audience than it was beforehand. In other words, he has made a plea to this audience for the persuasive definition of logic in terms of recent logic textbooks. In short, he has tried to convince us to use the term logic as synonymous with formal deductive logic. I do not think any disaster would follow if this stipulation were adopted, but if correct it would give a strange twist to Blair's argument. In any case, however, proposition (g), or (g'), emerges strengthened.

6. *The role of argument*. Finally, let us examine the fourth principle in Blair's list. It states that "the role of an individual argument in the larger setting of making a case will affect the norms relevant to its appraisal ... In other words, understanding the dialectical function of an argument—its role in responding to objections, counterarguments, or arguments for alternative positions—will be important to its assessment" (9). What Blair has in mind is arguments "designed just to establish a presumption ... to shift the burden of proof ... refuting an objection" (9), and the like.

Here I don't think I completely agree with Blair. It seems to me that such issues do not

---

[2] See, for example, Johnstone (1952; 1959; 1978) and chapter 7 above. Cf. Walton 1985, Eemeren and Grootendorst 1993, and Woods 1995.

escape the domain of deductive and inductive logic. On the one hand, if we are talking about case-building, where a particular argument is just one of several supporting a conclusion, then we are probably not in the domain of deduction, but in that of induction; that is, a particular argument is trying to show that the conclusion is likely to be true, and so questions of inductive correctness would become relevant. On the other hand, most and perhaps all of what Blair calls the role or design of an argument could be handled in terms of identifying its conclusion correctly. For example, if the argument is designed to just establish a presumption rather than the truth of a statement, then the conclusion should be stated in a form such as "we may presume that p," rather than "it is true that p." If the argument has the role of refuting an objection, then the conclusion would have to specify exactly what flaw is being attributed to the objection, rather than that the proposition being objected to is false.

However, I do not want to give the impression that Blair is completely wrong here. For even if we can reduce these "dialectical" issues pertaining to "case-building" to questions of the inductive or deductive correctness of arguments along the lines just sketched, we are dealing with special types of arguments. These are arguments such as what Eemeren, Grootendorst, and Snoeck Henkemans (1996, 17) call "multiple argumentation"; what Wellman and Govier would call conductive arguments of the third kind, as we saw in chapter 8 above; what I have called complex argumentation, e.g., in chapter 7 above; and what in this book I am generally calling meta-arguments or arguments about arguments. And such arguments are special both in the sense that they are usually not studied by deductive and inductive logicians, and that they deserve more careful study than they have received. Thus the spirit, if not the letter, of Blair's proposition (g), or (g'), is vindicated.

7. *Conclusion*. In summary, in this section I have tried to take seriously Blair's views on the logical and the philosophical study of argument by reflecting on them with the care they deserve. But since those views largely correspond to my own, I have mostly tried to *practice* some of what we preach. I have done so by taking Blair's views as constituting a "full-bodied" argument, reconstructing it in a nuanced manner, and evaluating it as much as possible in accordance with the principles elaborated by him for the philosophical study of argument. Hopefully, the result is a reinforcement of those views, as well as an exercise in the "philosophy of full-bodied argumentation." On a more critical note, my key point may be viewed as an *ad hominem* criticism of Blair's argument, *ad hominem* in the Whately-Johnstone sense (elaborated in chapter 7 above).

9.2 Eemeren on Fallacies as Strategic Derailments: *Tu Quoque Ad Hoc*?

In an essay entitled "More about Fallacies as Derailments of Strategic Maneuvering: The Case of *Tu Quoque*," Eemeren and Houtlosser (2003a)[3] make a contribution to the theory of fallacies. They revise the previous pragma-dialectical definition of a fallacy, according to which a fallacy is a violation of a rule for the reasonable or rational conduct of a critical discussion. The revised conception is that a fallacy is a derailment of strategic maneuvering with respect to two aims, the reasonable conduct of a critical discussion and the persuasion of an opponent.

1. *Agreement vs. disagreement*. This brief commentary is not the place for me to elaborate on my many agreements with the authors of this essay. For example, I agree that one of the best-known of the alternative accounts of fallacies (which here shall remain

---

[3] Here I limit myself to a critical examination of this essay, but it should be noted that the authors have discussed this topic in several other works. Cf. Eemeren (2009, 2010); Eemeren and Houtlosser (2002b, 2003b).

nameless) tends to be *ad hoc*; that it is extremely misleading (I would say self-contradictory) to claim that fallacies are not always fallacious; that to stress the traditional list of fallacies that have Latin labels is theoretically unproductive; and that the domain of persuasion is properly labeled rhetorical. After all, one might say that scholarly commentaries ought to be arguments; that arguments aim at the resolution of differences of opinion; and that therefore the present commentary should focus on some disagreement between myself and the authors.

On the other hand, whereas I see the fruitfulness of the conclusion just stated, I am not inclined to uncritically accept its supporting premises. At any rate, still less would I have space to elaborate, or even sketch, some alternative to their pragma-dialectical theory, such as the negative-evaluation approach to fallacies which I have advocated elsewhere (e.g., Finocchiaro 2005a, 128-47); and I feel that this practical temporal constraint must be obeyed even though there is a large amount of truth in Johnson's (2000a, 167-69) methodological requirement that an argument should include a dialectical tier, which (among other things) involves criticism of alternative positions, which in turn implies that the elaboration of alternative positions is part of a critical commentary.

2. *Interpretive summary*. To begin searching for a difference of opinion, let us recall that for pragma-dialectics, to use their own words, "fallacies are ... conceived as argumentative moves ... that violate the rules of the procedural model for conducting a critical discussion ... The reasons for finding fault with such moves are ... closely related with the general goal that is attributed to the discourse ... This general goal is resolving a difference of opinion by testing the acceptability of the standpoints at issue" (2).[4] Now, although the systematic theoretical treatments given in various dialectical approaches have much to recommend them, none of them has so far provided the comprehensive theory of fallacies that is desirable. This limitation applies to pragma-dialectics and hence serves to motivate the authors' present essay.

To be more specific, they point out that until now their theory had given no answers to the following three questions: (1) why fallacies are committed in the first place;[5] "(2) ... why a lot of fallacies can be so persuasive; (3) ... why fallacies do so easily go unnoticed" (p. 2). They now believe they can answer these questions, and the explanation is the same for all three phenomena. The explanation lies in the fact that arguers typically have other aims besides the canonical, pragma-dialectical one of resolving differences of opinion; that a crucially important additional aim is persuasion of the opponent; and that fallacies arise when such a secondary aim is improperly combined with the rules of critical discussion dictated by the canonical aim, or improperly interferes with their operation. They label the attempt to combine the two aims "strategic maneuvering," and a failure of such an attempt a "derailing" of strategic maneuvering. So conceived, fallacies become derailings of strategic maneuvering.

The authors then focus on a "specific type of strategic maneuvering that takes place in

---

[4] In this section, references to Eemeren and Houtlosser 2003a are given by just indicating the page numbers in parenthesis.

[5] Eemeren and Houtlosser (2003a) do not literally formulate this first question as I have, but instead speak of the problem of how one can formulate "adequate criteria for deciding in concrete cases univocally whether or not a certain rule has been violated" (2), that is "criteria that are needed to be able to check whether the rules are correctly applied in practice" (2). But when they start sketching their answer they say that "paying attention to the reasons a party may have in ordinary discourse for not complying with the rules because of the pursuit of other purposes, which may be at odds with the aims of a critical discussion, may lead to an explanation of why a violation of a rule for critical discussion is sometimes inevitable" (2). In short, their own answer does not really answer the question they literally ask, but the one I have attributed to them.

the opening stage of a critical discussion. In this type of maneuvering a party attacks the other party by pointing out a logical or pragmatic inconsistency between a starting point proposed by the other party and a starting point this party assumes on a different occasion" (4). Recall that for pragma-dialectics, the opening stage of a critical discussion is the second of four stages, the first being the so-called confrontation stage, the third the argumentation stage, and the fourth the concluding stage. Thus the opening stage is the one when the two parties, having already clarified what their difference of opinion consists in, try to agree on some procedural and material or substantive starting points, to be used at the argumentation stage. Procedural starting points include decisions on which party acts as protagonist or proponent (whose role is to defend a claim with reasons), and which as antagonist or opponent (whose role is to challenge the protagonist). Material or substantive starting points are propositions which the two parties agree to take for granted for the sake of the argument. From a purely pragma-dialectical point of view, the protagonist is not obliged to give a reason for not accepting a particular starting point. However, from a rhetorical point of view it is often desirable to do so. Generally speaking, such reasons are sometimes sound and sometimes unsound.

We have seen that one of these reasons may be an alleged inconsistency between what the protagonist is proposing now and what he proposed on some earlier occasion. When is such an allegation of inconsistency sound? The authors elaborate three conditions: the two propositions should be really inconsistent, according to some acceptable definition of inconsistency and some effective criteria for deciding it; the proposition which is attributed to the protagonist on some other occasion should be one that can be actually attributed to him, either explicitly on the basis of some avowed declaration, or implicitly on the basis of some action that contextually implies it; and the earlier occasion on which the protagonist advocated one of the two inconsistent propositions should be part of the same critical discussion in which he is now advocating the other proposition, where sameness of critical discussion is partly an empirical question and partly a theoretical question. These three conditions are apparently advanced as being individually necessary and jointly sufficient.

Another distinguishable element of the authors' account is their identification of the fallacy of *tu quoque* with an unsound rejection by the antagonist of a substantive starting point proposed by the protagonist, on the grounds that the latter's present proposition is inconsistent with one of his earlier starting points. On the other hand, cases where such allegations of inconsistency are sound do not constitute fallacies of *tu quoque*, but rather sound and successful cases of strategic maneuvering.

Finally, the authors provide an interesting and important historical illustration of strategic maneuvering involving an allegation of inconsistency. The case involves the rejection by William of Orange (commonly known as William the Silent) of the King of Spain's criticism that he (William) had violated the Treaty of Gaundt of 1576. William's defense consists primarily in pointing out that the King of Spain too had violated the treaty (cf. William 1581).

3. *Criticism*. If this interpretive summary is fair and accurate, then I would start asking myself questions such as the following. Regarding the last mentioned element (William of Orange's strategic maneuvering), I must say that I find it too sketchy. At the level of interpretation, not enough is said to make it clear that William's strategic maneuvering occurs at the opening, rather than at the argumentation, stage of the critical discussion; nor it is clear that the disputed proposition is a starting point instead of the main standpoint at issue, for it is not clear what is the main difference of opinion established at the confrontation stage. And at the level of evaluation, the authors do not tell us whether William's maneuvering is sound or unsound; such an evaluation and its justification would provide much needed elaboration of their own soundness conditions.

Another element of their essay is the interpretation of the *tu quoque* fallacy as an unsound strategic maneuvering at the opening stage of a critical discussion, consisting of a charge of inconsistency by the antagonist meant to justify a rejection of a starting point. I take this to be a theoretical definition of *tu quoque*. Here, I would have liked a justification of this definition by means of a theoretically-neutral description of the definiendum. In short, I feel the authors should have said more about what is ordinarily meant by *tu quoque*, so that their claim that it can be identified with the definiens they propose would have more empirical content and greater persuasiveness.

For example, as things stand, it is unclear why the definiens should not be identified also, or instead, with the *ad hominem* argument or fallacy, *ad hominem* in either the classical Galilean-Lockean sense,[6] or the modern abusive sense. Nor it is clear what the authors' pre-theoretic (i.e., pre-analytic *and* pre-evaluative) intuition of *tu quoque* is, and what it would tell them about whether or not the following are examples of *tu quoque*. These are arguments that were advanced by some during the debates that preceded the outbreak of the Iraq War in 2003: one was a defense of Saddam Hussein's regime from the charge that he was in material breech of various U.N. resolutions, by arguing that Israel was in material breech of many U.N. resolutions, and yet the USA did nothing about the latter violation; the other was a criticism by some Europeans of France's claim that the USA was wrong to try to unilaterally disarm the Saddam regime, by objecting that France's behavior within the European Union amounted to an attempt to unilaterally impose its will on the rest of the Union.

Continuing with my review of the various elements of this essay, there is little to find fault with in its identification and definition of the particular strategic maneuvering at the opening stage consisting of the antagonist's charge that the protagonist is being inconsistent. Still, one would have liked more of a theoretical motivation and contextualization for this particular strategic maneuvering. The authors show some awareness of this potential difficulty when they assert in a footnote that "it stands to reason that to provide a more refined inventory of the types of strategic maneuvering pertinent to the various stages, these stage-related 'local' aims need to be further specified" (p. 9, n. 13). My point is that one needs more clarification of, information about, and examples of the notions of strategic maneuvering, its derailment, and in general the interaction between the dialogical aim of dispute resolution and the rhetorical aim of persuasion.

My next question regards the conception of fallacies as derailments of strategic maneuvering. It is not clear to me whether the authors' thesis is that all fallacies are derailments of this sort, or that some are such derailments but some are violations of the pragma-dialectical rules pure and simple.

If the former is the case, that would seem to be a major revision of the original pragma-dialectical theory. Such a revision would be presumably justified by the facts of the case, the facts of argumentative practice. However, even if the revision is thus substantively justified by such evidence, one could ask how such a revision fares from a methodological (or meta-theoretical) point of view; that is, the point of view of what a revision indicates about the cognitive value of the original theory being revised and about the new version of the theory. For example, one might ask, was it ever justified for the old unrevised theory to have neglected the rhetorical element of argumentation, the fact that one of the aims of argumentation is persuasion? What are the prospects for the newly revised theory? That is, what reason is there to think that the addition of a second element is sufficient? Is it not

---

[6] For more on this notion, and its difference from the contemporary *ad hominem* fallacy, see, for example, Johnstone (1952; 1959; 1978), Eemeren and Grootendorst 1993, Woods 1995, Finocchiaro (2005a, 277-91, 329-39), and chapter 7 above.

obvious that one aim of argumentation is, as Alvin Goldman (1999, 131-60) has argued, epistemic and veritistic, and is one not entitled to predict that sooner or later the pragma-dialectical-cum-rhetorical version will have to find a way of coming to terms with the epistemic dimension of argumentation? And would not such additions and accretions to the pragma-dialectical theory give it a characteristic of *ad hocness*? Then, the previously dismissed theorists of fallacies could perhaps charge *tu quoque*, and however analyzed, it is unclear that this charge would be fallacious.

Let us now see what would follow if the thesis was that not all fallacies are derailments of strategic maneuvering, but some are simple violations of pragma-dialectical rules. Clearly the theoretical revision in this case would be smaller. One could say perhaps that the previous simpler analyses and definitions of some fallacies stand, but new more complex analyses of other fallacies are now formulated. Unfortunately this does not seem to be the case here because the *tu quoque* is given the simpler analysis in Eemeren and Grootendorst's (1992, 111-15, 212) work on *Argumentation, Communication, and Fallacies*. Moreover, I should report that Eemeren and Houtlosser's (2003a) present essay motivated me to read some other of their more recent works to which they refer,[7] and in one of these they assert explicitly and categorically that "all derailments of strategic maneuverings are fallacious and all fallacies can be regarded as derailments of strategic maneuvering" (Eemeren and Houtlosser 2002b, 142). So it does seem as if the revision is a major one.

4. *Conclusion.* In summary, there do seem to be points of disagreement between Eemeren and Houtlosser on the one hand and myself on the other. This may augur well for future argumentation, at least argumentation about argumentation. However, for the present occasion I believe that what I have done has been merely to work at the level of preliminaries for such argumentation, preliminaries which these pragma-dialectical theorists would probably label the confrontation stage. This is a label which I might accept if one meant a non-confrontational sense of "confrontation," which I believe is the pragma-dialectical sense of this word. More substantively and ambitiously, I should like to believe that I have suggested that Eemeren and Houtlosser's revised account of fallacies is vulnerable to the *tu quoque* charge of being *ad hoc*; for their previous pragma-dialectical conception was meant in part to be an alternative to the *ad hoc* treatment of traditional accounts, and yet the motivation for their revision seems to lack an adequate theoretical basis or systemic thrust.

9.3 Johnson on Anticipating Objections: *Caveat Emptor*?

In an essay on "Anticipating Objections as a Way of Coping with Dissensus," Johnson (2007a) discusses an important but neglected topic in the theory of evaluation. In the context of his dialectical theory of argument, he explores various normative principles which arguers may be obliged to follow.

1. *Arguing about objections?* In answer to the question, what are the arguer's dialectical obligations, Johnson (2007a) formulates three normative principles about the handling of objections. The first concerns strength: "the stronger the objection is, the stronger its claim on the arguer" (13).[8] The second concerns what Johnson calls proximity: "the closer it is to the arguer's position, the stronger its claim on the arguer" (13). The third

---

[7] At the time (2003), however, I was unable to check their references to Eemeren and Grootendorst's (2004) *Systematic Theory of Argumentation*, since this book was not yet available then.
[8] In this section, references to Johnson 2007a are given by just indicating the page numbers in parenthesis.

concerns what he calls salience: "the more salient the objection in the dialectical environment, the stronger its claim on the arguer to respond" (13). Combining these three principles and attempting to be clearer, I would say the following. An arguer has an obligation to anticipate objections, and the obligation is stronger for objections that are stronger, closer, or more salient.

In answer to the question, what is required for the successful dispatch of one's dialectical obligations, Johnson formulates another triad of normative principles. "First, the arguer must accurately and faithfully state the objection ... Second, the arguer must make an *adequate* response; i.e., must argue that the objection is not on target, does not really damage the argument" (14). And there is "a third requirement: the objection(s) anticipated must be appropriate" (14). Johnson calls these principles criteria of dialectical excellence and seems to suggest that they are individually necessary and jointly sufficient.

Now, these two sets of principles, for a total of six, seem to be Johnson's main conclusions in this essay; or we might say that he is advancing a six-part conclusion. However, if one looks for his justifying argument(s), one finds little more than the following. "The arguer take[s] the trouble to anticipate objections" (2) for two reasons; first because argument has "the goal of rational persuasion" (3); and second because argument "is an exercise in rationality in which the parties are interested in both the substance and the appearance of rationality" (3), that is, argument is an exercise in manifest rationality. And Johnson refers to his book (2000a) on the subject, since these assertions are obviously just the tip of an iceberg. At this point I could refer to the analysis and criticism of Johnson's arguments elaborated above (chapter 4). However, to leave it at that would be merely an invitation to stop our dialogue, which is the opposite of my intention here.

Thus, in the interest of further interchange, here I want to say that those arguments from Johnson's (2000a) book on *Manifest Rationality* pertain primarily and directly to the desirability of *responding* to objections, whereas in this essay he is talking about *anticipating* objections. The same applies to some of the arguments which Johnson gives in another essay, entitled "Responding to Objections" (Johnson 2007c). Thus, if we stress that in the present essay he is discussing the *anticipation* of objections, then I do not see that he has justified the six more specific claims described above. In short, Johnson's essay contains essentially no argument for these six claims.

However, it would be naïve to make too much of this objection to Johnson's essay. Even a theorist of argumentation must recognize that there are other things in the universe besides argumentation. Before argumentation can get started, a certain amount of preliminary work is needed, and such preliminary work involves definitions, distinctions, concept formation, description of approach, etc. Now, in most, indeed in almost all, of the present essay Johnson is engaged in such preliminaries. These should be judged on their own merits, not by means of criteria applying to other things. So it would be a kind of category mistake to object to this essay for its scarcity of argumentation. Let us then switch to that other kind of analysis.

2. *Appropriateness*. Let me begin by focusing on the last of Johnson's principles of dialectical excellence. In elucidating what he means by the appropriateness component of dialectical excellence, Johnson says that an appropriate response is one that deals with "a well-known and important objection (one that is salient [or] looms large in the dialectical environment" (14). So the third principle of his second set seems to duplicate the third one of the first set. That is, his six points really reduce to five, which could be summarized as follows: an arguer should anticipate objections that are strong, close, and salient, and the response should be accurate and adequate. We don't need to add that the response should be appropriate because that would only mean that the anticipated objections should be salient.

3. *Interpretation*. My next comment stresses an aspect of Johnson's first principle of dialectical excellence. It is the one that requires us to state the objection accurately and faithfully. Although the justification of this principle is another issue, I certainly accept the correctness and importance of this principle. But I want to point out that what is involved here is the *interpretation* of an objection, in a sense of interpretation that distinguishes interpretation from evaluation. In interpretation, one aims to understand what the objection is, what it is saying, what it is claiming. In evaluation, one is trying to determine whether the thing being evaluated (in this case the objection) is valid or invalid, relevant or irrelevant, right or wrong, and to what extent or in what sense it possesses these characteristics. Obviously, in the present situation, the evaluation comes later and is addressed in Johnson's second principle. So it should be clear that when anticipating an objection, the process of stating it accurately and faithfully is an interpretive activity.

Now, all this would hardly need saying were it not for the fact that there is a widespread tendency in informal logic, critical thinking, and argumentation theory to neglect the interpretation and understanding of arguments. Johnson seems to be part of this tendency. For example, in his book on *Manifest Rationality*, while elucidating what he means by the practice of argumentation, he says that "by this, I mean to refer to the sociocultural activity of constructing, presenting, and criticizing and revising arguments" (Johnson 2000a, 154). This does not seem to be merely an incidental remark.[9] Nor is it the case that this is merely a reflective pronouncement and that in fact he pays the proper attention to interpretation when working out the details of his theory of argument; this is not the case because there is not in fact a single chapter on the interpretation of argument, out of the twelve that make up the book on *Manifest Rationality*.

What I would say[10] is that an important part of argumentative practice is the interpretation of arguments, as distinct from their construction, presentation, evaluation, criticism, and revision. Therefore, concepts and principles of interpretation should be an important part of the theory of arguments. One of these principles is precisely Johnson's first principle of dialectical excellence.

4. *Pre-empting objections*. My fourth comment is an attempt to carry further the military metaphor introduced by Johnson and exploited by him to help him arrive at his formulation of the three principles of strength, proximity, and salience. He is well aware that the military metaphor does not constitute a justification but is merely a heuristic instrument, and that the results will have to be "translated out from the metaphor—a task for the future" (13), as he says. The notion I want to examine, however briefly, is that of *pre-empting* an objection as distinct from anticipating an objection, and by analogy to the notion of a pre-emptive war or pre-emptive attack.

Now, Johnson may be viewed as having already distinguished between responding to objections and anticipating objections. The first involves a response to something that already exists, to an objection that had already been advanced, whereas the second involves thinking in advance of objections that could be formulated and taking the necessary precautions. Responding to objections is analogous to waiting until one has already suffered a military attack and defending oneself from such an attack, including perhaps with counter-attacks. Anticipating objections is analogous to building up one's defenses before an attack has occurred, so that if and when it does occur one is well prepared to repulse it; anticipating objections is like building a Maginot line to defend France from a

---

[9] For example, the same elucidation appeared verbatim in the original version of Johnson's paper. However, as a result of my criticism, the formulation was later revised to include the terms 'interpreting' and 'analyzing'. See Johnson 2007a, 8 n. 25; 2007b, 2-3.

[10] Cf. Finocchiaro 1980a, 311-31; 2005, 14-15; and chapter 2 above.

German attack, although of course there is no reason why such anticipation should be as flawed or incomplete as the Maginot line, which did cover the French-German border, but was not extended to either the Belgian-French border or the Belgian-German border.

The twenty-first century has brought into prominence the notion of a pre-emptive war. In a pre-emptive war one attacks and destroys the threat before one has been attacked. One does more than wait for the attack and then respond to it; and one does more than anticipate the attack and prepare the response; one counter-attacks before the attack. What would that be like in rational and argumentative space?

It seems to me that the analogue would be the formulation of objections to alternative positions. One question which would then arise is whether one needs to consider only existing alternatives, or whether one should take into account potential alternatives. Another question would be how this differs from anticipating objections.

In his talk of both answering and anticipating objections, Johnson seems to consider three main things: formulating an objection to one's own position (the formulation being presented by someone else when one is merely responding, or by oneself when one is anticipating); criticizing or refuting the objection; and/or revising one's own position. Is any of this equivalent to the formulation and strengthening of objections to alternative positions, be they alternatives that have already been articulated or that could be articulated? If an alternative is taken to be a contradictory of one's own thesis, and if objections are reasons against, then objections to alternatives become reasons against contradictories, and these are reasons for one's own thesis. Thus the activity of pre-empting objections to one's own argument or conclusion seems more akin to a strengthening of the illative tier of one's own argument. This instance of dialectical excellence becomes then a case of illative excellence as well.

5. *Strengthening objections*. Fifthly, I want to focus on the notion of a strong objection, or to be more precise, the strengthening of an objection. Obviously, Johnson speaks of strong objections, in the context of both responding and anticipating. He does not seem to take into account that the strength of an objection is not always or necessarily a material or objective property; and that work may have to be done to make an objection stronger than it was or appears to be. To stress this fact, it may be useful to introduce the notion of *strengthening* an objection, as distinct that of a *strong* objection. The question then becomes whether arguers have the obligation to strengthen objections against their own position. Some famous arguers have thought so.

One of these was John Stuart Mill. In his essay *On Liberty*, in the context of his argument for freedom of discussion, one of his subarguments tries to show that knowledge of objections is necessary if one wants to properly understand and appreciate the reasons supporting one's own conclusions (cf. chapter 10 below). His key point is that in almost all subjects outside of mathematics, "truth depends on a balance to be struck between two sets of conflicting reasons ... [and so] he who knows only his own side of the case, knows little of that" (Mill 1951, 128). Now, referring to such conflicting reasons, or objections (as we might call them), Mill holds that the arguer "must know them in their most plausible and persuasive form; he must feel the whole force of the difficulty ... [for] else he will never really possess himself of the portion of truth which meets and removes that difficulty" (Mill 1951, 129). In other words, "that part of truth which turns the scale ... is [n]ever really known, but to those who have attended equally and impartially to both sides, and endeavoured to see the reasons of both in the strongest light" (Mill 1951, 129-130). Applied to the present context, I take this to be an eloquent plea for an additional dialectical obligation for arguers: whether responding to or anticipating objections, before refuting them an arguer should not only interpret objections accurately and fairly, but present them in their strongest possible light.

Another famous advocate and practitioner of this principle was Galileo. In the winter of 1615-1616 he spent several months in Rome to defend the Copernican theory of the earth's motion from various objections, especially those based on Scripture, but also those based on the laws of physics, on astronomical observation, and on epistemological principles. Here is how a witness of these discussions described them in correspondence with an official from out of town:

Your Most Illustrious Lordship would really like Galileo if you heard him argue, as he often does, surrounded by fifteen or twenty people who launch cruel attacks against him, now about one thing and now about another. But he is so well fortified that he is amused by them all; and although he does not persuade them, on account of the novelty of his opinion, nevertheless he proves the invalidity of most arguments with which his enemies try to bring him down. On Monday in particular, at the house of Mr. Federigo Ghisilieri, his arguments were astonishing; what I liked most was that, before answering the contrary reasons, he amplified and strengthened them with new grounds of great plausibility, so that after he destroyed those reasons, his opponents would appear more ridiculous.[11]

In saying that the opponents were made to look all the more ridiculous, this observer seems to draw a conclusion that is the opposite of the correct one; for Galileo's amplification and fortification of the opposing reasons would make the opponents emerge as intelligent and reasonable persons, who happened to be wrong. But this is not the proper occasion to elaborate.[12] In any case, these references to Mill and Galileo are not meant to be arguments from authority, but rather they are made here primarily to suggest an additional reason why the argumentative practice of these thinkers would be well worth studying. Argumentation theorists, especially dialectically oriented ones, could find fruitful examples, formulations, and justifications of the principle that it is sometimes desirable for arguers to strengthen objections to their own arguments before responding to them.

6. *Defending alternatives.* This leads to my sixth comment, which involves a related principle or technique. That is, sometimes it may be a good idea to defend an alternative position from some objections, before one formulates others more damaging ones against it. By so doing, one displays what might be called open-mindedness and fair-mindedness, and to that extent one might perhaps be able to construct an argument in favor of this technique. Another line of argument could perhaps be gleaned, again, from John Stuart Mill's argument for freedom of discussion. His third subargument involves considerations about avoiding one-sidedness and appreciating partial truths (cf. chapter 10 below). His key point is that in arguments about human affairs, the most common situation is one "when the conflicting doctrines, instead of being one true and the other false, share the truth between them; and the nonconforming opinion is needed to supply the remainder of the truth, of which the received doctrine embodies only a part" (Mill 1951, 140).

However, for this principle, I must admit that I do not know of good examples from argumentative practice. It might be useful to undertake a systematic search. I have occasionally engaged in this technique, as the beginning of this section illustrates (see also chapter 5.2 above). But I leave the problem for future investigation.

7. *Virtues vs. obligations.* Finally, the last two principles provide the occasion for exploring a distinction that might prove useful in a dialectical theory of argument. I shall label it the distinction between dialectical obligations and dialectical virtues. Johnson may have this in mind with his talk of dialectical adequacy versus dialectical excellence,

---

[11] My translation from Galilei 1890-1909 12: 226-27; cf. Santillana 1955, 112-113. Here I have followed Motta's (1993, 612) transcription of the last clause of the first sentence as being about *cosa* rather than *casa*, which is the reading in Galilei 1890-1909, 12: 226.

[12] But cf. Finocchiaro (1980a, 114-15; 2005a, 409-29; 2009a; 2010b, xiii-xliii; and 2011b).

although in his essay (Johnson 2007a) he introduces the notion of excellence primarily as a synonym of adequacy when the latter is meant in a broad sense, needed because the notion adequacy has both a narrow and a broad meaning and is thus ambiguous.

Here I would begin by admitting that I am hesitant to say that an arguer has an *obligation* to *strengthen* objections against his own argument and to *defend an alternative* position from some objections. I would be more comfortable in saying that it is a good thing for an arguer to do these things. In other words, if and to the extent that arguers do these things, that adds extra value to their arguments, at least on some occasions; but if one does not do them, I am not sure that not doing them is in itself a flaw or fault. In Johnson's language, one might say that following these two practices is a sign of dialectical excellence; not following them does make one fall short of excellence, but not short of dialectical adequacy, if and to the extent that the other more basic principles, pertaining to adequacy, are satisfied. In my own language, it is a virtue, but not an obligation or duty, to do them. Of course, this is a claim in argumentative or rational space, and so the virtues, obligations, and duties in question are cognitive or epistemic ones, not ethical or moral.

8. *Conclusion.* In the context of Johnson's dialectical theory of argument evaluation, I began with an appreciation of several of his principles about the obligations to consider, answer, and anticipate objections. But then I attempted to follow this perspective further than he does himself, by exploring the importance of some norms involving the interpreting, pre-emptying, and strengthening of objections. These further norms may appear to be counterintuitive, unconventional, self-destructive, or paradoxical, but such difficulties can perhaps be moderated by regarding those norms as methodological virtues, rather than duties.

9.4 Conclusions: Varieties of Self-referential and *Ad Hominem* Arguments

As suggested at the beginning of this chapter, a self-referential argument is one that can or should be interpreted or evaluated in terms of principles contained in it, that is, relatively explicitly contained in it. The chapter has examined three examples of self-referential arguments. Let us now reflect on this exercise.

I interpreted Blair's argument partly in terms of his notion of a "full-bodied" argument, and evaluated it in terms of his principles regarding the arguer, the audience, and the argument's role or function. I feel the interpretation was sufficiently enlightening, and the evaluation sufficiently positive, as to result in a strengthening of Blair's argument. My evaluation also contained some negative criticism, but this criticism can be viewed as an instance of *ad hominem* argumentation in the Whately-Johnstone sense, and thus as theoretically instructive and constructive.

I interpreted Eemeren and Houtlosser's argument as a defense of their revised conception of fallacies generally, and of the *tu quoque* specifically, in terms of their notion of a derailment of strategic maneuvering; and I evaluated their argument primarily in terms of a principle which they use to criticize traditional accounts as *ad hoc*. The result was that their revision could be also criticized as being *ad hoc* vis-à-vis the original pragma-dialectical conception of fallacies as violations of the rules of critical discussions aimed as the rational or reasonable resolution of disagreements. My criticism can be viewed as a *tu quoque* charge of *ad hocness*, and it is hoped that its value from a meta-theoretical or meta-argumentative point of view will make up for its provocative potential. Furthermore, provocation and self-reference aside, my critique illustrates, once again, the Whately-Johnstone notion of *ad hominem* argument.

I interpreted Johnson's discussion as a preliminary clarification of several principles about the role of objections in argumentation and its evaluation. Accordingly, I defended his discussion from the potential objection that it lacked supporting (meta)argumentation. Then, I made a clarification regarding his notion of *appropriate* objections. In the main thread of my critique, I made some further clarifications by extending his discussion of the consideration, answering, and anticipation of objections to a discussion of the interpretation, pre-empting, and strengthening of objections. Although such an extension seems natural and very much in the spirit of Johnson's account, perhaps it follows that spirit and/or letter too far, and so it is unclear whether he would welcome it. If not, then besides continuing the discussion further at the level of arguing about objections, I could also defend my critique as an *ad hominem* argument (à la Whately-Johnstone) at the next higher level.

In sum, the self-referential interpretation of arguments is always desirable and instructive, whenever it can be carried out. The same applies to the self-referential evaluation of argumentation. Moreover, insofar as such evaluation is positive, the object argument would seem to acquire extra strength. Insofar as the self-referential evaluation is negative, the justification of the latter would seem to reduce to *ad hominem* meta-argumentation (again, in the Whately-Johnstone sense of *ad hominem*).

PART III
FAMOUS META-ARGUMENTS

Chapter 10
Mill on Liberty of Argument

In part i of this work, I elaborated an approach to logic and argumentation theory that studies arguments in a manner characterized as pragmatic, comparative, empirical, historical, naturalist, and both normative and descriptive. It corresponds to Toulmin's idea of an applied logic, as well as to the historical-textual approach which I have advocated previously. Besides elaborating such an approach for the study of arguments in general, I also proposed a novel application to the domain of meta-arguments, i.e., arguments about arguments.

In part ii, I followed this approach and implemented this proposal by studying a number of theoretical meta-arguments, namely arguments advanced by logicians and argumentation theorists in favor of their own claims about arguments. Accordingly, the last several chapters have had a double relevance. One has been the meta-argumentative aspect of the arguments considered and the meta-argumentative perspective of my point of view. The other has been the substantive topics and issues examined, which have ranged from definitions of the concept of argument to methods of argument criticism, deep disagreements, conductive arguments, and self-referential arguments.

In accordance with that earlier project, it is now time to examine some famous meta-arguments, namely meta-arguments which have been historically influential and have become classics, and which regard topics that are intrinsically important, universally significant, and perennially interesting. These too can be anticipated to have a double relevance, methodological and substantive. We begin with John Stuart Mill's argument for liberty of thought and discussion in the second chapter of his book *On Liberty*.

10.1 Introduction

Mill's essay *On Liberty* needs little introduction. Its classic status in moral, social, and political philosophy is widely acknowledged. However, its pertinence to logic and argumentation theory is little known, although this relevance has been suggested explicitly by Hans Hansen and implicitly by some other authors.[1] Inspired by Hansen's suggestion, this chapter aims to explore this connection.

The relevance stems from the second chapter of Mill's essay entitled "of the liberty of thought and discussion." That chapter consists of 44 paragraphs. Although these were not numbered by him and so do not bear numbers in any editions I have used, I suggest we number them sequentially in order to keep track of passages carefully and simplify references. In fact, in this chapter quotations from Mill's text will be referenced by giving the paragraph number in parenthesis. However, because of the large number of quotations, when I quote in close proximity several passages that come from the same Millean paragraph, I will reference only the first quotation and thus spare readers' eyes from redundant distractions.

---

[1] Feyerabend (1981, 1:139-41, 2:65-71; 1999, 212-16); Hansen (1997; 2002, 271; 2005; 2006); Riley 1998; Rosen 2006.

Mill on Liberty of Argument

The reconstructions given below also utilize the systematic numbering system first introduced in chapter 2 for keeping track of various claims in argumentation, and for indicating their place in the network that makes up the macrostructure of arguments. The numbers are given in brackets at the beginning of the sentences or clauses that express the various claims, and such numbers should not be confused with the numbers in parenthesis which are usually given at the end of quoted sentences, and which designate paragraph references to Mill's text. Although this may add considerable encumbrance to parts of the exposition, I find it an efficient way of indicating explicitly, in the course of long and complex arguments, how the various propositions relate to each other and to Mill's text, respectively. Here such numbers are usually inserted without comment and intended to be an aid in the reconstruction; but this is not meant to exclude more analysis later, for their analysis is intimately related to the justification of the accuracy of the reconstruction and to the understanding and evaluation of Mill's argument.

Recall that the key idea of that systematic numbering system is that if a given claim is labeled [n], then the linked premises that directly support it are labeled [n1], [n2], [n3], etc.; and if [nm] is also part of another subargument, then the linked premises directly supporting it are labeled [nm1], [nm2], [nm3], etc. Moreover, when [n] is directly supported by two or more *independent* sets of premises, then letters may be used to distinguish one set from another, e.g., [na1], [na2], [na3], ..., [nb1], [nb2], [nb3], ..., [nc1], [nc2], [nc3], etc. However, another convention will be used below: sometimes the propositions in a given subargument will be numbered by using first the number of the Millean paragraph in which they occur, and then adding digits or letters to the various propositions of the macrostructure in accordance with the usual rules.

In presenting my reconstruction of Mill's arguments, I also found useful to occasionally utilize without explanation the technical terminology of *illative* and *dialectical component* or *tier* of an argument. These terms are taken from the earlier discussion (in chapter 4), where they were adapted them from Johnson (2000a), who in turn adapted them from Blair (1995). Briefly, the illative component of an argument consists of the network of premises that are intended to provide supporting reasons for the conclusion. The dialectical component of an argument refers to the arguer's defense of the conclusion or supporting reasons from objections, and consists of replies to objections (together with a statement of the objections).

10.2 Discovering Truth and Appreciating Fallibility

The first strand (par. 1-20) of Mill's argument consists of an illative component of supporting reasons and a dialectical component of replies to five objections. The illative component is the following.

"[1] The peculiar evil of silencing the expression of an opinion is, that [11] it is robbing the human race" (par. 1) of the opportunity to determine for oneself whether the opinion is true or false; and [12] to deprive others of this opportunity amounts to "an assumption of infallibility" (par. 3), for [121] "assumption of infallibility ... is the undertaking to decide that question for *others*, without allowing them to hear what can be said on the contrary side" (par. 11); but [13] the assumption of infallibility is highly undesirable because [13a] it contradicts the principles of corrigibility, open-mindedness, and epistemological modesty. [13a1] The principle of corrigibility asserts that "the source of everything respectable in man ... [is] that ... he is capable of rectifying his mistakes" (par. 7); [13a2] the principle of open-mindedness states that the capacity to rectify one's mistakes derives from the willingness and ability to seek for, listen to, and learn from criticism (par. 7); and [13a3] the principle of epistemological modesty claims that what is attainable is not absolute certainty but rather provisional knowledge or approximate truth, which consists of beliefs that are open to criticism and have survived criticism (par. 8). Finally, the assumption of infallibility is undesirable because [13b] it has produced serious errors, such as the [13b1] condemnation of Socrates and [13b2] of Jesus and [13b3] Marcus Aurelius's persecution of Christianity.

After this illative tier, let us go on to the dialectical component of objections and replies. One objection claims that suppression of discussion does not assume infallibility but rather only the right and willingness to act upon one's own best judgment (par. 5). Mill's reply to this objection is that suppression of opinion presupposes that one is not permitting the potential refutation of one's own best judgment, i.e., that one has the right to prevent the potential refutation of one's own best judgment; this is much more than the right to act upon one's own best judgment since the latter is merely the right to presume that one's own judgment is true until it has been refuted (par. 6).

Another objection (par. 9) is that the argument for free discussion should not be "pushed to an extreme," i.e., applied to the case of doctrines that are certain; that is, doctrines that are certain need not be open to criticism. Mill replies that a doctrine which is regarded as certain is often a doctrine for which *we* are certain that it is true; but even if we are certain, the doctrine is not and we should not be certain, "while there is any one who would deny its certainty if permitted, but is not permitted."

A third objection (par. 10) argues that some opinions should be immune to criticism because they are highly socially-useful (not because they are true), and to question them would undermine social stability or peace (although it would not undermine intellectual honesty or rigor); hence, some suppression of opinion does not assume infallibility, and so it is right.

Mill replies to the third objection as follows (par. 10). This objection presupposes infallibility with regard to the proposition that the opinion under discussion is socially-harmful, although admittedly not with regard to the proposition that this opinion is false; but we can never be absolutely certain of the proposition that the opinion is harmful; so this proposition should be open to criticism; but if this proposition is open to criticism, then it is wrong to suppress the proposition that the opinion is harmful; and so it is wrong to suppress discussion of the opinion. Moreover, there are connections between the truth and the usefulness of an opinion; "the truth of an opinion is part of its utility ... no belief which is contrary to truth can be really useful"; thus, to question the truth of an opinion is in part to question its usefulness, and those who would not allow questioning of its usefulness would not allow questioning of its truth; hence, as long as one allows questioning of the truth of an opinion, one must allow questioning of its usefulness.

Fourthly, one could object "with Dr. Johnson ... that persecution is an ordeal through which truth ought to pass, and always passes successfully" (par. 15). One reply to this objection is that the persecution of truth does not mean the persecution of the propounders of new truths; these persons should not be treated "as the vilest of criminals" (par. 16); thus, free discussion should be allowed; and so the persecution ordeal for truth does not imply the suppression of free discussion. A second reply is that it is false that "truth always triumphs over persecution" (par. 17): for example, Protestant reformers were successfully suppressed some 20 times before Luther; then even after Luther, they were successfully suppressed "wherever persecution continued ... in Spain, Italy, Flanders, the Austrian Empire" (par. 17); and "Christianity might have been extirpated in the Roman Empire" (par. 17) if persecutions had been more constant; so, the correct thing to say is that truth triumphs sooner or later.

Fifthly, one could object (par. 18) that suppression of some opinions could be allowed as long as one does not put dissidents ("heretics") to death but punishes them with lesser penalties, as was the practice in England in the 1850's. Mill replies that "unhappily there is no security ... that the suspension of worse forms of legal persecution ... will continue" (par. 19). Moreover, he points out that "the chief mischief of the legal penalties is that they strengthen the social stigma. It is that stigma which is really effective" (par. 19) and "which makes this country not a place of mental freedom" (par. 19). Furthermore, "though we do not now inflict so much evil on those who think differently from us as it was formerly our custom to do, it may be that we do ourselves as much evil as ever by our treatment of them" (par. 19); for this milder legal but mostly social treatment results in a state of "peace in the intellectual world" (par. 19); and "the price paid for this sort of intellectual pacification is the sacrifice of the entire moral courage of the human mind" (par. 19); and such intellectual cowardice is disastrous (par. 20) for dissidents and conformists alike, and for "great thinkers ... as much [as] ... average human beings" (par. 20). By contrast, controversy and mental freedom have characterized three great ages that have produced all kinds of mental and institutional improvements: the immediate aftermath of the Reformation in Europe; "the latter half of the eighteenth century" (par. 20) in Continental Europe; and Germany at the time of Goethe.

10.3 Considering Objections and Appreciating Reasons

After this examination of the harm which the suppression of discussion causes to the discovery of whether an opinion is true, "let us now pass to the second division of the argument" (par. 21), which considers the harm done to an opinion that is known to be true (par. 21-33). Mill's key claim here is that "however true it may be, if it is not fully, frequently, and fearlessly discussed, it will be held as a dead dogma, not a living truth" (par. 21); in other words, "unless it is suffered to be, and actually is, vigorously and earnestly contested, it will ... be held in the manner of a prejudice, with little comprehension of its rational grounds. And not only this, but ... the meaning of the doctrine itself will be in danger of being lost, or enfeebled, and deprived of its vital effect on the character and conduct" (par. 43). In short, [21] without freedom of discussion, several harmful consequences follow: primarily, true opinions will be held without a proper understanding and appreciation of [a] their supporting reasons and of [b] their practical or emotional meaning. If this is so, then it would obviously follow that [2] suppression of discussion is undesirable, although here Mill does not explicitly draw this conclusion. Instead he focuses on justifying his key claim. This justification advances two subarguments, corresponding respectively to the two main parts of the key claim. Before elaborating the illative component of the first subargument, Mill begins with the following preliminary point (par. 22).

Even if true, opinions which we are unable to defend from objections are "apt to give way before the slightest semblance of argument" (par. 22), and so they will likely be rejected "rashly and ignorantly" rather that "wisely and considerately"; but such rejection might be taken to mean that we could not properly hold onto those opinions; it follows that even if true, opinions that cannot be defended from objections cannot be held very steadily or for very long. The connection with freedom of discussion is that without such freedom it is likely that people will be unable to defend their true opinions from objections, and so [21c] without freedom of discussion we would not properly hold onto true opinions.

More importantly, Mill claims that, "assuming that the true opinion abides in the mind, but abides as a ... belief independent of ... argument—this is not the way in which truth ought to be held by a *rational* being. This is not *knowledge* of the truth. *Truth*, thus held, is but one superstition the more, accidentally clinging to the words which enunciate truth" (par. 22, my ital.). In short, we may say that [21a1] holding a true opinion without argument amounts to holding it without a proper understanding and appreciation ("knowledge") of the reasons why it is true. In fact, as Mill argues in the next paragraph, [21a11] "truth depends on a balance to be struck between two sets of conflicting reasons" (par. 23), whereas for him [21a12] holding a true opinion without argument seems to mean that "a person assents undoubtingly to what they think true, though he has no knowledge whatever of the grounds of the opinion, and could not make a tenable defence of it against the most superficial objections" (par. 22). Implicit in this last proposition is Mill's concept of argument; for if we ask why it is that holding a true opinion without argument means holding it without knowing its supporting reasons and without being able to defend it from objections, the answer is clearly that [21a121] an argument just is an attempt to justify a conclusion by giving reasons in support of it or defending it from objections. The connection with freedom of discussion is that obviously [21a21] without freedom of discussion people will not ask each other why they hold the opinions they do and will not advance objections when they happen to disagree, and hence [21a2] without freedom of discussion even true opinions will be held without argument, and therefore finally [21a] without freedom of discussion true opinions will be held without a proper understanding and appreciation of their supporting reasons.

As I shall discuss later, Mill is here obviously presupposing a dialectical conception of argument.[2] If this commitment to dialectics is implicit, his practice of it is explicit; for, as he did before in the first strand of his argument, he now goes on to defend the illative tier of his second strand from objections.

The first objection (par. 23) is that the example of geometrical demonstration shows that an understanding of supporting reasons is enough and that there is no need to examine objections.

Mill replies (par. 23) that although this is true of mathematics, it is not so of other subjects ranging from natural philosophy and forensic oratory "to morals, religion, politics, social relations, and the business of life"; in all such subjects, [{23}[3], or 21a11] "truth depends on a balance to be struck between two sets of conflicting reasons." This view of truth means partly that [{23}1] "he who knows only his own side, knows little of that"; for [{23}11] "if he is ... unable to refute the reasons on the opposite side; if he does not so much as know what they are, he has no ground for preferring either opinion." That is, [{23}111] the evaluation of supporting reasons is always comparative; good reasons for an opinion are always reasons that are better than the reasons for the opposite; the evaluation of supporting reasons presupposes the evaluation of objections. Besides this conception of reason assessment, Mill advances a view that amounts to a formulation and a justification of the principle of charity: [{23}2] the arguer "must know them [objections] in their most plausible and persuasive form; he must feel the whole force of the difficulty"; for [{23}21] "else he will never really possess himself of the portion of truth which meets and removes that difficulty"; that is, [{23}211] "that part of truth which turns the scale ... is [n]ever really known, but to those who have attended equally and impartially to both sides, and endeavoured to see the reasons of both in the strongest light."

Another objection (par. 24) goes like this. Knowledge of objections to a particular opinion may be important for experts and professional "philosophers and theologians," but not for "common men" and "simple minds"; for these it is enough to know that *the experts*

---

[2] For the relevant sense of dialectics, cf. Johnson 2000a; Hansen 2002; and chapter 4 above.

[3] Here and elsewhere braces are used to treat numbers larger than single digits as a unit rather as an ordered pair resulting from the application of the standard labeling system introduced in chapter 2 above. That is, here '23' designates the initial label for a particular conclusion, rather that the third premise supporting proposition 2.

know, understand, and can answer the objections. This may be labeled the objection from the elite-mass distinction, especially since Mill himself uses these two terms in his reply (par. 25).

Mill replies to this second objection as follows (par. 25). If common people are to be in a position to know "that all objections have been satisfactorily answered," then some experts must be free to present and defend objections, so that others can satisfactorily answer them; but if there is such a freedom of discussion among experts, "in the present state of the world, it is practically impossible" to prevent this freedom from extending to the common people as well; therefore, [21d] without freedom of discussion, common people could not come to know that the objections to an opinion have been answered.

The second subargument of Mill's second strand of his overall argument tries to justify the conclusion that [21b] without freedom of discussion, which generates objections, a true opinion loses its practical or emotional meaning: in "the absence of free discussion" (par. 26), a true opinion has no "influence on the character"; it ceases being "a vivid conception and a living belief"; and "the shell and husk only of the meaning is retained, the finer essence being lost."

This loss is shown by [21b1] the history of ethical and religious doctrines (par. 27): at the beginning their adherents hold their views with "meaning and vitality" because these views constantly need to overcome objections and difficulties; but after a creed becomes established "it almost ceases to connect itself with the inner life of the human being" because it does not have to overcome objections any longer.

[21b11] The history of Christianity provides an example. Nowadays most Christians pay lip service to most Christian doctrines but have "no feeling which spreads from the words to the things signified" (par. 28). On the other hand, early Christians "had a much livelier feeling of the meaning of their creed than they have had since" (par. 29); and strict Christians today have such lively feelings only with respect to the peculiar distinguishing characteristics of their sect, rather than the common beliefs of all Christians sects.

Besides moral and religious doctrines, [21b2] "the same thing holds true, generally speaking, of all traditional doctrines—those of prudence and knowledge of life" (par. 30) that have reached the status of truisms, proverbs, or common sayings: although one cannot "truly learn" the "full meaning" of such a doctrine except by experience, it can be understood better and more deeply if one is "accustomed to hear it argued *pro* and *con*."

However, it could be objected (par. 31) that Mill's argument seems to presuppose a conception of truth, knowledge, and rationality such that absence of consensus is a necessary condition for truth and knowledge; that "as soon as mankind have unanimously accepted a truth, ... the truth perish[es] within them."

Mill replies (par. 32) that it is inevitable, indispensable, and good that human progress should lead to an increase of uncontested doctrines, to an increase of consensus. But progress has some drawbacks, the key one of which is "the loss of so important an aid to the intelligent and living apprehension of a truth, as is afforded by the necessity of explaining it to, and defending it against, opponents." This drawback can and should be minimized if we "provide a substitute for it; some contrivance ...," so to speak.

This substitute (par. 33) can be found partly by re-appropriating "Socratic dialectics" and "the school disputations of the Middle Ages," partly by cultivating the art of "negative logic" and "negative criticism," and partly by encouraging and respecting dissent when it happens to exist. For "in any but the mathematical and physical departments of speculation ... no one's opinions deserve the name of knowledge, except so far as he has either had forced upon him by others, or gone through of himself, the same mental processes which could have been required of him in carrying on an active controversy with opponents."

Synthesizing the two subarguments, as well as the illative and dialectical components of Mill's second strand, we may this. [2] Suppression of discussion is undesirable because [21] without the freedom of discussion of true opinions, many harmful consequences are likely to follow: we will be unable [21a] to properly understand and appreciate the reasons supporting them (par. 23); [21b] to experience their full practical or emotional meaning (par. 26-30); [21c] to hold on to those opinions in the first place (par. 22); and [21d] to come to know that the experts can satisfactorily answer the objections (par. 24-25). It is true that [22] one can raise various objections to freedom of discussion: that [22a] holding true opinions without arguments is enough (par. 22); that [22b] knowing only the supporting reasons is enough (par. 23); that [22c] mere knowledge *that* the objections can be answered by the experts is enough (par. 24-25); and that [22d] Mill's own conception seems to have the paradoxical consequence that absence of consensus is a necessary condition for knowledge. But [23] these objections are all wrong: [231] they presuppose an unsound conception of truth, knowledge, and rationality; in particular, [2311] regarding the fourth objection, the correct thing to say is that absence of consensus is a natural precondition for acquiring knowledge, as well as an artificial but effective instrument for learning or re-learning it, for teaching or disseminating it, and for expanding or deepening it.

10.4 Avoiding One-sidedness and Appreciating Partial Truths

So far Mill has discussed the case when we are trying to discover whether an opinion is true and the case when it is believed to be true, and he has examined explicitly the harmful effects of suppressing discussion, and implicitly the beneficial effects of freedom of discussion. Next he takes up the more usual case when an opinion is partly true and partly false (par. 34-39).

"One of the principal causes which make [3] diversity of opinion advantageous" (par. 34) is that [31] the most common case is "when the conflicting doctrines, instead of being one true and the other false, share the truth between them; and [32] the nonconforming opinion is needed to supply the remainder of the truth, of which the received doctrine embodies only a part." Now, the last case is the most common because "[311] in the human mind, one-sidedness has always been the rule, and many-sidedness the exception." On the other hand, the nonconforming opinion is needed because [321] "so long as popular truth is one-sided, it is more desirable than otherwise that unpopular truth should have one-sided assertors too; such being usually the most energetic, and the most likely to compel reluctant attention to the fragment of wisdom which they proclaim as if it were the whole."

For example (par. 35), [321a1] in the eighteenth century, when elites and masses alike were infatuated with the refinements of civilization and modernity, Rousseau came along to extol the virtues of a simple life and of the state of nature and to bemoan "the enervating and demoralizing effect ... of artificial society." And [321a2] Rousseau's doctrine was "salutary" because [321a21] it corrected the one-sidedness of the prevailing views even though "they were nearer to [the truth]; they contained more of the positive truth, and very much less of error."

Another example comes from politics (par. 36). [321b] In this domain a conservative and a progressive or reformist party "are both necessary elements of a healthy state of political life." For [321b1] "truth, in the great practical concerns of life, is ... a question of the reconciling and combining of opposites"; but [321b21] "very few have minds sufficiently capacious and impartial to make the adjustment with ... correctness"; and so [321b2] such reconciliation and combination can only be made "by the rough process of a struggle between combatants fighting under hostile banners." [321b3] In this struggle, each party "derives its utility from the deficiencies of the other; but it is in great measure the opposition of the other that keeps each within the limits of reason and sanity." Now, the reason why truth is a synthesis of opposites is that [321b11] in oppositions such as the following, we need to give their due to both elements: "to democracy and to aristocracy, to property and to equality, to co-operation and to competition, to luxury and to abstinence, to sociality and individuality, to liberty and discipline."

It could be objected (par. 37) that "*some* received principles ... are more than half-truths. The Christian morality, for instance, is the whole truth on that subject."

Mill replies (par. 37) by distinguishing several senses of this phrase. [{37}1] If Christian morality means that of the New Testament, then it is not the whole truth because [{37}11] the New Testament needs to be supplemented by parts of the Old Testament and to be interpreted in the context of Greek and Roman morality. [{37}2] If Christian morality means the theological morality articulated by the Church Fathers, then "it is, in many important points, incomplete and one-sided" because, for example, [{37}21] "its ideal is negative rather than positive; passive rather than active ... Abstinence from Evil, rather than energetic Pursuit of Good"; [{37}221] "it holds out the hope of heaven and the threat of hell," and so [{37}22] it has "an essentially selfish character"; [{37}23] "it is essentially a doctrine of passive obedience; it inculcates submission to all authorities found established"; and [{37}24] "duty to the State ... is scarcely noticed or acknowledged."

Mill's reply continues as follows (par. 38). [{37}3] If Christian morality means "the doctrines and precepts of Christ himself," then "many essential elements of the highest morality are among the things which are not provided for, nor intended to be provided for, in the recorded deliverances of the Founder of Christianity." For example, "I much fear that [{37}31] by attempting to form the mind and feelings on an exclusively religious type ... there will result ... a low, abject, servile type of character, which ... is incapable of rising to or sympathising in the conception of Supreme Goodness."

The final conclusion of Mill's reply to this objection is implicit but obvious, namely that [{37}, or 321c1] Christian morality is only a partial truth. However, he does explicitly add (par. 38) that [321c2] "it is not necessary that in ceasing to ignore the moral truths not contained in Christianity men should ignore any of those which it does contain. Such prejudice, or oversight, when it occurs, is altogether an evil; but it is one from which we cannot hope to be always exempt, and must be regarded as the price to be paid for an inestimable good. [321c21] The exclusive pretension made by a part of the truth to be the whole, must and ought to be protested against; and if a reactionary impulse should make the protestors unjust in their turn, this one-sidedness, like the other, may be lamented, but must be tolerated." Thus, [321c] the example of Christian morality also supports the claim that [321] it is desirable to counteract one partial truth with another.

Mill ends his discussion (par. 39) of partial truth with considerations analogous to those with which he began it. [{39}11] Freedom of discussion will not put an end to sectarianism, partisanship, and one-sidedness because [{39}1111] most people have a "narrow capacity" of mind and lack impartiality or judiciousness, and so [{39}111] most truths will always be advocated one-sidedly. However, because [{39}1211, or {39}1111] most people are so narrow-minded, [{39}121] "truth has no chance but in proportion as every side of it ... not only finds advocates, but is so advocated as to be listened to"; thus, [{39}12] freedom of discussion is the only means of coping with the problem of partisanship. It follows that [{39}1] "not the violent conflict between parts of the truth, but the quiet suppression of it, is the formidable evil"; and so [{39}, or 3] diversity of opinion is a formidable good.

The chapter ends with a discussion (par. 44) of a general objection. This final objection is that freedom of discussion is proper only if an opinion is expressed with temperance and fairness. Mill replies that, first, there is no practical test for determining whether an opinion is being expressed with fairness and temperance. Second and more importantly, even when it is clear that an opinion is being expressed with unfairness (e.g., with sophistry, distortion, and concealment), this unfairness is not a good reason for preventing its expression because it "is so continually done in perfect good faith, by persons who are not considered ... morally culpable; and still less could law presume to interfere." Third, with regard to intemperate expression (e.g., "invective, sarcasm, personality"), it would be a good idea to denounce its occurrence on both sides, though not by legal means; but there is an asymmetry between a prevailing and an unpopular opinion, insofar as the former can be criticized effectively only "by studied moderation of language, and the most cautious avoidance of unnecessary offence," whereas "unmeasured vituperation employed" against an unpopular opinion is effective; so, "it is far more important to restrain this employment of vituperative language" when directed against unpopular opinions than against prevailing ones.

10.5 Conclusions: Structure, Contents, and Epistemology

The numbering system discussed in chapter 2 and utilized in this chapter has obviously been an aid in the reconstruction of Mill's long and complex series of arguments. But the numbers assigned to the various propositions are also the tip of an analytical iceberg that could be studied at great length.

One general interpretive issue is the question of how many distinct arguments or subarguments Mill is advancing. Part of this question hinges on how one regards a reply to an objection, whether such a reply is regarded as the dialectical component of an argument that also has an illative tier or as a distinct subargument in its own right. A good example of a distinct subargument is Mill's reply to the first objection in the second strand of his argument (par. 23); but note that this reply can be grafted onto the illative tier by identifying the main conclusion of the reply (proposition 23) with a key premise of that illative tier (proposition 21a11). Overall, in accordance with the reconstruction above, we can count replies to ten objections, five in the first subargument, three in the second, one in the third, and one general.

There is also the question of whether Mill's overall argument subdivides into three main strands, with the second one in turn subdivided into two main substrands, or whether it is best to view it as consisting of four main strands. The former interpretation corresponds to the way Mill presents his argument in the course of his exposition and to my reconstruction above; the latter corresponds to a suggestion made by Mill himself when he summarizes his main argument at the end of the chapter (par. 40-43). Ultimately this question reduces to the question of the relationship between the two main parts of the second strand, of how independent of each other they are viewed. My labeling above presumes that each part is independent of the other insofar as the conclusions of the two parts (propositions 21a and 21b) are largely independent of one another; but note that they also get combined into the key claim of the second strand, which I have labeled proposition 21.

A similar question arises with regard to the main strands (whether three or four) of the overall argument. How do they relate? The reconstruction above indicates that the final conclusion of the first strand speaks of "[1] the peculiar evil of silencing the expression of an opinion" (par. 3); that is, it claims that the suppression of discussion is evil. The final conclusion of the second strand, implicit in Mill's own exposition, I have formulated as the proposition that [2] suppression of discussion is undesirable. And the final conclusion of the third strand tries to "make [3] diversity of opinion advantageous" (par. 34); that is, it claims that diversity of opinion is advantageous. Unless these three claims are plausibly interpreted as synonymous or equivalent, we would have three different arguments and not three strands of the same argument. But my reconstruction above has already ensured that they are seen as three strands, for the differences in these three propositions are essentially semantical or linguistic.

In fact, in this context the terms evil and undesirable may be regarded as synonymous to one another and as antonyms of the term advantageous. The diversity of opinion mentioned in the conclusion of the third strand may be taken to include not only the mental thinking of an opinion but also and especially its linguistic or public expression; and the opinions in question are not only opinions which Mill labels popular, prevailing, majoritarian, or conformist but also opinions which he labels unpopular, dissenting, minoritarian, or nonconformist; finally, once conflicting opinions are included, we have to include not only their expression but their discussion. So, in talking about diversity of opinion, the conclusion of the third strand is really talking about freedom of expression. Finally the suppression of discussion of which the conclusions of the first two strands speak is obviously the opposite of freedom of discussion. Thus, all three propositions are saying, negatively formulated, that suppression of discussion is undesirable, or positively expressed, that freedom of discussion is desirable.

On the other hand, while the conclusion of each strand is essentially the same, the key reasons are different, in the sense that they provide independent support. And since we have already seen that the second strand contains two independent subarguments, Mill is really giving us a total of four main reasons why freedom of discussion is desirable. They pertain, respectively, to the appreciation of the fallibility, of the supporting reasons, of the practical meaning, and of the partiality of one's opinions.

An additional interesting question is whether Mill's argument, or parts thereof, is a conductive argument. It would seem so. For the several supporting reasons are, and are intended to be, independently, cumulatively, convergently, and nonconclusively supportive of the claim on the desirability of freedom of discussion; and the various objections are obviously contrary reasons; and Mill's discussion of these objections is clearly an acknowledgment of their seriousness. It is true, of course, that he criticizes each one of the objections, so that ultimately he does not think that any of them have any real force. However, this criticism only means that he is justifying his balance of considerations claim, rather than leaving it unsupported, as a mere judgment call.

A good example of conductive argumentation is the second strand of Mill's argument. In fact, let us look at the summative reconstruction at the end of section 10.3 above. The pro-reasons are propositions 21a, 21b, 21c, and 21d; and the con reasons are propositions 22a, 22b, 22c, and 22d. Moreover, proposition 21 can be viewed as the pro-reasons acknowledgment claim, proposition 22 as the con reasons acknowledgment claim, and proposition 23 as the balance-of-considerations claim. Furthermore, these three claims are obviously linked, as obviously as the pro-reasons are independently supportive of or positively relevant to proposition 2, and the con reasons are independently disconfirming or negatively relevant. This analysis is in accordance with the general one I proposed earlier (chapter 8.7), on the basis of my analysis of Brooks's health-care argument.

Much more could be done to continue the analysis of the reconstructed argument from the point of view of argumentation theory, for the argument is complex and full of theoretically interesting features, and the considerations just made barely scratch the surface. However, here they will to suffice because there are some other points I want to discuss. In fact, the substantive content of Mill's argument is as interesting as its inferential structure.

First of all, the main conclusion is itself a claim in argumentation theory for it could be construed as the claim that *freedom of argument* is desirable. We have already seen that the main conclusion is essentially the claim that freedom of discussion is desirable, and it is obvious that by discussion Mill means not only the mere expression of opinions (whether popular or unpopular), but their advocacy and justification, their support and defense, in short argument. Now, this move from discussion to argument is warranted if one defines an argument as an attempt to justify a conclusion by supporting it with reasons or defending it from objections. And we have seen that Mill presupposes this conception of argument; indeed it appears as a latent or tacit premise in the first part of the second strand of his main argument, specifically as proposition 21a121. Such an assumption represents a second substantively interesting feature of Mill's argument.

However, note that I am not saying that Mill is supporting, defending, or arguing for this conception of argument, for obviously this tacit proposition is unsupported in this chapter of *On Liberty*; rather he is merely assuming or utilizing it. What he is arguing *for* is freedom of argument, whereas this conception of argument is what he is arguing *from* (among other things). Nevertheless, the fact that Mill accepts such a conception of argument is important, and I would elaborate this importance as follows (summarizing the relevant discussion from chapter 4 above).

This conception of argument is a key theoretical definition in an approach to argumentation that may be labeled dialectical. To be sure there are several varieties of dialectical approaches. For example, in regard to the definition of argument, the Millean conception may be called a "moderately" dialectical definition, insofar as it makes the dialectical tier of defense from objections a sufficient but not necessary condition for argument; some recent scholars[4] have advanced this type of definition. Others advance a "strongly" dialectical definition that makes the reply to objections a necessary but not sufficient condition; for example, Johnson (2000a, 168) defines an argument as an attempt to persuade someone of the truth of a conclusion by supporting it with reasons *and* defending it from objections. Some advocate a "hyper" dialectical conception that makes the defense from objections both a necessary and a sufficient condition for an argument; for example, the pragma-dialectical school defines an argument as "an attempt to meet the critical reactions of an antagonist, that is, to take away anticipated objections and doubt."[5] And then there are the many "nondialectical" definitions, for which the dialectical tier is neither sufficient nor necessary; they define an argument in terms of only an illative core, namely a set of propositions whose purpose is that the premises are intended to provide support, evidence, or reasons for the conclusion.[6]

There are at least two reasons for placing Mill in the moderately dialectical category of this taxonomy. First, at the beginning of the first part of the second strand of his argument (par. 22), Mill apparently equates believing something true *without* knowing supporting reasons *and without* knowing how to defend it from objections with believing it *without* argument; he does *not* equate lack of argument with lack of supporting reasons *or* lack of replies to objections; hence, believing with argument must be believing with *either* supporting reasons *or* replies to objections. Secondly, in the same context (par. 23), Mill asserts that in everyday life, "three-fourths of the arguments for every disputed opinion consist in dispelling the appearances which favor some opinion different from it."

---

[4] Goldman 1999, 131; Reed 2000, 1. Cf. chapter 4 above.

[5] Rees 2001, 233. Cf. Finocchiaro 1980a, 419; Eemeren and Grootendorst 1992, 73; Snoeck Henkemans 1992, 179; Eemeren, Grootendorst, Jackson, and Jacobs 1993, 12, 14; and cf. also chapter 4 above.

[6] Note that the classification scheme elaborated in this paragraph does *not* yield a classification of arguments but of approaches to argumentation theory, or theories of argumentation, or definitions of argument, or conceptions of argument.

A third substantively relevant element of Mill's discussion is his mention of "Socratic dialectics" (par. 33). This is an explicit recognition that there exists a dialectical tradition in argumentation and that he regards himself to be part of that tradition. However, this reference is more than just a self-reflective dialectical pronouncement. For it occurs in the context of Mill's reply (par. 33) to the objection that he seems to be committed to what might be called the dissensus theory of truth, namely the paradoxical and almost self-contradictory view that disagreement about the truth value of an opinion is a necessary condition of its truth. Mill's reply (cf. proposition 22d1 above) amounts to saying that dissensus is not part of the conception or meaning of truth or knowledge, but a key part of the way by which are arrive at the truth and acquire knowledge. Now, let us replace the talk of dissensus by talk of argument, which would be an easy thing to do in light of the dialectical definition of argument. Then we get the thesis that argument is a key method in the search for truth and the acquisition of knowledge. And this is another interesting and important claim in argumentation theory.

Again this sketch of this three-fold substantive relevance of Mill's chapter to argumentation theory will have to suffice here, but the third just-mentioned thesis brings us to one further issue worthy of discussion; that is, the relationship between epistemology and argumentation theory. Let us begin by focusing on the following aspect of Mill's argument.

The first strand of Mill's argument grounds the desirability of free discussion on what I have called the principles of [13a1] corrigibility, [13a2] open-mindedness, and [13a3] epistemological modesty. These three principles could be further subsumed under the principle of fallibility since the "assumption of infallibility" (as Mill phrases it) conflicts with those principles. That first argument reduces to claiming that free discussion is good because it is in accordance with such principles, or alternatively, suppression of discussion is bad because it violates them.

In the second strand of Mill's argument, its main part grounds the desirability of free discussion on what I have called a conception of truth, knowledge, and rationality. The major elements of this conception are principles such as the following (par. 23): that [{23}, or 21a11] "truth depends on a balance to be struck between two sets of conflicting reasons"; that [{23}1, or 21a111] one "who knows only his own side, knows little of that," which may be called the principle of comparative reason-assessment; and that [{23}2, or 21a112] one "must know them [objections] in their most plausible and persuasive form; he must feel the whole force of the difficulty," which I have called a principle of charity. This part of the second strand reduces to claiming that suppression of discussion violates these principles, or alternatively, that these principles require freedom of discussion.

The third strand of Mill's argument grounds the desirability of free discussion on principles such as the following. One is that "[311] in the human mind, one-sidedness has always been the rule, and many-sidedness the exception" (par. 34); this might be called the human predicament of partisanship. Another is that [{39}121] "truth has no chance but in proportion as every side of it ... not only finds advocates, but is so advocated as to be listened to" (par. 39); this principle might be called the dialectical antidote to partisanship. A third is the claim that [321c21] "the exclusive pretension made by a part of the truth to be the whole, must and ought to be protested against; and if a reactionary impulse should make the protestors unjust in their turn, this one-sidedness, like the other, may be lamented, but must be tolerated" (par. 38). Again, Mill's argument reduces to claiming that these principles require freedom of discussion, or alternatively that the suppression of discussion disregards them.

Such claims might be called epistemological principles, if we define epistemology as the study of the nature of knowledge and related concepts such as truth, rationality, reason assessment, fallibility, corrigibility, open-mindedness, and impartiality. Now, let us see what follows if in Mill's main conclusion we replace the talk of discussion by that of argument, and we rephrase it as the claim that freedom of argument is desirable, which is a claim in argumentation theory; then it seems that he is basing a conclusion in argumentation theory on epistemological principles, and this might be taken to suggest that argumentation theory has to be based on epistemology, and so perhaps that the latter has some kind of priority over the former. In turn, this might be taken to lend support to those scholars who advocate an "epistemic" approach to argumentation theory.[7]

On the other hand, the evaluation of the methodological principles of these epistemically oriented scholars would require that one look at the arguments they advance, and so it would be an exercise in argumentation theory, or more precisely in argument analysis. Indeed, even before evaluating them, one would want to make sure one understood them fairly, and such interpretation would be another exercise in argument analysis. Thus, to be in the position to say that the epistemic approach to argumentation theory is correct would in some important sense presuppose argument analysis, and ultimately argumentation theory; and this suggests that argumentation theory has some kind of priority over the theory of knowledge. This claim, in turn, is along the lines of Toulmin's approach. For as we saw in chapter 1, according to Toulmin, epistemology is the branch of logical theory (i.e., argumentation theory) that studies a particular class of arguments, namely arguments for knowledge claims. Now, in the same spirit we can think of the class of arguments justifying *epistemological* claims, and suggest that the study of this related class of arguments would be another, related branch of logical theory.

Thus, the real issue is that of the proper relationship between logic and argumentation theory on the one hand, and epistemology and the theory of knowledge on the other. One next step in this investigation would be to study the details of the methodological principles of the advocates of the epistemic approach to argumentation theory, in order to understand what they mean; in particular whether they mean something similar to the claim that epistemology is prior to argumentation theory; and in any case to assess whether their methodological principles are correct. Obviously that is a future task that does not fall within the scope of the present investigation. However, to anticipate or preview that further investigation, one other consideration is worth making.

Going back to Mill's argument, let us recall that its argument-theoretical conclusion was based on epistemological premises. But now let us ask: How does Mill, or how could one, check the correctness of these premises? Is there any better way than to examine the arguments that can be advanced to justify them? In doing so, should we not use the dialectical conception of argument, and so ask for critical objections as well as supporting reasons? And in searching for, formulating, interpreting, and evaluating such justifying arguments, would not argumentation theory be immensely helpful? Is there any other discipline, field, or branch of learning that would be equally helpful? Could epistemology itself be such a field? If that were so, would we not be trying to justify epistemological principles by means of epistemological principles? Would there not be some circularity in that?

These are not simply rhetorical questions. In fact, not all circularity is vicious, but rather some is virtuous; thus, the epistemological circularity just mentioned may be of the virtuous kind, or may be so arranged with some care.[8] Moreover, insofar as such

---

[7] Biro and Siegel 1992; Feldman 1994; Freeman 2005; Goldman (1986, 82; 2003); Govier (1987, 271-92; 1992); Lumer (2005a; 2005b); Pinto (1994; 2001, 21-31).
[8] For details on this topic, see Alston 2005, 191-210; Bergmann 2004; Goldman 2003; Sosa (1997; 2011, 140-57); Van Cleve 1979.

epistemological circularity is problematic, the problem may be no more serious than the following similar circularity affecting logic and argumentation theory: the justification of principles of argumentation also requires concrete arguments, which in turn presuppose principles of argument; thus, if one is not careful one may fall into a vicious circle.

Thus, again, the proper way to formulate the problem is to ask for the proper relationship between argumentation theory and the theory of knowledge, whether either is prior to the other, or whether they are mutually reinforcing.

10.6 Summary

The second chapter of Mill's *On Liberty* has been reconstructed as a long, nuanced, and complex argument for freedom of discussion. The argument consists of three subarguments, each possessing illative and dialectical components. The illative component is this. Freedom of discussion is desirable because, first, it enables us to determine whether an opinion is true, whereas its denial amounts to an assumption of infallibility; second, it improves our understanding and appreciation of the supporting reasons of true opinions, and our understanding and appreciation of their practical or emotional meaning; third, it enables us to understand and appreciate every side of the truth, given that opinions tend to be partly true and partly false and people tend to be one-sided. The dialectical component consists of replies to ten objections, five in the first subargument, three in the second, one in the third, and one general.

This reconstruction has been presented not only as textually accurate, but also as logically explicit and transparent. For this reason, it may be viewed as a contribution to argumentation theory, at least if we regard the interpretation of arguments as an important part of argumentation theory, and if the latter is practiced in accordance with the applied-logic or historical-textual approach. But Mill's argument (as reconstructed) is itself a contribution to argumentation theory because its main conclusion, a principal premise, and a key claim may be construed as claims in argumentation theory; indeed Mill's theoretical claims represent a good example of the dialectical approach in argumentation theory. Thus, Mill's argument is a meta-argument, and my analysis of it is a case study in the meta-argumentation project—the study of meta-arguments proposed in chapter 1 above, and being carried in this book.

Finally, Mill's argument is also significant because it raises in a vivid and striking manner the issue of the viability of an epistemic or epistemological approach to argumentation theory. Additionally, it suggests the reverse Toulminian possibility of an argument-theoretical approach to epistemology, and more generally the question of the relationship between epistemology and argumentation theory.

Chapter 11
Mill on Women's Liberation

We have seen in the last chapter that Mill's discussion of freedom of thought in his essay *On Liberty* is not only a famous meta-argument, but also almost a theoretical meta-argument for the simple reason that the topic can be easily seen to be freedom of argument. Thus, that meta-argument represents a good transitional case between part ii and part iii of this book. In any case, the added value was welcome and enhanced the significance of that case study. However, Mill also published other works whose content and classic status makes them equally famous and equally argumentative, but whose lessons are perhaps less theoretical and more practical. From the point of view of our meta-argumentation project that is not a drawback, but almost an advantage since we want to study meta-arguments of many different kinds and from many different contexts. Mill's book on *The Subjection of Women* provides one such example.

11.1 Introduction

Mill's essay on *The Subjection of Women* is well-known as an important contribution to the cause of women's liberation and to feminist theory. Some scholars also recognize the essay as one of Mill's "finest pieces of argument" (Okin 1988, v). However, when it comes to understanding, assessing, and appreciating the details and nuances of Mill's argument, we find little beyond the point that it consists of two main parts: the first aims "to disarm his opponents" (Okin 1988, xi) by showing that "there is absolutely no reason to suppose that [women] are not" (Ryan 1997, xxxix) the equals of men; the second part articulates a number of reasons in favor of women's liberation and equality. On the other hand, Hans Hansen (2006) has moved the analysis of Mill's argumentation in general to a higher and more sophisticated level, but I believe even he has missed the crucial meta-argumentative component of this particular Millean work.

Mill begins his essay with a clear and explicit statement of his aim. It is, he says, to argue "that the principle which regulates the existing social relations between the sexes—the legal subordination of one sex to the other—is wrong in itself, and now one of the chief hindrances to human improvement; and that it ought to be replaced by a principle of perfect equality, admitting no power or privilege on the one side, nor disability on the other" (par. 1).[1] For the purpose of our discussion, a number of approximations and conventions will be useful.

Mill's conclusion has two equivalent aspects: that the principle of the subordination of women (to men) is wrong; and that the principle of equality of women (to men) is right. That is, in this context the principle of subordination and the principle of equality are taken as opposites. Moreover, for short, we may drop the term "principle," and speak more simply of subordination or equality. Furthermore, since Mill also uses the term subjection as synonymous to subordination, we shall follow him in treating these two terms as equivalent. Thus, arguments for the equality of women are simultaneously arguments against subjection; arguments against equality are simultaneously arguments for subjection; and objections to one are simultaneously arguments for the other.

Despite the clarity and explicitness of his statement of the conclusion, Mill does not go

---

[1] References to chapter 1 of Mill's *Subjection of Women* (e.g., Mill 1988; 1997) will be given by indicating the paragraph rather than page number.

on immediately to elaborate supporting reasons or replies to objections. The first supporting reason does not come until chapter 2 of his book, where he tries to show that the subordination of women is wrong because it produces considerable evil and harm in marriage. The second reason comes in chapter 3, where he argues that subjection is wrong because of its harmful effect in the public sphere of employment and citizenship. And a third positive reason is found in the fourth and last chapter, where he suggests that subordination is wrong because its abolition would result in considerable benefits and advantages in marriage, social relations, and the psychological well being of women. These reasons may be regarded as the illative tier of Mill's overall argument.

Mill articulates these reasons after he tries to refute the objection that the subordination of women is right because it corresponds to the nature of men and women. His reply to this objection is found at the end of chapter 1 (par. 17-24), and it may be regarded as the dialectical tier of his overall argument.

However, at the beginning of chapter 1, immediately after his clear and explicit statement of the conclusion, Mill engages in a number of methodological or metacognitive reflections (par. 2-17) required by the special features of the situation at hand. These considerations are the focus of this chapter since they are best viewed in terms of the notion of meta-argument.

11.2 Argument Unnecessary and Counterproductive?

To begin with, in this case the conclusion to be justified conflicts with deep feelings and strong emotions. But, Mill asserts, "so long as an opinion is strongly rooted in the feelings, it gains rather than loses in stability by having a preponderating weight of argument against it" (par. 2). If we accept this principle, it would seem to follow that argumentation is going to be not only ineffectual, but counterproductive.

Additionally, Mill is acutely aware of the fact that the conclusion he advocates conflicts with "almost universal opinion … universal usage" (par. 3) and "established custom" (par. 4). But, he fears, "a cause supported on the one hand by universal usage, and on the other by so great a preponderance of popular sentiment, is supposed to have a presumption in its favour, superior to any conviction which an appeal to reason has power to produce in any intellects but those of a high class" (par. 3). If this principle is correct, it would follow that the question is settled and there is nothing to argue about.

But are these principles correct? Do we have to accept these two principles which ground these skeptical, misological conclusions? That is, here we have two arguments about the role of argumentation in assessing the subjection of women: they advance conclusions that limit the role of argumentation in some ways; and they use premises some of which are factual and uncontroversial, but some of which are themselves principles of argumentation. It should come as no surprise that one may decide to question these principles themselves.

In fact, according to Mill, although these principles cannot be summarily dismissed, they are one-sided and too extreme. In his own words, "I do not therefore quarrel with them for having too little faith in argument, but for having too much faith in custom and general feeling" (par. 4). We need a more balanced and multi-faceted principle. Mill formulates it as follows: "established custom, and the general feeling, should be deemed conclusive against me, unless that custom and feeling from age to age can be shown to have owed their existence to other causes than their soundness, and to have derived their power from the worse rather than the better parts of human nature" (par. 4). That is, custom and feeling should not be allowed to prevail unconditionally or simpliciter; but rather they can prevail if and only if they cannot be shown to derive from causes that are logically unsound and morally questionable. In particular, if one can show that the custom and feeling which favor the subjection of women derive from logically or morally questionable causes, then they will lose their presumptive force and the issue can be argued about on its merits. This is turn would open the door to argumentation.

However, although this balanced and many-sided principle thus rescues argumentation, it also places a very heavy burden on the arguer. For in this situation, Mill is saying that he has to do something above and beyond what is required in normal argumentation. What is normally required is already subject to a very high standard.[2] In Mill's own words, "before I could hope to make any impression, I should be expected not only to answer all that has ever been said by those who take the other side of the question, but to imagine all that could be said by them ... and besides refuting all arguments for the affirmative [i.e., the subjection of women], I shall be called upon for invincible positive arguments to prove a negative [the denial of subjection]" (par. 3). I interpret Mill to be saying that to justify a claim C, the normal requirements are (1) to answer all objections that have already been advanced against the claim by its opponents; (2) to invent potential objections and answer them; (3) to advance reasons supporting the claim; and (4) to defend these supporting reasons from objections. These requirements could also be interpreted in terms of the illative tier and the dialectical tier of an argument. But these tiers and these requirements are not enough in certain cases, for, in Mill's words, "even if I could to all this, and leave the opposite party with a host of unanswered arguments against them, and not a single unrefuted one on their side, I should be thought to have done little" (par. 3). The additional requirement is the one mentioned above.

This is not just an additional requirement that increases the burden of proof, but a new *kind* of requirement. For it does not seem to involve the examination of evidence for and against the truth of the claim in question, but rather an examination and critique of the causes that produced the prevailing belief with which the claim conflicts. Let us call this additional requirement *causal undermining*. In fact, the main argument in chapter 1 of Mill's essay fits this scheme and performs this function. Let us analyze the details of that argument.

11.3 Causal Undermining

One key claim is that "the inequality of rights between men and women has no other source than the law of the strongest" (par. 5). For, Mill speculates, this inequality "arose simply from the fact that from the earliest twilight of human society, every woman (owing to the value attached to her by men combined with her inferiority in muscular strength) was found in a state of bondage to some man" (par. 5). This speculation, in turn, is in accordance with (and supported by) the fact that history shows "how entirely, in former ages, the law of superior strength was the rule of life; how publicly and openly it was avowed" (par. 7). From that initial origin, the subordination of women was kept in existence until our present age by the fact that it generally happens that "institutions and customs which never had any ground but the law of force, last on into ages and states of general opinion which never would have permitted their first establishment" (par. 8).

With this argument (par. 4-8), Mill is stating and supporting his causal-historical explanation of the subjection of women. The root cause is claimed to be the so-called law

---

[2] My account may be compared and contrasted to the one given by Hansen (2006, 100-102). He also speaks of Mill's "standard," and interprets it in terms of a tripartite structure that corresponds to what I am calling the normal requirements but does not take into account exceptional situations like the present one.

of the strongest, that is, the superior physical strength of men. But he is clear that this by itself does not amount to showing that the subjection of women is wrong;[3] to think so would be tantamount to committing the so-called genetic fallacy, to use present-day terminology. Rather, this argument is merely one step of the main meta-argument of chapter 1 trying to show that, in accordance with the nuanced and balanced principle stated above, custom and feeling do not provide a conclusive presumption in favor of the subjection of women and the issue must be argued on its own merits. That principle stipulated a second condition besides the causal explanation, namely a critique of the logical soundness and moral propriety of this cause. Accordingly, Mill tries to show that this cause is neither. He does this dialectically so to speak, namely by replying to possible objections.

The first objection is that the law of the strongest, when applied to the relations between men and women, is natural. In Mill's words, one could object that "a comparison cannot fairly be made between the government of the male sex and the forms of unjust power ... adduced in illustration of it, since these are arbitrary, and the effect of mere usurpation, while it ... is natural" (par. 9). Mill's illustrations had been slavery, absolute monarchy, and military despotism (par. 8).

Mill's reply is that this objection presupposes a flawed concept of what is natural. In his own words, for those who raise this objection, "unnatural generally means only uncustomary, and ... everything which is usual appears natural. The subjection of women to men being a universal custom, any departure from it quite naturally appears unnatural" (par. 9). At this stage in the discussion, Mill does not say what the correct conception of natural is, and whether or not the subjection of women is natural in that correct sense, and why. Later (par. 18-24), he has more penetrating things to say about that issue. Here his main point seems to be that to claim that the subordination of women stemming from the law of the strongest is natural, says nothing more than that it is customary and universal. But he has already agreed that the subjection of women is an established custom and universal practice; indeed his predicament derives from this fact together with the principle that universal custom creates a presumption stronger than reason or argument. His way out of this predicament is not to deny this fact, but to deny this principle, and replace it by a more nuanced principle. In short, the question is not whether the subjection of women is customary, but whether it is right or natural, and to argue that it is natural because it is customary is to beg the question.

Another objection brings us closer to the question of moral propriety. It is that the application of the law of the strongest to the relationship between men and women is accepted by women. In Mill's words, one could object that "the rule of men over women differs from all ... others in not being a rule of force: it is accepted voluntarily; women make no complaint, and are consenting parties to it" (par. 10).

Mill replies to this objection by saying that as a matter of fact, many women do not find their state of subjection acceptable. This is shown by the fact that in England there had been petitions to Parliament for women's suffrage; in the United States there was an organized party favoring women's liberation; and similar developments had occurred in France, Italy, Switzerland, and Russia. Secondly, "many more women ... silently cherish similar aspirations; but ... are ... strenuously taught to repress them as contrary to the properties of their sex" (par. 10). Thirdly, "no enslaved class ever asked for complete liberty at once" (par. 10). Fourthly, there would be many more complaints by women if complaints were not a major cause for their being mistreated. Finally, "all causes, social

---

[3] "I have not yet shown that it is a wrong system: but everyone who is capable of thinking on this subject must see that even if it is, it was certain to outlast all other forms of unjust authority" (par. 8).

and natural, combine to make it unlikely that women should be collectively rebellious to the power of men" (par. 11).

In this reply, Mill seems to be admitting the relevance and probative force of the objection, but denying its truth. In this case, to deny that women find their subjection acceptable, is equivalent to asserting that they find it unacceptable. And in turn this assertion would lend probative weight to the counter-conclusion that the subjection, and the root case that produced it, are morally questionable.

With this critique, Mill has undermined the initial presumption in favor of the subjection of women due to custom and feeling, for he has now satisfied both conditions stipulated in his balanced guiding principle. That is, Mill has now completed his main meta-argument. This meta-argument starts with the premise that "established custom, and the general feeling, should be deemed conclusive ... unless that custom and feeling from age to age can be shown to have owed their existence to other causes than their soundness, and to have derived their power from the worse rather than the better parts of human nature" (par. 4). He then goes on to argue that the universal custom and feeling owe their existence to the law of the strongest, and this claim provides the second premise of the main meta-argument. Additionally, he argues that the law of the strongest is not really natural but customary, and it is endorsed by women not freely but under duress; and this critical claim provides a third premise. The conclusion is that, in his own words, "the preceding considerations are amply sufficient to show that custom, however universal it may be, affords in this case no presumption, and ought not to create any prejudice in favor of the arrangements which place women in social and political subjection to men" (par. 12). This meta-conclusion in turn opens the door to ground-level argumentation: "the question should not be considered as prejudged by existing fact and existing opinion, but open to discussion on its merits, as a question of justice and expediency" (par. 17).

11.4 Predictive Extrapolation

However, in this particular case, Mill thinks he can argue that there is a presumption against subjection (i.e., in favor of equality). In his words, "but I may go farther, and maintain that the course of history, and the tendencies of progressive human society, afford not only no presumption in favor of this system of inequality of rights, but a strong one against it; and that, so far as the whole course of human improvement up to this time, the whole stream of modern tendencies, warrants any inference on the subject, it is, that this relic of the past is discordant with the future, and must necessarily disappear" (par. 12).

Mill's argument in support of this claim is a historical prediction or predictive extrapolation: he detects a long-lasting historical trend; he claims that this trend is a good thing; he predicts that the trend will continue; but the subjection of women conflicts with this trend; and he concludes that the subjection of women will eventually disappear.

The trend is described and evaluated by Mill as follows: "the modern conviction, the fruit of a thousand years of experience, is, that things in which the individual is the person directly interested, never go right but as they are left to his own discretion; and that any regulation of them by authority, except to protect the rights of others, is sure to be mischievous" (par. 13). Here we have what may be called the principle of individual freedom, and Mill seems to be saying that this principle has been found to be good through historical experience and empirical observation. However, he also gives a deeper reason for the soundness of this principle, namely that "in all things of any difficulty and importance, those who can do them well are fewer than the need, even with the most unrestricted latitude of choice: and any limitation of the field of selection deprives society of some chances of being served by the competent, without ever saving it from the incompetent" (par. 14).

Mill's prediction that this trend will continue is less explicit, but it may be found in his call to action meant to ensure that the principle continues to be followed and is expanded to the case of women. He urges us that "if this principle is true, we ought to act as if we believed it, and not to ordain that to be born a girl instead of a boy, any more than to be born black instead of white, or a commoner instead of a nobleman, shall decide the person's position through all life" (par. 14).

However, as things stand, i.e., until the principle is expanded to this new area, "the disabilities, ... to which women are subject from the mere fact of their birth, are the solitary examples of the kind in modern legislation" (par. 15). This discrepancy between the subjection of women and the principle of individual freedom "raises a *prima facie* presumption on the unfavorable side, far outweighing any which custom and usage could in such circumstances create on the favourable; and should at least suffice to make this ... a balanced question" (par. 16).

In what sense is this a meta-argument? Its final conclusion is that the principle of individual freedom provides a presumption against the subjection of women. Its premises are the claims that this principle is widely practiced, generally beneficial, and specifically inconsistent with the subjection of women.

Let us compare this argument with the previous one, aimed at the causal undermining of the universal belief in subjection. The conclusion of that meta-argument was that established custom and general feeling do not provide a presumption is favor of the subjection of women. Its premises were that such custom and feeling are the result of the law of the strongest, and that this law is logically and morally questionable. There the ground-level argument was this: the subjection of women is right because it is in accordance with almost universal custom and feeling. That meta-argument was trying to show that this ground-level argument is inferentially unsound, at least in the sense of deductively invalid.

What is the ground-level argument in the case of the predictive extrapolation? Perhaps it is this: the subjection of women is wrong because it conflicts with the principle of individual freedom. If so, what is the predictive extrapolation saying about it? It seems to be claiming that this argument is inferentially sound, at least to some degree.

Thus, the causal undermining and the predictive extrapolation seem to constitute two complementary meta-arguments. They are complementary in the sense that one is retrospective and the other prospective. But they are also complementary insofar as from each Mill seems to want to draw the same further conclusion, namely that the issue should be argued about on its merits, by means of the presentation, defense, and criticism of supporting reasons and critical objections. In other words, the causal undermining and the predictive extrapolation are distinct subarguments of the same longer, more complex argument advocating the value of merit-based argumentation on the question at hand. Moreover, the two arguments are meta-arguments insofar as they are, respectively, about two identifiable ground-level arguments.

11.5 Conclusion: Meta-argumentative vs. Dialectical vs. Illative Tiers

But now a problem emerges. If this analysis is accepted, is not the causal-undermining meta-argument merely a negative evaluation of the corresponding ground-level argument about the presumptively decisive role of custom and feeling? If so, it would be a normal part of the dialectical tier of Mill's argumentation. And is not the predictive-extrapolation meta-argument merely a positive reason in support of Mill's main conclusion that subjection is wrong? In this case, it would be a normal part of the illative tier of Mill's argumentation. However, as we saw above, Mill claims that with these preliminary considerations he is doing something above, beyond, distinct from, necessary for, and prior to what is normally required in argumentation.

My solution of this difficulty is along the following lines. The distinction between the meta-level and the ground-level of argumentation is similar to the distinction between the illative and the dialectical tiers of argumentation. It is not a dichotomy or material difference such that a piece of reasoning belongs to one of these tiers in virtue of properties which it may possess or lack in a physical sort of way. Rather the distinction is a conceptual difference such that a piece of reasoning may or may not be analyzed from the point of view of one or more of these tiers depending on the purpose at hand.

Let me begin by elaborating this point with regard to the distinction between dialectical and illative tiers. This issue first arose in chapter 4, as a possible objection to the moderately dialectical definition of argument from the point of view of the hyper dialectical definition, and then again in chapter 5, with regard to the justification of the hyper dialectical definition. The latter discussion is perhaps worth repeating here.

Given any claim, one could always raise the question, what reasons if any there are in support of the claim. This question may be regarded as the prime or minimal objection to any claim. If one anticipates it, one constructs the illative tier and gives the supporting reasons even before the objection has actually been raised. Or one can wait until after the objection has been explicitly raised. In either case, the illative component can be interpreted as a part of the dialectical tier.

However, I do not think this establishes the exclusivity, or even the primacy, of the dialectical tier, as claimed by the proponents of pragma-dialectics. For in a way analogous to how one attempts to interpret the illative component in terms of the dialectical tier, one can try to do the reverse and reinterpret the dialectical tier in terms of the illative component.[4]

Consider an argument whose illative component consists of premises P-1 through P-n and conclusion C. And suppose the argument also has a dialectical tier with objections O-1 through O-k, respectively answered by replies R-1 through R-k. Now consider the conjunction of an objection and its corresponding reply, (O-j & R-j), or some appropriately reworded phrasing of it that might be needed for grammatical propriety. It seems to me that such a conjunction would constitute a reason supporting the conclusion C. It would be like saying that one reason for accepting the conclusion is that if one objects to it in such and such a way, such an objection would be incorrect; or collectively considered, one reason for accepting conclusion C is that all objections against it fail, i.e. that there are no objections to it. In other words, an objection to a conclusion C may be seen as a reason against it, a reason for claiming not-C; and if a reason R for not-C is a bad reason, then the claim that R is a bad reason for not-C, may be seen as a reason for C. Of course, such a reason would not be a conclusive reason, and to claim such conclusiveness would be to commit a damaging version of the fallacy from ignorance. But we are clearly dealing with reasons that, however strong, fall short of conclusiveness, and for such cases the explicit refutation of an explicit objection may be viewed as a supporting reason.

The upshot of these considerations is that while the presentation of supporting reasons may be regarded as a reply to a weak or minimal objection, the refutation of objections may be regarded as a weak or minimal supporting reason. There thus seems to be a symmetry between the illative and the dialectical tiers.

A similar symmetry seems to exist between the meta-argumentative level and the illative and dialectical tiers. First of all, as mentioned earlier, the dialectical tier can be

---

[4] Here I am repeating a discussion that has already arisen in chapters 4.7 and 5.3..

easily viewed as meta-argumentation insofar as it consists of replies to arguments against the conclusion C, and hence as arguments about arguments against C. Next, the illative tier consists of positive reasons, premises P-1 through P-n, supporting conclusion C; but the explicit statement of such an argument may be viewed as an implicit claim that this argument (P-1, ..., P-n, so C) is correct; the latter claim is a meta-argumentative assertion about a ground-level argument; and the considerations that made that original ground-level argument cogent are implicitly reasons supporting the meta-argumentative assertion.

The two points made so far involve viewing the dialectical tier and the illative tier in terms of the meta-argumentative tier.[5] Can we now do the reverse? Can we start with a meta-argument and view it in terms of illative and/or dialectical tiers. I believe we can. In fact, the two meta-arguments from Mill reconstructed above provide good examples of this process, which may be called the "grounding" of the meta-level. The causal-undermining meta-argument can be seen as a criticism of the objection by Mill's opponents based on the alleged probative conclusiveness of established custom and general feeling. On the other hand, the predictive-extrapolation meta-argument can be seen as a roundabout statement of a useful, if inconclusive, preliminary argument by Mill, grounding his conclusion that subjection is wrong on the fact that it contradicts the principle of individual freedom.

Thus, the meta-argumentative tier may be viewed as the meta-level reflection of the dialectical and illative tiers, and the latter two tiers as the ground-level embodiment of the meta-argumentative tier. However, just as the symmetry between the illative and dialectical tiers does not amount to a conflation or confusion of these two roles, the symmetry between the meta and ground levels ought not to lead to their conflation. Both symmetries may be viewed as refinements contributing to a better understanding of the relationship among the meta-argumentative, the illative, and the dialectical tiers of argumentation.

11.6 Summary

This chapter has focused on the first part of chapter 1 of Mill's *Subjection of Women*, where he engages in some metacognitive reflections, before going on to reply to a key objection in the rest of that chapter and to articulate three supporting reasons in the other three chapters. Mill begins by formulating two objections that seem to suggest that the subjection of women is a special topic where argumentation is counterproductive or superfluous. He answers these two objections by rejecting the two principles of argumentation on which they are based and replacing them by a more balanced and nuanced principle.

This principle places on him the heavy burden of causally undermining the universal belief in the subjection of women, to pave the way to arguments and objections on the merits of the issue. In accordance with this principle, he argues that the subjection of women derives from the law of the strongest, but that this law is logically unsound and morally questionable, and hence that custom and feeling provide no presumption in favor of the subjection of women. Additionally, Mills thinks that in this case he can also make a predictive extrapolation; accordingly he argues that there is a presumption against subjection based on the principle of individual freedom, which is widely accepted,

---

[5] This thesis may be taken to correspond to the account advanced by Boella, Gabbay, van der Torre, and Villata (2009), in which they attempt to "show how to use meta-argumentation as a general methodology for modeling argumentation" (p. 50). I hesitate to make the comparison, because (as noted in chapter 1.3) my approach is very different from theirs, which is formal, mathematical, aprioristic, and oriented toward artificial intelligence and computer science; but I have no doubt that there is a substantive similarity, that the similarity is important, and that it deserves further reflection.

generally beneficial, and inconsistent with subjection.

This predictive extrapolation and the earlier causal undermining constitute two complementary meta-arguments. Nevertheless, they may also be viewed, respectively, as the criticism of an objection, and the statement of a supporting reason, and hence as elements of the dialectical and illative tiers, rather than as a distinct meta-argumentative part of the overall argument. The chapter ends with an attempt to solve this difficulty, by stressing that there is a symmetry between meta and ground levels analogous to the symmetry between illative and dialectical tiers, and that meta-argumentation may be seen as a component of argumentation distinct from but related to the illative and dialectical components.

Chapter 12
Hume on Intelligent Design

We now come to a topic which may be said to have been anticipated since the first chapter, when I gave an ostensive definition of argumentative reality by quoting, among other examples, Aquinas's fifth argument for the existence of God. In fact, argumentation about God's existence represents a very robust strand of the history of thought, and so it is not surprising that it has generated its share of meta-argumentation. Recall that Aquinas's fifth argument makes some claims about design being present in nature and tries to show the need for the existence of an intelligent designer responsible for the presence of that design. Now, the most famous, significant, and instructive critique of the theological design argument is perhaps David Hume's *Dialogues Concerning Natural Religion*. Hume's critique thus represents a classic, famous meta-argument, and a natural candidate to be examined in the present investigation. Moreover, the study of Hume's *Dialogues* is very promising also because the subject matter continues to be especially important and relevant, as the ongoing controversy about "intelligent design" indicates.

12.1 Introduction: Barker's Meta-argumentative Analysis

There is another reason why the study of the critique of the design argument in Hume's *Dialogues* is especially promising. That is, it so happens that there already exists an insightful discussion of the meta-argumentative aspects of Hume's critique, by Stephen Barker (1989).[1] His meta-argumentative analysis is so relevant that before undertaking my own, I shall reconstruct Barker's account and then revise it appropriately.

In what follows in this section, whether or not I reproduce a direct quotation, it is to be understood that the views being expressed and stated are those in Barker's (1989) essay, unless I indicate otherwise.

The logic of argumentation in Hume's *Dialogues* has not been adequately appreciated. Although Kant's criticism of the teleological argument for the existence of God may have been more influential than Hume's, Hume did it first, and in any case, "the incisiveness of Hume's treatment far outshines that of Kant on this matter" (173).[2] Although the argument for design can be formulated as a deductive argument, for example Aquinas's "fifth way," such formulations are ultimately question-begging. "It is a distinctive merit of Hume's approach to the argument from design that he carefully avoids having Cleanthes formulate it as a deductive argument" (174).

Hume has Cleanthes formulate the argument as follows:

Look round the world, contemplate the whole and every part of it: you will find it to be nothing but one great machine, subdivided into an infinite number of lesser machines, which again admit of subdivisions to a degree beyond what human senses and faculties can trace and explain. All these various machines, and even their most minute parts, are adjusted to each other with an accuracy which ravishes into admiration all men who have ever contemplated them. The curious adapting of means to ends, throughout all nature, resembles exactly, though it much exceeds, the productions of

---

[1] O'Connor 2001 is also valuable, although his focus is on other aspects of Hume's argument. I have consulted a few other accounts of Hume's argument (e.g., Aiken 1948; Smith 1947, 57-76, 97-123), but found them unhelpful.

[2] In this section, references to Barker 1989 will be given in parenthesis by means of just the page number(s)

human contrivance—of human design, thought, wisdom, and intelligence. Since therefore the effects resemble each other, we are led to infer, by all the rules of analogy, that the causes also resemble, and that the Author of nature is somewhat similar to the mind of man, though possessed of much larger faculties, proportioned to the grandeur of the work which he has executed. By this argument *a posteriori*, and by this argument alone, do we prove at once the existence of a Deity and his similarity to human mind and intelligence. [Hume, *Dialogues*, II-5][3]

First, note that this is an *inductive* argument, in the following sense: "(I) ... such an argument is not deductively valid, and the person advancing it does not claim that it is so; the arguer claims merely that the premises increase significantly the probability of the conclusion. (II) The argument takes as its premises empirical propositions which are to be known by observation. (III) The conclusion is a proposition whose empirical content goes beyond that of the premises" (176). Moreover, it is an argument *by analogy*, in the sense that it has this structure: a, b, c, etc., have the property F and the property G; n has the property F; so, probably n has property G. Here, a, b, c, etc., are the known cases, of which there could be only one; n is the new case analogous to them and sharing property F; the conclusion infers that n also shares the property G; and the word "probably" indicates that the premises are claimed to merely increase the reasonableness of believing the conclusion. In short, "Cleanthes presents this as an inductive argument by analogy" (175).

In this type of argument, "its strength does not just depend upon its logical form" (176). Moreover, "we do not have available any 'decision procedure' to guide us ... judgments about the strength of an inductive argument by analogy cannot be made in that mechanical way" (177). However, that does not mean that we are "falling back here upon the idea that it is by dogmatic appeal to untestable intuitions that we decide" (178). Rather, we must take into account evaluative factors such as the following: the similarities among the known cases; the dissimilarities among them; the similarities between the known cases and the new case; the dissimilarities between them; and the scope of the conclusion. And then we must engage in "reasoning about arguments by analogy" (178).

Hume's *Dialogues* is highly instructive for this purpose because Cleanthes and his opponent Philo do engage in "a fruitful exchange" (178) that makes some "genuine intellectual progress" (178). But they do this not by allegedly deductive appeals to logical form, decision procedures, or infallible intuitions. "Cleanthes and Philo carry out their discussion in the *Dialogues* in another manner. Throughout, they employ reasoning by analogy" (179).

For example, consider what Cleanthes is doing when he defends his argument by mentioning that the "steps of a stair are plainly contrived, that human legs may use them in mounting ... [and] human legs are also contrived for walking and mounting" (Hume, *Dialogues*, II-9). "What Cleanthes is doing is to urge that there is an important analogy between two specimen arguments; let us call them A and B. A is the argument about the stairs ... the other argument (argument B) ... is a special case of the argument from design ... Cleanthes, in this part of his discussion, is putting forth an argument by analogy (we may call it argument C) ... which ... affirms that B must be a good argument because of its strong analogy to A, which we already recognize to be good" (179-80).[4]

This analysis is just an interpretation and does not mean that one is "endorsing

---

[3] In this chapter, references to Hume's *Dialogues* will be given in parenthesis by a sequence of a roman numeral and an arabic numeral, the first indicating the part number (as designated by Hume himself) and the second indicating the paragraph number within a given part. This particular passage corresponds to Hume (1947, 143; 1948, 17).

[4] This is reminiscent of Woods's (meta)argument by parity of reasoning, for which see Woods and Hudak 1989 and chapter 6.3 above.

Cleanthes's meta-argument concerning the status of his initial argument. Actually, Cleanthes's meta-argument is quite a bad argument, because B is not similar enough to A. Thus argument C does not succeed in defending argument A against the charge of being bad" (180).

"However, through the central portions of the *Dialogues* it is Philo who most extensively uses reasoning by analogy in order to reason about how Cleanthes's argument is to be evaluated. Philo compares Cleanthes' argument from design to a wide variety of other inductive arguments by analogy ... both strong and weak. Of course, his conclusion is ... that Cleanthes' argument is very weak" (180). That is, "on the one hand, Philo cites examples of strong arguments ... However, he urges, there is little analogy between Cleanthes' argument and these strong arguments ... Furthermore, Philo cites a rich variety of other possible inductive arguments by analogy that we can recognize as very weak ... Philo claims that Cleanthes' argument is very like [these], and therefore is weak too" (180).

Finally, let us ask, "what kind of reasoning by analogy is it that Cleanthes and Philo are employing in their meta-arguments?" (181). Such meta-arguments by analogy are non-inductive because some of their premises, as well as their conclusion, are non-empirical. That is, when we are evaluating an "argument like C, insofar as it deals with cases that are merely possible rather than actual, the similarities and differences referred to in [the evaluative factors] will not be empirically observed; they will be discerned by reflection. A second aspect in which argument C is less empirical than A and B concerns the empirical content of the conclusion. In A and B the empirical content of the conclusion does go beyond that of the premises, giving the conclusion a predictive aspect; while in C this is not the case, for its conclusion is a proposition about logical force and has no empirical content" (182).

In short, Barker's thesis is that Hume's *Dialogues* is a (1) powerful (2) non-inductive (3) meta-argument (4) by analogy, claiming that the design argument is an (5) inductive ground-level argument by analogy, which is (6) weak because (7) it is similar to many other arguments by analogy that are obviously weak and (8) it is dissimilar from many other arguments by analogy that are obviously strong. And the theoretical and methodological lesson is that one can and should employ non-inductive meta-arguments by analogy in evaluating inductive ground-level arguments by analogy.

12.2 Multiplicity of Barker's Claims and of Hume's Meta-arguments

As previously mentioned, my reconstruction of Barker's reconstruction of Hume's *Dialogues* will now be tested by actually examining this work. To this we now turn.

Let me begin by saying that I obviously agree that Philo's critique of Cleanthes's design argument is a meta-argument (Barker's claim no. 3). I also agree that Philo's meta-argument is powerful (claim no. 1) and non-inductive (claim 2). However, I reserve judgment on whether Philo's meta-argument is (primarily) an argument *by analogy* (claim 4). Similarly, there is no question that Cleanthes's design argument is inductive and ground-level (which are the first two parts of claim 5), but I am not sure that it really is an argument *by analogy* (the third part of claim 5). Finally, Barker is partly right in claiming (nos. 6, 7, and 8) that the content of Philo's meta-argument is that the design argument is weak because of its similarity to many other weak arguments and its dissimilarity to many other strong arguments; but there are other important parts of Philo's critique which argue for different flaws.

In other words, Barker has reconstructed only one strand of Philo's meta-argument, the strand relating to arguments by analogy. In the context of a study of such arguments, such a choice is of course justified. However, my stress is on meta-argumentation, and from this point of view much more needs to be done. Nevertheless, to make my task more manageable, as well as more focused, I too will need to delimit my task and make some simplifying assumptions.

That is, a complete analysis of the meta-argumentation in Hume's *Dialogues* would have to include several other meta-arguments: the introductory discussion of the meaning of skepticism and its connection with natural religion (part I); Cleanthes's defense of the design argument (part III); Cleanthes's criticism of Demea's *a priori* argument for the existence of God (part IX); Philo's criticism of Demea's same argument (part IX); Philo's argument that the controversy over the nature of God is a "verbal dispute" (part XII); and Philo's criticism of the argument that theism provides the only foundation of morality (part XII). However, these meta-arguments are beyond the scope of this chapter, and instead I shall focus only on what I call Philo's critique of Cleanthes's design argument. My choice is motivated by the fact that this is the most inclusive, comprehensive, complex, and instructive meta-argument in the *Dialogues*.

12.3 Cleanthes's vs. Philo's Ground-level Argument

Before criticizing the design argument, Philo finds the occasion to amplify and reformulate Cleanthes's statement of it. The context is provided by Demea's objection that only *a priori* arguments are proper, and it is "extravagant" that "the proofs of a Deity fall short of perfect evidence" (II-10), as is the case with Cleanthes's *a posteriori* argument.

Philo replies (II-12-13) that the only thing we could know *a priori* about the universe is that it must be free of contradictions. However, only experience tells us what particular properties the universe has and which causes produce which effects. Then he elaborates Cleanthes's argument as follows:

Order, arrangement, or the adjustment of final causes, is not of itself any proof of design, but only so far as it has been experienced to proceed from that principle. For aught we can know *a priori*, matter may contain the source or spring of order originally within itself, as well as mind does; and there is no more difficulty in conceiving that the several elements, from an internal unknown cause, may fall into the most exquisite arrangement, than to conceive that their ideas, in the great universal mind, from a like internal unknown cause, fall into that arrangement. The equal possibility of both suppositions is allowed. But, by experience, we find (according to Cleanthes) that there is a difference between them. Throw several pieces of steel together, without shape or form, they will never arrange themselves to compose a watch. Stone and mortar and wood, without an architect, never erect a house. But the ideas in a human mind, we see, by an unknown, inexplicable economy, arrange themselves so as to form the plan of a watch or house. Experience, therefore, proves that there is an original principle of order in mind, not in matter. From similar effects we infer similar causes. The adjustment of means to ends is alike in the universe, as in a machine of human contrivance. The causes, therefore, must be resembling. [II-14]

Cleanthes's original argument may be reconstructed as follows:[5]

---

[5] The numbering system for the propositions in my argument reconstructions is the one presented in chapter 2 above; it is a variation of the system presented by various authors when they discuss the representation of complex argument by means of structure diagrams in the shape of either tree branches or tree roots (e.g., Angell 1964, 369-93; Scriven 1976, 41-43; Finocchiaro [1980a, 311-31; 2005a, 39-41]; Eemeren and Grootendorst 1984, 87-93; Eemeren, Grootendorst, and Kruiger 1984, 17-36; Freeman 1991). The key idea is that if a given claim is labeled *n*, then the premises that directly support it are labeled *n1*, *n2*, *n3*, etc.; and if *nm* is also part of another subargument, then the premises directly supporting it are labeled *nm1*, *nm2*, *nm3*, etc.

[11]   the universe is similar to a machine;
[12]   machines are produced by human intelligent design (HID);
[1]   so, the universe was produced by a cause similar to HID.

On the other hand, Philo's elaboration amounts to the following:

[211] when material things are ordered and organized (OO), they are produced by HID;
[212] similar effects have similar causes;
[21]   (so, when material things are similar to OO, they are produced by causes similar to HID);
[22]   the universe is similar to an OO material thing;
[2]   so, the universe was produced by a cause similar to HID.

Here proposition 21 is placed in parenthesis because it is an intermediate conclusion not explicitly stated in the passage.

Let us compare and contrast these two versions. It is obvious that the two formulations have identical conclusions. Moreover, premise 11 corresponds to premise 22, with the difference that the latter is more general than the former. Similarly, premise 211 is a generalization of premise 12.

Next, the principle of similarity of effects and causes is explicitly stated in Philo's version (proposition 212), but is merely mentioned and implicitly used in Cleanthes's formulation. I have omitted it from my reconstruction of Cleanthes's argument in order to make his version look like an argument by analogy, as it is interpreted by Barker and as suggested explicitly by Cleanthes himself. Despite such talk of analogy and all the talk of similarity, however, it seems to me that Cleanthes's argument is not really an argument by analogy. The reason is that although Cleanthes bases his conclusion on the similarity between the two analogues and on the fact that one of the analogues (machines) is known to have a special property, what he concludes is not that the other analogue (the universe) has the *same* property, but rather that it has a *similar* property. In other words, in arguments by analogy one reasons from the fact that two things are known to be similar in certain respects to infer that they are probably also similar in another additional special respect, where similarity means sharing the *same* properties, not *similar* ones; here the similarity is a relationship applying to the individual entities and consists of their sharing some properties, it is not a relationship applying to the properties. A third way of saying this is to point out that in reasoning by analogy one is arguing from the fact that two things share some properties to the conclusion that they share another special property, not to the conclusion that they possess two additional similar properties.

Philo's version of the argument makes all this clear, since it does not even look like an argument by analogy. Instead, it seems to be an argument from generalizations to a particular case. However, it is not a deductive universal instantiation because the generalizations (propositions 211, 212, and 21) are not meant to be strict universal generalizations, but empirical claims advanced as being true for the most part, or in typical cases, or as a matter of likelihood; and so the conclusion is not meant to follow necessarily but with some probability. This type of argument corresponds to what Toulmin (1958, 109, 131) has labeled "quasi-syllogism." Other philosophers[6] have called this type of argument "statistical syllogism," a term that may be adopted as long as one is not misled by the word

---

[6] Barker 1957, 70; Hempel 1965, 55; W. Salmon 1984, 94-97; M. Salmon 2002, 112-15.

"statistical" or the word "syllogism."[7] The simplest case of such an argument has the form: a high percentage of F's are G's; a is F; so, a is G. The last step of Philo's argument (from 21 and 22 to 2) is a variant of this form. On the other hand, the first step of that argument (from 211 and 212 to 21) is obviously not of that form, but rather is a type of inductive argument from empirical generalizations to empirical generalizations.

12.4 Philo's Constructive and Critical Meta-arguments

12.4.1 Empirical Explicitness

Philo's design argument (argument 2), as it stands, is of course a ground-level argument and not a meta-argument. However, he makes two initial claims about argument 2, and that is how some meta-argumentation arises in this context. First, Philo claims that [M1] argument 2 makes explicit the empirical and inductive credentials of argument 1, and the justification of this claim is that [M11] argument 1 can be amplified and strengthened in the manner done in the last several paragraphs above. Here, then, is the first (constructive) meta-argument that may be attributed to Philo.

12.4.2 Empirical Exclusivity

Second, Philo also claims that [M2] the design argument (argument 2 or some variant of it) is the only serious argument that can be advanced to try to support the existence or nature of God. The reason for this is that [M21] assertions of God's existence and of his intelligent nature are statements of fact, but [M22] "all inferences ... concerning fact are founded on experience, and [M23] all experimental reasonings are founded on the supposition that similar causes prove similar effects, and similar effects similar causes" (II-17). And here we have another constructive meta-argument, which appears to be deductively valid and deductive, although it could be questioned by questioning the truth of the epistemological principles asserted in its final premises.[8]

Despite the presence of these two constructive and appreciative meta-arguments, the main point of this passage of the *Dialogues* (II-11-17) is to formulate ground-level argument 2, so that it can be the focus of the main thread of critical meta-argumentation in the rest of the book. To this we now turn.

12.4.3 Hasty Generalization

Philo's first criticism of the design argument concerns the premise that when material things are ordered and organized, they are produced by human intelligent design (proposition 211). To begin with (II-18), it should be noted that this proposition is not universally and absolutely true since, for example, the physical forces of attraction and repulsion produce order and organization in material things but do not involve human intelligent design. Furthermore (II-19), the justification of this proposition 211 would

---

[7] Scriven (1976, 205) discusses this type of argument at some length, but does not coin a term for it, describing it as "reasoning from facts about a whole population to conclusions about the individual members of the population."

[8] The textual presence of this meta-argument and its role in the main thread of the *Dialogues* is also confirmed by the fact that part VII contains a good recapitulation that begins with this meta-argument: "since no question of fact can be proved otherwise than by experience, the existence of a Deity admits not of proof from any other medium" [VII-3].

consist of a generalization argument from some to all, in which the premises would be claims about order and organization such as machines, watches, houses, and ships; and such a subargument would be deductively invalid.

These two points are essentially correct, but should not be taken to have much critical import because it is understood, and has already been stressed, that the design argument is an inductive argument and that this premise is an empirical generalization of the statistical or typical sort. Instead, these two points should be understood as a clarification and introduction for the next.

That is (II-20), experience reveals that nature is extremely diversified on earth, and so it is unlikely that it just repeats itself in the universe as whole. Thus, insofar as watches and ships are typical examples of order and organization on earth, they are probably not typical of order and organization in the universe in general. It follows that the argument from cases of terrestrial order and organization to material things in the universe at large is a hasty generalization. In short, Philo's meta-argument here is that [M3] premise 211 is unjustified because [M31] it is supported by a subargument that is a hasty generalization.

12.4.4 Fallacy of Composition

A similar difficulty affects the premise that the universe is similar to an ordered and organized material thing (proposition 22). To see this, it must be clarified that although from the point of view of how this proposition is utilized in the design argument, it is a particular claim about an individual entity (the universe), from the point of view of how it is arrived at, the proposition is a kind of generalization. That is, in argument 2, this proposition (22) is combined with a generalization (proposition 21) to arrive at the final conclusion (proposition 2), by an inference of the statistical-syllogism type. Even in argument 1, the corresponding claim (proposition 11) is meant to be a statement about particular cases, about the similarity between two particular things (the two analogues), so that by analogy one can attribute to one of them (the universe) the property which is already known to belong to the other. However, if we examine the support for this claim about the universe (proposition 22), we see (II-5) that it consists of claims about various parts of the universe. In short, the subargument supporting proposition 22 (and the one for proposition 11) is an argument from parts to whole.

But is this subargument correct? Philo claims that it is not, that it commits the fallacy of composition. His meta-argument here is that [M4] proposition 22 is unjustified because [M41] its supporting subargument commits the fallacy of composition. Philo does not use this technical term, but his language leaves no doubt: intelligent design "is an active cause by which some particular parts of nature ... produce alterations in other parts. But can a conclusion, with any propriety, be transferred from parts to the whole? Does not the great disproportion bar all comparison and inference?" (II-18).

12.4.5 Second Hasty Generalization

But this is not all, for there is a third analogous difficulty (II-21). The difficulty is analogous insofar as it again consists of a type of hasty generalization. But now the difficulty involves the inference from propositions 22 and 21 to 2. The point is that whereas premise 22 is a claim about how order and organization *are* produced, in the sense of *presently* or *recently* produced, the conclusion (proposition 2) is a claim about how the universe *was* produced at some time in the *extremely remote* past. This inference presupposes the principle that "the operations of a world constituted, arranged, and adjusted, can with any propriety be extended to a world which is in its embryo state, and is advancing towards that constitution and arrangement" (II-21). And Philo seems to think that this is a gratuitous extrapolation (II-22-23). Here the meta-argument is that [M5] the step from propositions 22 and 21 to 2 is illegitimate because [M51] it involves an inference from the present to the remote past.

12.4.6 Misapplication of Generalization

Furthermore, there is another difficulty with the same step of the design argument. Here (II-24-28), the discussion is more complex because Philo's criticism elicits an interesting response from Cleanthes. Philo's objection may be reconstructed as the following meta-argument: [M611] the universe is a unique entity, and so [M61] it is not covered or subsumed by the generalization formulated in proposition 21; [M612] the latter is a generalization about ordinary material things, whereas [M613] to apply it to the universe by combining it with proposition 22 it would have to be a generalization about universes or universe-like entities, specifically about the origin of universes; thus [M6] the inference to proposition 2 from 21 and 22 is illegitimate.

Cleanthes replies that "to prove by experience the origin of the universe from mind is not more contrary to common speech than to prove the motion of the earth from the same principle. And a caviller might raise all the same objections to the Copernican system which you have urged against my reasonings. Have you other earths ... which you have seen to move?" (II-25). This reply seems to question the truth of proposition M613: it is not true that to infer a conclusion about the origin of the universe one needs a generalization about how universes originate, any more than to infer a conclusion about the motion of the earth one needs a generalization about how earths move.

Philo counter-replies (II-26-27) that in astronomy, one could see other "earths," namely the moon, the planets, and the satellites of Jupiter and observe that they move; but one also had to establish the similarity between these bodies and the earth. In fact, "if we peruse Galileo's famous *Dialogues* concerning the system of the world, we shall find that that great genius, one of the sublimest that ever existed, first bent all his endeavours to prove that there was no foundation for the distinction commonly made between elementary and celestial substances" (II-27).

By way of evaluation, I would say that although it is true that one of Galileo's arguments for the earth's motion around the sun was the inductive argument by analogy sketched by Philo, this was neither his only argument, nor his strongest; for example, a better argument was the explanatory argument that the earth's revolution around the sun provided the best explanation of the annual cycle of sunspot motion observed across the solar disk.[9] And if we include the further development of science after Galileo, Philo's claim is disconfirmed further; for example, Isaac Newton's systematization of celestial mechanics provided a proof for the earth's motion based on the law of universal gravitation and the fact that the center of mass of the sun-earth system lies inside the surface of the sun.

It follows that in this particular exchange, Cleanthes seems to prevail over Philo. Cleanthes's reply is essentially a non-inductive meta-argument by analogy. It is cogent and remains unaffected by Philo's counter-meta-argument. Thus it basically succeeds in undermining proposition M613.

However, this is not as damaging as it may sound for the purpose of the criticism embodied in Philo's meta-argument 6. For although Philo does assert proposition M613, and although it is effectively undermined by Cleanthes, Philo's present criticism of the design argument does not really need this proposition. In fact, if we delete proposition

---

[9] Cf. Galilei 1997, 232-33; Finocchiaro 1980a, 246-49, 129-30.

M631 from meta-argument 6, we are left with a cogent criticism. For the rest of Philo's meta-argument may be interpreted as claiming that the last step of the design argument (21, 22, so 2) has a flaw reminiscent of the fallacy of stereotyping; that is, the misapplication of a statistical or typical generalization to an individual case to which it should not be applied, either because this case is not typical or because other things are known about the case that should be taken into account.[10]

12.4.7 Objection from Divine Simplicity

To find the next criticism of the design argument, we must go to part IV of the *Dialogues* (IV-2-3). There it is Demea who raises an objection to the argument's conclusion, even before Philo raises some others. Demea argues that [M7] proposition 2 is problematic since [M71] it implies that God's mind is similar to the human mind; but [M72] this consequence is untenable because [M721] the human mind is "a composition of various faculties, passions, sentiments, ideas" (IV-2), whereas [M722] such composition is not "compatible with that perfect immutability and simplicity which all true theists ascribe to the Deity" (IV-2).

For the record, it should be stated that Cleanthes replies that such theists are really atheists without knowing it, because a simple mind in their sense is a contradiction in terms.

12.4.8 Infinite Regress

At this point Philo intervenes (IV-5-14) to advance another criticism—an infinite-regress objection.[11] The focus is again the conclusion of the design argument, the proposition (number 2) that the universe was produced by a cause similar to human intelligent design. "It is not easy ... to see what is gained by this supposition ... we are still obliged to mount higher in order to find a cause of this cause" (IV-6); for from the point of view of *a priori* reason, "a mental world or universe of ideas requires a cause as much as does a material world or universe of objects" (IV-7), whereas from the point of view of our experience we cannot "perceive any [important] difference in this particular between these two kinds of world, but [find] them to be governed by similar principles, and to depend upon an equal variety of causes in their operations" (IV-8). "How, therefore, shall we satisfy ourselves concerning the cause of that being whom you suppose the Author of nature ... ?" (IV-9). This question generates a vicious infinite regress, which it is best to stop at the very beginning, so as "never to look beyond the present material world" (IV-9). For if we "say that the different ideas which compose the reason of the Supreme Being fall into order of themselves" (IV-10), this is meaningless, or else it is equally plausible "to say that the parts of the material world would fall into order of themselves" (IV-10). Nor can one say that it is unreasonable to expect that "if I assign a cause for any event" (IV-13) then I should also "assign the cause of that cause, and answer every new question which may incessantly be started" (IV-13); for the nature of explanation is such that "naturalists indeed very justly explain particular effects by more general causes, though these general causes themselves should remain in the end totally inexplicable, but they never surely thought it satisfactory to explain a particular effect by a particular cause which was no more to be

---

[10] Scriven's (1976, 205-10) discussion of this type of argument and the corresponding fallacy is still instructive. See also Schauer 2003; Jones 2010, 179-98.

[11] For an account of infinite-regress arguments in general, see Gratton 2010, which also contains a different analysis of this particular argument.

accounted for than the effect itself" (IV-14).

These remarks can be reconstructed as the following meta-argument. [M8] The conclusion of the design argument (proposition 2) leads to an infinite regress because [M81] if the material universe was produced by an intelligent cause, then that cause must have been produced by some other cause, and [M82] so on. But [M83] this infinite regress cannot be stopped by saying that the intelligent cause of the material universe does not need a cause or caused itself, because [M831] the latter is no more plausible than saying that the material universe does not need a cause or caused itself. [M84] Nor can the infinite regress be stopped by appealing to the logic of explanation and pointing out that it is the piecemeal process of accounting for one effect in terms of a cause, without having to explain everything in the process; for [M841] such a piecemeal process works properly only when we are explaining a particular effect in terms of a general cause, not in terms of a particular cause.

12.4.9 Occult-Quality Objection

Another one of Philo's criticisms is contained in a paragraph (IV-12) interspersed within the critique just examined. It too targets the conclusion of the design argument (proposition 2). The meta-conclusion is simply the claim that [M9] the ground-level conclusion lacks any explanatory power. For [M91] to explain the order and organization of the material universe as caused by God is like explaining the nourishing effects of bread in terms of its "nutritive faculty"; but [M92] in the latter case, "this subterfuge was nothing but the disguise of ignorance" (IV-12); and the former case is similar, for [M911] "when it is asked what cause produces order in the ideas of the Supreme being, can any other reason be assigned ... than that it is a *rational* faculty, and that such is the nature of the Deity?" (IV-12); and [M912] this is an answer that could have been given in explaining the order of the material universe.

I shall label this last critique the occult-quality criticism or objection. It is obvious that it consists of a meta-argument by analogy and that the meta-argument is deductive. It does not share this analogy form with the infinite-regress objection or Demea's divine-simplicity objection, although all three of these meta-arguments try to find fault with the conclusion of the design argument, considered individually and somewhat independently of the supporting reasons.

12.4.10 Divine Infinity, Perfection, and Unity

In the next group of criticisms (part V), Philo focuses on a previously unexamined aspect of the design argument, i.e., the premise that similar effects have similar causes (proposition 212). His aim is not to refute or undermine this principle, but rather, having accepted it, to derive some consequences that are problematic for theists like Cleanthes.

The first consequence is the finiteness of God, which is inconvenient given that theists usually hold that God is infinite. The argument here is that since [311] "the cause ought only to be proportional to the effect, and [312] the effect, so far as it falls under our cognizance, is not infinite" (V-5), it follows that [31] the cause of the universe is finite; but [32] God is (usually conceived as) the cause of the universe; therefore, [3] God is finite.

The second consequence is the imperfection of God, which contradicts the theist claim that God is perfect. Here Philo argues that since [411] "there are many inexplicable difficulties in the works of nature which ... perhaps, will be insisted on as new instances of likeness to human art" (V-6), [41] the Author of nature is imperfect; therefore, [4] God is imperfect.

Third, there is a problem with the theist idea of "the unity of the Deity" (V-8). Since [511] usually "a great number of men join in building a house or ship" (V-8), the proper conclusion to draw would be that [51] "several deities combine in contriving and framing a world" (V-8); so, [5] there is not just one God but several; and [52] this conclusion does not violate the principle that we ought not "to multiply causes without necessity" (V-9) because [5211] it has not yet been proved that one god is sufficient, and hence [521] we don't know that the other gods are unnecessary.

As they stand, these three arguments are ground-level. That is why I have numbered them without the prefix 'M' but rather in the sequence of ground-level arguments, of which the first two were Cleanthes's and Philo's versions of the design argument. It is unclear whether or how they could be transformed into meta-arguments. The issue would be to ask what these arguments show about the design argument, that is, what further conclusion one could derive. Perhaps the connection could be worked out as follows.

Argument 2 could be extended to add some further conclusion that would explicitly use the notion of God. Indeed it is obviously intended to be so extended. The first further conclusion to infer immediately from proposition 2 would be that [2.a] there exists something that is the cause of the universe. Now, adding the premise that [2.b (=32)] God is (usually conceived as) the cause of the universe, it would follow that [2.c] God exists.

Let us now extend argument 3. It concludes that [3] God is finite. But [3.a] if God exists, he is infinite, because [3.b] God is (also usually conceived as) infinite; therefore, [3.c] God does not exist; therefore, [3.d] the further conclusion of the design argument (proposition 2.c) is false. In short, [M{10}] the design argument (argument 2) is incoherent, because [M{10}1] from one of its crucial premises, proposition 212, one can plausibly infer the denial of its intended further conclusion, proposition 2c. And this is a meta-argument.[12] This meta-argument would obviously be further instantiated by the obvious extensions of Philo's arguments involving perfection and unity.

One final point should be noted about these three ground-level arguments involving infinity, perfection, and unity, which can be extended in the manner just indicated to become instances of this incoherence meta-argument. They are *ad hominem* arguments in the sense of Henry Johnstone, Whately, Locke, and Galileo.[13] That is, they are arguments that criticize the original argument by deriving a conclusion not acceptable to the proponents of the original argument from premises accepted or acceptable to them. In this sense such *ad hominem* arguments point out an incoherence in the position (or set of beliefs) of an opponent.

12.4.11 The Organism Objection and the Fallacy of Incomplete Evidence

The criticism of the design argument next moves to a lengthy discussion (parts VI-VII) of a topic which I shall call the organism analogy objection. Philo begins with a clear and succinct statement, which as it stands is a ground-level argument and will be numbered 6 in this chapter's sequence:

---

[12] Note that I am labeling it M-ten and symbolizing it by using braces to denote the number of the meta-argument and of its corresponding main conclusion, thus treating '10' as the tenth numeral rather than as a sequence of two single digits.

[13] Needless to say, this concept of *ad hominem* argument should not be confused with the contemporary meaning, according to which it is the fallacy of criticizing an argument by criticizing the motives or character of the person advancing the argument. Cf. Finocchiaro 2005a, 277-91, 329-39; and chapter 7 above.

If we survey the universe, so far as it falls under our knowledge, it bears a great resemblance to an animal or organized body, and seems actuated with a like principle of life and motion. A continual circulation of matter in it produces no disorder; a continual waste in every part is incessantly repaired; the closest sympathy is perceived throughout the entire system; and each part or member, in performing its proper offices, operates both to its own preservation and to that of the whole. The world, therefore, I infer is an animal; and the Deity is the *soul* of the world, actuating it, and actuated by it. [VI-3]

Against this, Cleanthes makes a cogent criticism. That is, the analogy between the world and an animal is flawed because the world has "no organs of sense; no seat of thought or reason; no one precise origin of motion and action" (VI-8). The analogy is better with a vegetable, from which one could not infer the conclusion about the soul of the world.

Philo seems to accept the cogency of this criticism, for at the beginning of the next chapter he reformulates his objection so as to take into account the vegetable possibility besides the animal possibility. This reformulation has an explicit meta-argumentative character, and so it will be called meta-argument {11} (eleven) in our sequence:

[M{11}11] If the universe bears a greater likeness to animal bodies and to vegetables than to the works of human art, it is more probable that its cause resembles the cause of the former than that of the latter, and [M{11}1] its origin ought rather to be ascribed to generation or vegetation than to reason or design. [M{11}] Your conclusion, even according to your own principles, is therefore lame and defective. [VII-1]

Note that it is obvious from the context that Philo is assuming and implicitly asserting the antecedent of the conditional premise M{11}11, namely the proposition that [M{11}12] the universe *does* bear a greater likeness to animals and vegetables than to human artifacts. Note also that this is a weaker claim than the corresponding premise of the original argument (6), but that this weakening makes it more tenable. It should be noted too that Philo strengthens this claim in another respect, namely insofar as he is now (with M{11}12) saying that the universe is *more* similar to an organism, and not merely that the similarity is *great*, as stated in the original argument (6).

The explicit meta-argumentation derives from the content and import of the meta-conclusion (M{11}) that the conclusion of the design argument is "lame and defective." If this argument had stopped short of that, with proposition M{11}1, then we would have a ground-level argument whose conclusion (M{11}1) would be the denial of the conclusion of the design argument. Such a shortened argument, if advocated as correct, would be a refutation of the design argument by way of refuting its conclusion. However, this would presuppose the correctness of the shortened argument, and Philo obviously does not hold this presupposition; for the shortened argument is logically analogous to the design argument 2, and so its logical worth is the same as that of argument 2, which Philo has argued is very little. Thus, he is not proposing the shortened argument in order to advocate its conclusion (M{11}1), rather he is merely considering the ground-level segment in order to point out that it and the design argument have equal worth (because they have identical form), and since they cannot both be good (because their conclusions contradict each other), they must both be bad.

This interpretation is also suggested by the explicit wording of the meta-conclusion M{11}. For it does not say that the conclusion of the design argument is false, which would follow from the assertion of proposition M{11}1. Philo's saying that that conclusion is "lame and defective" suggests primarily that there is an inferential or logical flaw in the design argument. That is, that conclusion does not follow. And the reason for that is that if it did follow, then the conclusion of the animal similarity objection, and of the vegetable similarity objection, and of the organism similarity objection would also follow, and that would lead into contradiction. The flaw in each such argument is that it focuses on one similarity to infer a property correlated with that similarly, but it neglects other similarities with are correlated with conflicting properties.

Here we seem to have an example of the violation of what some logicians have called the requirement of complete, or total, evidence.[14] This is a principle of evaluation that applies to inductive arguments that have the form of a statistical syllogism (or quasi-syllogism): a high percentage of F's are G's; a is F; so probably a is G. The principle stipulates that to be inductively correct the premises of such argument must embody all available relevant evidence. For suppose that a is also an H; and suppose that a high percentage of H's are not-G's; then by violating the requirement of complete evidence, we can give two arguments both with true premises, but with one conclusion claiming that a is probably G and the other conclusion that a is probably not G. A good example would be: a high percentage of Swedes are not Roman Catholic, and Petersen is a Swede, so Petersen is not a Roman Catholic; but a high percentage of visitors to Lourdes are Roman Catholics, and Petersen is a visitor to Lourdes, so Petersen is a Roman Catholic.[15]

In other words, one defect of the design argument is that it focuses only on the order and organization in the universe and the corresponding similarity with human artifacts, while ignoring the features that make it similar to an animal, vegetable, or organism. Then it derives a conclusion from the incomplete evidence on which it focuses. Such selectivity is illegitimate, for all available evidence should be taken into account. If that is done, it is unclear that anything can be inferred. The design argument thus commits the fallacy of incomplete evidence.

12.4.12 Necessity of Partial Order and Inference to the Best Explanation

A new type of criticism is elaborated by Philo in part VIII. He begins by entertaining the Epicurean hypothesis that the universe is eternal but composed of a finite number of particles (VIII-2). He also hypothesizes that some of these particles are in motion, and that the total quantity of motion is conserved (VIII-5). Then he argues that even if such a universe were ultimately chaotic, some semblance of order would be necessarily produced (VIII-6-9); for in a temporally infinite universe, every possible arrangement of a finite number of particles would be repeated infinitely many times, this being a necessary consequence of the concept of infinity (VIII-2).

For example, "whenever matter is so poised, arranged, and adjusted, as to continue in perpetual motion, and yet preserve a constancy in the forms, its situation must of necessity have all the same appearance of art and contrivance which we observe at present" (VIII-6). In other words, after "the universe goes on for many ages in a continued succession of chaos and disorder" (VIII-8), it is "possible that is may settle at last, so as not to lose its motion and active force ... yet so as to preserve an uniformity of appearance, amidst the continual motion and fluctuation of its parts" (VIII-8). Again, "an animal could [not] subsist unless its parts were so adjusted" (VIII-9); and more generally, "if it were not so [ordered], could the world subsist?" (VIII-9). So even a chaotic world "must ... pass through new positions and situations till in great but finite succession it fall, at last, into the present or some such order" (VIII-9).

---

[14] Barker 1957, 77; Carnap 1950, 211; Hempel 1965, 163-67; M. Salmon 2002, 117-18; W. Salmon 1984, 94-97.

[15] I have adapted this example from Toulmin 1958, 131-33, 139-40, and Hempel 1965, 55-56.

Philo admits that the "foregoing hypothesis is so far incomplete and imperfect" (VIII-11), but so are all other attempts to explain the order and organization in the universe, including that advanced by the design argument. And so his conclusion is not that the observable cosmological order is to be explained as the result of such necessity, but rather that "a total suspension of judgment is here our only reasonable resource" (VIII-12).

In other words, Philo mentions his Epicurean explanation argument not to advocate its conclusion, but rather to undermine the design argument. That is, in this criticism, Philo is interpreting the design argument as (having an aspect of) an inference to the best explanation; then he constructs another argument with the same crucial premise (proposition 22) but inferring an alternative explanation; then he admits that the two explanations are on a par; and so he concludes that the design argument is explanatorily weak, namely has the flaw that the inferred explanation is no better that the alternative one. The meta-argumentative reconstruction would be as follows: [M{12}] the design argument is explanatorily weak because [M{12}1] it may be interpreted as an inference to the best explanation, but [M{12}2] the inferred explanation is not the best since [M{12}21] it is possible to construct another inference to the best explanation, the [M{12}211] Epicurean explanation argument, and [M{12}22] the conclusion of the design argument is no better an explanation than the conclusion of the Epicurean argument.

Finally, it should be noted that this criticism of the design argument provides a good concrete illustration of the explanatory disconnection type introduced earlier (chapter 2).

12.4.13 The Argument from Evil

We now come to the criticism and the meta-argument connected with the argument from evil. The discussion begins with Philo and Demea joining forces to describe and elaborate the existence, quantity, and variety of evil, pain, and misery in the world: "a perpetual war is kindled among all living creatures ... hunger ... fear, anxiety ... anguish" (X-8); "the stronger prey upon the weaker ... the weaker, too, in their turn often prey upon the stronger" (X-9); "man, it is true, can, by combination, surmount all his *real* enemies and become master of the whole animal creation; but ... what new enemies does it not raise to us? ... man is the greatest enemy of man" (X-11-12); "though these external insults ... from animals, from men, from all the elements which assault us form a frightful catalogue of woes, they are nothing in comparison of those which arise within ourselves, from the distempered condition of our mind and body" (X-13); finally, "not satisfied with life, afraid of death ... we are terrified, not bribed to the continuance of our existence" (X-17). It should be added that even if (as Cleanthes objects) "health is more common than sickness; pleasure than pain; happiness than misery" (X-31), still (as Philo notes) "pain ... is infinitely more violent and durable. One hour of it is often able to outweigh a day, a week, a month of our more common insipid enjoyments" (X-32).

Such considerations make it undeniable that [711] there is evil in the world, and this paves the way for the statement (X-25) of the argument from evil as such: [712] if God were both able and willing to prevent evil, then there would be no evil; so, [71] God is either not able or nor willing to prevent evil; but [72] if God were willing but not able, then he would not be infinitely powerful; and [73] if God were able but not willing, he would not be benevolent; so, [7] God is either not infinitely powerful or not benevolent. It also follows that [7.a] if God is infinitely powerful, then he is not benevolent.

How does this argument affect the design argument? And what is the meta-argumentative import of this argument?

To answer these questions, let us note that argument 7 is a ground-level argument. But it is not the only version of the argument from evil which we find in this context. Another one is argument 8: "[8111] His power, we allow, is infinite; [811] whatever he wills is executed; but [812] neither man nor any other animal is happy; therefore, [81] he does not will their happiness. [8211] His wisdom is infinite; [821] he is never mistaken in choosing the means to any end; but [822] the course of nature tends not to human or animal felicity; therefore, [82] it is not established for that purpose ... [8] In what respect, then, do his benevolence and mercy resemble the benevolence and mercy of men?" (X-24). In other words, [8.a] if God is infinitely powerful and infinitely wise, then his benevolence and mercy do not resemble human benevolence and mercy; or [8.b] if God is infinitely powerful, then his benevolence does not resemble human benevolence. This further conclusion, proposition 8.b, may be viewed as a weakening of the further conclusion of the previous argument, proposition 7.a.

And then there is the following version, argument 9: "Why is there any misery at all in the world? Not by chance, surely. From some cause then. It is from the intention of the Deity? But he is perfectly benevolent. Is it contrary to his intention? But he is almighty ... these subjects exceed all human capacity, and ... our common measures of truth and falsehood are not applicable to them" (X-34). This argument may be reconstructed as follows: [91] there is evil in the world; [92] the cause of evil cannot be chance; [93] the cause is not in accordance with God's will because [931] he is benevolent; [94] the cause is not contrary to his will because [941] he is almighty; so, [9] these subjects are beyond human comprehension. To draw a further conclusion and make it fit the progression suggested from 7.a to 8.b, one might say that [9.a] if God is infinitely powerful, then his benevolence is beyond human comprehension; that is, [9.b] if he is infinitely powerful, then we cannot know anything about his benevolence.

To move in the direction of meta-argumentation, we must note that Philo's opinion of the worth of these arguments is very high: "through the whole compass of human knowledge there are no inferences more certain and infallible than these" (X-24); and "nothing can shake the solidity of this reasoning, so short, so clear, so decisive" (X-34). Let us call this judgment meta-proposition thirteen-one, or [M{13}1]. To this judgment we may add the analysis of the argument from evil, in its three versions of ground-level arguments, 7, 8, and 9; here the claim would be that [M{13}2] arguments 7, 8, and 9 are good examples of inductive arguments that undermine the moral attributes of God, such as benevolence. From these two premises, Philo infers that [M{13}] the attempt to justify the benevolence of God from experience is hopeless (X-35); that is, there can be no good inductive arguments justifying the benevolence of God.

However, although we do have a meta-argument here, its connection with the design argument is problematic. In fact, this meta-argument is not *directly* about the design argument. The indirect connection is that people who accept the design argument and its conclusion also accept that God is benevolent, which is undermined by the argument from evil. So the argument from evil undermines an additional belief held by the proponents of the design argument. On the other hand, everyone can accept the distinction between the *natural* or *cognitive* attributes of God, such as intelligent design, and the moral attributes, such as benevolence and mercy. In other words, the argument from evil is an argument *against* divine benevolence. Its strength does undermine the possibility of good arguments *for* divine benevolence, but not the possibility of good arguments for *divine intelligence*.

12.4.14 The Moral-Indifference Argument

The next and last meta-argument in our sequence also involves a discussion of the problem of evil. The discussion begins (XI-1) with Cleanthes conceding the main point of the preceding meta-argument, but the conclusion he draws is to abandon attributing infinity to God and to conceive of him as finite. That is, we ought to abandon attributing infinity to God because the existence of evil not only does not support (human-like) goodness for an infinite God, but also is incompatible with it. In other words, inductive arguments not only do not justify that an infinite God is good, but rather justify the opposite, that an infinite God is *not* good; and so we ought to conceive of God as finite. A finite God is compatible with evil. In fact, in that case evil would be explained as follows: "a less evil may then be chosen in order to avoid a greater; inconveniences be submitted to in order to reach a desirable end; and, in a word, benevolence, regulated by wisdom and limited by necessity, may produce just such a world as the present" (XI-1).

For Philo, however, this revision opens the door to another difficulty. Although a finite God is *not incompatible* with the phenomena of evil in the world, his goodness is still *not justifiable* from those phenomena. The logic of the situation seems to be the following: "if the goodness of the Deity (I mean a goodness like the human) could be established on any tolerable reasons *a priori*, these phenomena, however untoward, would not be sufficient to subvert that principle, but might easily, in some unknown manner, be reconcilable to it. But ... as this goodness is not antecedently established but must be inferred from the phenomena, there can be no grounds for such an inference while there are so many ills in the universe, and while these ills might so easily have been remedied" (XI-12).

Regarding the moral attributes of a finite God, the only correct argument Philo is willing to allow is one that justifies the *moral indifference* of the deity:

Look round this universe. What an immense profusion of beings, animated and organized, sensible and active! You admire this prodigious variety and fecundity. But inspect a little more narrowly these living creatures, the only beings worth regarding. How hostile and destructive of each other! How insufficient all of them for their own happiness! How contemptible or odious to the spectator! The whole presents nothing but the idea of a blind nature, impregnated by a great vivifying principle, and pouring forth from her lap, without discernment or paternal care, her maimed and abortive children! [XI-13]

From these facts, there is only one consequence that follows:

The true conclusion is that the original Source of all things is entirely indifferent to all these principles, and has no more regard to good above ill than to heat above cold, or to drought above moisture, or to light above heavy. [XI-14]

The reason is that

There may *four* hypotheses be framed concerning the first causes of the universe: that they are endowed with perfect goodness; that they have perfect malice; that they are opposite and have both goodness and malice; that they have neither goodness nor malice. Mixed phenomena can never prove the two former unmixed principles; and the uniformity and steadiness of general laws seems to oppose the third. The fourth, therefore, seems by far the most probable. [XI-15]

This is obviously a ground-level argument, and so let us number argument ten in our list. It is important enough that it deserves a name; let us call it the moral-indifference argument.

But what does the moral-indifference prove about the design argument? And where is the meta-argumentation?

It is certainly striking that the moral-indifference argument begins with the same exact words as (Cleanthes's version of) the design argument (II-5). It is equally striking that the first part of the moral-indifference argument quoted above (XI-13) is an inductive generalization argument, like the longest and initial part of Cleanthes's argument. The difference is, however, that whereas earlier Cleanthes's intermediate conclusion was that the universe is similar to a machine, now Philo's intermediate conclusion is that the universe is indifferent to human misery or happiness. Moreover, whereas Cleanthes's next step of his argument appeared to be analogy, and concluded that the cause of the universe is similar to human intelligent design, here Philo proceeds by an explanatory step (inference to the best explanation) and concludes that the cause of the universe is morally indifferent (and hence utterly dissimilar from human moral sensibility). However, these two respective conclusions are not inconsistent with each other since the former pertains to the natural or cognitive attributes of the deity, whereas the latter pertains to the moral attributes. Thus, the moral-indifference argument cannot be interpreted as implying a charge that the design argument violates the requirement of complete evidence, and still less as a criticism of refutation by logical analogy, or a criticism by parity of reasoning.

On the other hand, theists, who accept the similarity of God to human intelligent design (the conclusion of the design argument), would reject his moral indifference (the conclusion of the present argument), since they obviously attribute to him the moral attribute of benevolence. Moreover, although the conclusion of the two arguments are not strictly inconsistent, the question arises of how likely it is that God resembles man regarding the natural attributes of intelligence and design if he does not resemble man regarding the moral attributes of goodness and benevolence. This can only be considered to be suggestive, and so we have the following weak presumptive meta-argument: [M{14}] the conclusion of the design argument is unlikely to be true because [M{14}1] we can formulate the moral-indifference argument, whose conclusion asserts the moral indifference of God, and [M{14}2] this argument is very strong, and [M{14}3] its conclusion renders it unlikely that God is similar to man in regard to intelligent design.

12.5 Conclusion: The Manageable Complexity of Meta-argumentation

We have seen that the design argument tries to justify (proposition [2]) the existence of an intelligent human-like cause of the universe from the premises that [211] organized systems are produced by human intelligence, that [212] similar effects have similar causes, and that [22] the universe is similar to an organized system. The argument has at least two steps or subarguments: in step one, from the first two premises one infers the intermediate proposition that [21] things similar to organized systems are produced by causes similar to human intelligence; and in the second step, this proposition is combined with the third premise to arrive at the final conclusion.

We have also seen that such a reconstruction of the design argument amounts to a strengthening and amplification (by Philo) of the argument first advanced by Cleanthes. Such a reconstruction makes more explicit its empirical and inductive character and its superiority to any *a priori* deductive argument for God's existence and about his nature.

Then Hume elaborates his criticism of the design argument. He criticizes the first premise (proposition 211) when he objects (in meta-argument M3) that it is improperly justified by means of a hasty generalization. He faults the third premise (proposition 22) by objecting (in meta-argument M4) that its supporting argument commits the fallacy of composition. He questions the inference in the second, main subargument (from propositions 21 and 22 to 2) for several reasons: because it embodies a hasty generalization from the present to the remote past (meta-argument M5); because it misapplies a generalization, proposition 21, to an individual case, proposition 22 (meta-argument M6); because it violates the requirement of complete evidence by overlooking the organism analogy (M{11}); and because it is an inadequate inference to the best explanation that overlooks the necessity of some partial and temporary order (M{12}). Hume criticizes the conclusion by arguing that its content contradicts the notion of divine simplicity (M7); that it generates an infinite regress of causes (M8); that it embodies an occult-quality pseudo-explanation (M9); and that it is basically inconsistent with the second premise (proposition 212), which plausibly implies that God is finite, imperfect, and multiple, whereas the conclusion presupposes the opposite (M{10}).

This is a long and varied list of criticisms. However, if we keep in mind the content of the claims and the structure of the reasoning of the design argument, all those criticisms follow into a simple and elegant pattern. That is, while agreeing with one premise (proposition 212 about the similarity of causes and effects), Hume criticizes the argument by questioning the tenability of the other two premises, the inference to the conclusion, and the content of the conclusion, including its logical coherence with the uncontroversial premise.

Furthermore, we have seen that the last two criticisms, pertaining to the moral attributes of God, do not really fit into this pattern. They are primarily arguments against divine benevolence: the argument from evil is against the existence of an infinite and benevolent God; the moral-indifference argument is against the existence of a finite morally-sensible God. And their corresponding meta-arguments are that no good inductive argument can justify the benevolence of an infinite God, and that no good inductive argument can justify the moral sensibility of a finite God. These affect the design argument only insofar as its proponents would also attribute benevolence or moral sensibility to God.

12.6 Summary

In short, Hume's critique of the design argument in the *Dialogues* is indeed, as Barker claimed, a powerful non-inductive meta-argument. However, the main line of argument is not a meta-argument by analogy but rather a complex meta-argument, consisting of two main parts, an interpretive constructive part and an evaluative critical part. The interpretive meta-argument claims that the design argument is an inductive ground-level argument; but not that it is an argument by analogy; rather that it is complex argument, consisting of three premises and two subarguments, one of which subarguments is an inductive generalization, while the other one is a statistical syllogism. The critical meta-argument argues that the design argument is weak because two of its three premises are justified by inadequate subarguments; because its main inference embodies at least four flaws; and because the conclusion is in itself problematic for at least four reasons. Finally, the design argument is indirectly undermined by two powerful ground-level arguments justifying conclusions that are in probable or presumptive tension with the conclusion of the design argument, while admittedly not in strict contradiction with it.

Chapter 13
Galileo on the Motion of the Earth

Our next and last example of a famous meta-argument is Galileo's critique of Aristotle's geostatic argument from vertical fall. Neither the original Aristotelian argument, nor the Galilean critique, are completely new topics in this book. Aristotle's argument from vertical fall was mentioned in the passage which I quoted earlier (chapter 1.1) as part of my ostensive definition of argumentative reality. And Galileo's critique was used (in chapter 2.5) to illustrate some of the elementary principles of argument interpretation and evaluation. Similarly, other Galilean examples have been previously used on several occasions: Galileo's critique of the observational argument for heavenly unchangeability was also used is that same chapter for the same purpose. Then in chapter 3, I used the ship analogy argument against the earth's motion as an example of ground-level argumentation; Galileo's critique of it as an example of argument analysis; and Galileo's simplicity argument for terrestrial rotation as an example of self-reflective argumentation. Such examples were, or were meant to be, relatively unproblematic because their function was primarily illustrative.

On the other hand, in the discussion of conductive arguments, there was another Galilean example with a somewhat more complex function (chapter 8.9). The example was Galileo's whole main argument in his *Dialogue on the Two Chief World Systems, Ptolemaic and Copernican*. This example was partly illustrative, since it was presented as an instance of a conductive argument. But recall that I had to give an argument to show that the argument in the *Dialogue* was indeed conductive; and my own argument turned out to be itself conductive. So that example had in large part a theoretical function.

Now in this chapter, the Galilean critique of vertical fall has not only the illustrative function of providing an instance of a famous meta-argument, and not only a theoretical function, but also a methodological function. In other words, it provides not just an illustration of a meta-argument, but primarily a case study of the approach advocated in this book and a test case of some theoretical claims to be elaborated presently.

13.1 Introduction: A Galilean Case Study and Test Case

Before undertaking our examination of this Galilean meta-argument, it will be useful to review our theoretical and methodological motivation. Galileo's argument is an important and significant one, from the points of view of the Copernican Revolution and of his own scientific achievements, which were so crucial and pivotal that he came to be regarded as one of the founding fathers of modern science. I have selected and will be studying this argument by employing the applied-logic or historical-textual approach, being sensitive to its pragmatic, comparative, empirical, historical, naturalist, and normative-cum-descriptive requirements. This will involve the careful observational description of this segment of argumentative reality, to ensure that my own conclusion and claims are robustly grounded in it. Since Galileo's argument is a meta-argument, it also constitutes a famous meta-argument in accordance with our meta-argumentation project. This means that Galileo's argument will also be treated in part as a piece of logical theorizing, and so we will be on the lookout for possible theoretical lessons.

Although these lessons cannot be fully anticipated, nevertheless there are some guiding ideas or questions that will be showing us the way in the observational description of this material and in drawing conclusions therefrom. There are three main guiding ideas. One is the just mentioned methodological question of the possibility or viability of doing argumentation theory as meta-argumentation. Another idea is the substantive-theoretical problem of what is meant by begging the question or *petitio principii*, and whether there are more than one version of it; the reason for this is simply that, as we shall soon see, Galileo's criticism of the Aristotelian argument from vertical fall is that it begs the question.

A third guiding question is the problem of the relationship between what Perelman would call demonstration and argumentation, but which I shall call the relationship between logic and rhetoric. For I start with argumentation as the domain of reality under investigation. Then, I take it to have (at least) two main components or aspects: one pertains to reasoning, demonstration, and relationships among truths; the other pertains to persuasion, acceptance, and relationships among beliefs. And I label the latter rhetoric, and the former logic; this is a narrow or restricted sense of the word logic, not to be confused with the broader meaning, such as when one speaks of logical theory, equating it with the general theory of argument, adapted from Toulmin. To avoid confusion, here one could speak of the rational instead of the logical, and of the problem of the relationship between reason or rationality and rhetoric or persuasion.[1]

A few other introductory remarks are in order, to provide some historical background and context.[2] Galileo included the critique of Aristotle's argument from vertical fall in the *Dialogue on the Two Chief World Systems, Ptolemaic and Copernican* (1632). This book contains Galileo's mature synthesis of astronomy, physics, and methodology. Its basic structure is that of a critical examination of all the arguments for and against the motion of the earth. He argues that the arguments in favor are much stronger than the arguments against, and in that sense the Copernican geokinetic theory is more likely to be true that the Ptolemaic, Aristotelian, geostatic theory. This work also happens to be the book that occasioned the trial and condemnation of Galileo by the Inquisition in 1633; he was convicted of being a suspected heretic for holding the astronomical and physical thesis that the earth moves, as well as the methodological and hermeneutical principle that Scripture is not a authority in physics and astronomy. The book is written in the form of a dialogue among three characters: Salviati, named after a deceased friend of his, represents the Copernican geokinetic side; Simplicio, named after the sixth-century Aristotelian commentator Simplicius, represents the Aristotelian, Ptolemaic, geostatic side; and Sagredo, named after another deceased friend, represents an educated intelligent layman who does not know much about the topic, but is interested, and wants to listen to both sides, and make up his mind as a result of critical reasoning.

The passage in question occurs at the beginning of the "Second Day" of the *Dialogue*, where the focus is the mechanical or physical arguments concerning the earth's daily axial rotation.[3] This comes after a "First Day" that discusses the similarities and differences between the earth and the heavenly bodies; these similarities and differences provide the basis of various arguments for and against the earth's location in the heavenly region among other heavenly bodies. There follows a "Third Day," which concentrates on the

---

[1] This is, and ought to be, reminiscent of what Eemeren and Houtlosser call the problem of the relationship between "dialectic and rhetoric" (2002a) or the phenomenon of "*strategic manoeuvering* ... the need to balance a resolution-oriented dialectical objective with the rhetorical objective of having one's position accepted" (2003b, 179). However, needless to say, the problem of the relationship between rationality and persuasion is not new, but goes back, for example, to Aristotle, Pascal 1657-1658/1974, and Feyerabend 1975; for a recent wide-ranging account, see Johnson 2009.
[2] For more details, see Galilei 2008, and the annotated bibliography there (pp. 37-43).
[3] Galilei 2008, 222-26; cf. the Appendix to this chapter, below.

astronomical evidence for and against the earth's annual revolution around the sun. Finally, there is a "Fourth Day" in which Galileo advances a particular argument in favor of the earth's combined motion of daily axial rotation and annual heliocentric revolution; the argument tries to show that the combination of these two terrestrial motions provides the best explanation of the occurrence and various properties of the phenomenon of the tides.

It should be kept in mind that the Galilean passage under consideration examines only one strand of the argumentation contained in the Aristotelian passage quoted earlier (chapter 1.1), namely the part at the end of the passage involving vertical fall. The other strands are criticized by Galileo in earlier parts of his book. In particular, the Aristotelian passage begins with a statement of the so-called argument from natural motion, which Galileo criticizes in the first part of the First Day of the *Dialogue*;[4] and the middle of the Aristotelian passage formulates the empirical argument for the coincidence of the centers of the earth and of the universe, which Galileo refutes later in the First Day.[5]

13.2 Observational Description of the Galilean Argumentation

We are now ready to undertake the observational description of the sample of argumentative reality reproduced in the passage quoted in the Appendix. Galileo begins with the following reconstruction of the argument from vertical fall:

(1) if the earth were moving (with daily axial rotation), then the place of ejection of a body thrown vertically upwards would move eastward along with the earth and the body would be left behind and fall some distance westward of that place;

(2) so, if the earth were moving, bodies would not fall vertically;

(3) but bodies do fall vertically;

(4) so, the earth does not move.

Next Salviati asks Simplicio how he knows that bodies fall vertically. Simplicio answers that we know it from observation, namely because they are *seen* to fall vertically. Here, Simplicio's subargument is the following:

(5) bodies are seen to fall vertically;

(3) so, bodies do fall vertically.

Now comes a crucial step. Salviati gets Simplicio to recognize that:

(6) *if* the earth were moving *and* bodies were *seen* to fall vertically, then they would actually *not* be falling vertically.

Later, I will give an explanation and justification of this claim. For now, staying closer to the text, from this claim Salviati, with Simplicio's endorsement, infers that:

(7) if one does not assume that the earth stands still, then one cannot conclude that bodies would actually fall vertically from the mere fact that they are seen to fall vertically.

The justification of (7) with (6) is repeated later, in Sagredo's speech. Hence we must regard as a partial restatement of (6) the reason for (7) given by Simplicio:

(8) if the earth moved, then bodies would actually *not* be falling vertically.

That is, in making Simplicio say this, Galileo means that if the earth moved, then (since bodies are seen to fall vertically) they would actually not be falling vertically. Finally, from (7) Salviati reaches his next conclusion to the effect that:

(9) the Aristotelian argument from vertical fall begs the question by assuming the very same thing it is trying to prove.

For, in inferring (3) from (5), the Aristotelian argument is concluding that bodies would actually fall vertically from the mere fact that they are seen to fall vertically. In light of (7),

---

[4] Galilei 1967, 9-32; cf. Finocchiaro 1980a, 32-34, 349-53.
[5] Galilei (1967, 32-38; 1997, 83-90); cf. Finocchiaro (1980a, 32-34, 353-56; 1997, 83-90); cf. also Arnauld and Nicole 1996, 190.

it follows that it is being assumed that the earth stands still (does not move), which is the argument's conclusion (4).

After this explanation of the nature and seriousness of the paralogism in the Aristotelian argument, Sagredo, with the help of Simplicio, objects to this criticism, especially to the justification of proposition (7). The objection is that (6), the reason for (7), hides a contradiction in its antecedent and that a new argument for the motionlessness of the earth can be formulated:
- (10) if the earth moved, then it would be impossible that bodies should be *seen* to fall vertically;
- (5) but bodies are *seen* to fall vertically;
- (4) so, the earth does not move.

And the justification for proposition (10) is that:
- (11) if the earth moved, then if bodies were seen to fall vertically, they would simultaneously have two motions—toward and around the center; and
- (12) it is impossible to have two such motions.

As Salviati implicitly admits, this revised argument is not open to the same objection as the original one. In particular, proposition (11) is an immediate consequence of of (6). What Simplicio is doing here is to try to turn the tables around. Whereas Galileo had used (6) to show the question-begging character of the original Aristotelian argument, Simplicio uses (6) to formulate a revised argument not beset by the same difficulty. Of course, his doing so is an implicit admission that the original argument was indeed begging the question.

However, Salviati notes that the worth of the revised argument depends wholly on the alleged impossibility of motion both toward and around the center, proposition (12). He finds three things wrong with this claim:
- (13) if Aristotle had proposition (12) in mind when giving the original argument, he should have explicitly said so (since it is an important part of the argument);
- (14) the mixed motion mentioned in (12) is not only not impossible, but necessary (something which Galileo promises to demonstrate later);
- (15) Aristotle himself would probably not accept (12), since he allows that kind of mixed motion when he admits that fire and the upper atmosphere move up by nature and rotate by participation.

Proposition (13) is obviously true, but it is not clear in what sense, if any, it makes the *revised* Aristotelian argument faulty. This question will be pursued below.

As regard proposition (14), it is clear how the revised Aristotelian argument would be affected: it would have a false premise, and thus be unsound, although it would remain formally valid. It is also clear that today we know that (14) is physically true, in the sense that such mixed motion is possible, though not necessarily "necessary." However, the relevant critical question would be whether (14) is true *in the context* of Galileo's discussion, that is, whether he does give the promised argument, and whether it is contextually correct. He does fulfill the promise, at least in part, in his argument dealing with the falling of bodies on a moving ship, and I believe that argument is basically correct.[6] This issue will be further pursued later.

Proposition (15) is partly true, because Aristotle does allow mixed motion for fire.[7] But whether it is completely true, i.e., whether Aristotle would consequently allow mixed motion for falling bodies is questioned by Simplicio on the ground that:
- (16) there is an important difference between the two situations: fire particles are so light that they could easily be carried along by the rotating air, but falling rocks are so heavy that they could not.

However, Salviati's response (not given in this passage) would be the following. This

---

[6] Galilei 2008, 226-33. Cf. Finocchiaro (1980a, 389-91; 1997, 159-70; 2010b); and chapter 3.2 above.
[7] Aristotle, *Meteorology*, book 1, chapter 7, 344a9-14.

Galileo on the Motion of the Earth 223

objection is misconceived because, for Aristotle, fire particles are supposed to be carried along not by the rotating air but by the rotating lunar sphere, which is made of the element aether; and compared to fire particles, this element is even rarer (less dense) than air is compared to rocks. Thus, Salviati's third criticism of the revised Aristotelian argument seems cogent; its effect is to make the revised argument seem un-Aristotelian.

The statement of proposition (16) is followed by a discussion of the ship experiment as possible evidence for (12). This further discussion need not concern us here, but will be pursued below, to the extent that it is relevant to the issues raised in this passage.[8]

13.3 Merits and Defects of Ground-arguments and Meta-arguments

So far my main concern has been with the argumentation as found in the Galilean text. Such Galilean argumentation is of course meta-argumentation, in the sense that it is about the argument from vertical fall stemming from Aristotle. Such meta-argumentation is both interpretive and evaluative, that is, aimed at understanding what the original ground-level argument was, as well as assessing its worth. However, so far we have just scratched the surface. Let us now delve a little deeper, starting with some evaluative issues, involving questions of what Toulmin calls merits and defects.

Let us consider more carefully proposition (6): if the earth were moving and bodies were seen to fall vertically, then they would actually not be falling vertically. I have already mentioned that this proposition is crucial, and that both sides seem to agree about its truth. Let us see how this truth can be made evident. For this purpose we need some visual aids, such as the following.

In Figure 13.1, the semicircles on the left side represent portions of the earth's equator, and the structures labeled AB and A'B' represent towers on the earth's surface. The parts of the figure on the right side are simply highly magnified representations of the situations on the left, so that the earth's surface appears flat because the distance involved (BB') is very small. The top part (a) of the figure represents a motionless earth, whereas the middle part (b) and the bottom part (c) represent the earth undergoing axial rotation from west to east (or clockwise), as suggested by the arrows. In parts (b) and (c) of the figure, each situation exhibits two different positions of the tower: the *unprimed* position (AB) represents the tower's position at the beginning of the experiment of dropping a rock from the top of a tower and letting it fall freely; the *primed* position (A'B') represents the tower's position at the end of the experiment when the fallen rock has reached the ground. Along the vertical wall of the tower, and between towers, there are solid and dotted lines: solid lines represent apparent vertical fall, namely the trajectory as seen by an observer standing near the tower; and dotted lines represent actual vertical fall, namely what would be happening in reality (or at least from the fixed vantage point of an extra-terrestrial observer). The lines (whether solid or dotted) between towers are drawn both as straight slanted lines and as parabolic slanted lines; here the main point to note is that they are both slanted, and although the parabolic representation is more accurate, this refinement plays no role in this discussion. In part (a) of the figure representing a motionless earth, apparent and actual vertical fall coincide. In part (b), representing a situation where the earth is rotating and apparent vertical fall is experienced on earth, the actual path is slanted or nonvertical. In part (c), representing a situation where the earth is again rotating but where actual vertical fall is taking place, a terrestrial observer would experience apparently slanted or nonvertical fall. Thus, part (b) of the figure shows that proposition (6) is true; and part (c) shows the truth of

---

[8] But, again, for details, see Galilei 2008, 226-33. Cf. Finocchiaro (1980a, 389-91; 1997, 159-70; 2010b); and chapter 3.2 above.

the related claim that: if the earth moved and bodies actually fell vertically, then they would be seen not falling vertically, but rather in a slanted path.

Figure 13.1

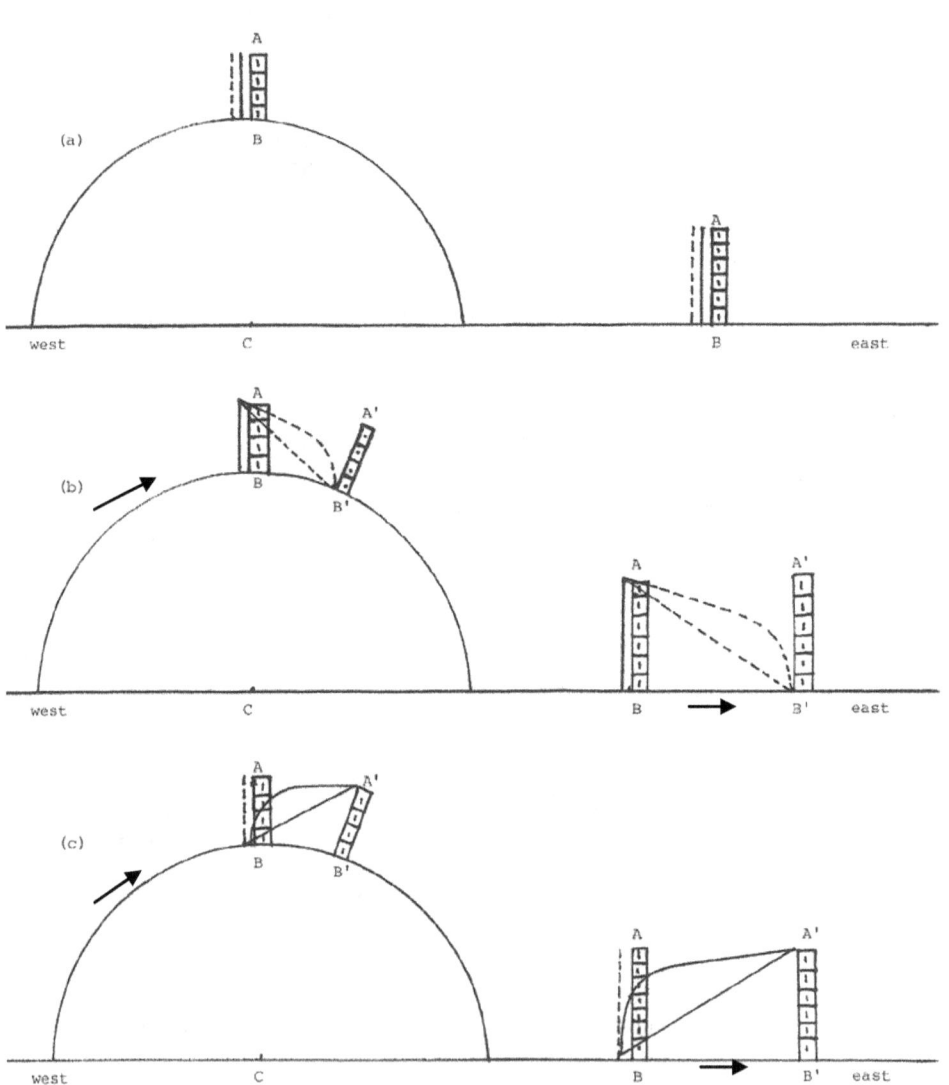

Another issue raised by Galileo's meta-argument is whether and how the revised Aristotelian argument is faulty because it is a revision of the original one. The revised argument begins by grounding the immobility of the earth directly and explicitly on *apparent* vertical fall, which is an unquestionable observational fact; so it might be called the argument from apparent vertical fall. This argument is quickly elaborated to make it dependent on the crucial premise about the alleged impossibility of motion being both around and toward the center, or "mixed," so to speak. The argument is formally valid, although questionable insofar as this premise is questionable.

The issue is, what does it matter if Aristotle did not have the revised argument in mind when he gave the original one. Is it really improper to revise the original argument from actual vertical fall? The impropriety, if any, is clearly not a formal (logical) one, any more than question begging is. It has to do with the way the (revised) argument originates. The Aristotelian (Sagredo acting as one) thinks of it when he clearly sees that if one infers actual vertical fall from apparent vertical fall, as he is, then one is assuming that the earth stands still. Now, this was implied by the crucial hypothetical that if the earth moves, then if bodies are seen to fall vertically they do not actually fall vertically. So we might say that the Aristotelian makes the switch when he realizes the damaging consequences of that crucial hypothetical. What he does at this point is to draw from it a consequence that suits him, namely:

(11) if the earth moved, then if bodies were seen to fall vertically, they would be moving with "mixed" motion.

How does this consequence suit him? In the sense that according to it, the motion of the earth implies mixed motion for falling bodies, which he then quickly denies. The propriety of the revision then depends on the propriety of this denial, more specifically on whether or not the denial is arbitrary. It does not depend on whether this denial is true (absolutely or contextually), which relates to the second possible deficiency of the revised argument discussed by Galileo.

The denial would be arbitrary if it were arbitrarily introduced, and it would be arbitrarily introduced if it were un-Aristotelian, which in one sense it is. Hence the premise on which the argument depends seems arbitrary. The concept which this defect suggests is that of an *ad hoc* argument. Such an argument could be regarded as a kind of question-begging argument, one where a conclusion is supported by means of a premise which is as much in need of support as the conclusion. This kind of argument is referred to by Galileo in the passage, and it is described as "an attempt to prove the unknown by means of what is equally unknown." The reference is made during the analysis of the original Aristotelian argument, when both sides assume and agree that the original argument is not an instance of this kind of question begging, although that shared assumption leads to the realization that the original argument is question-begging in the sense of circular.

This kind of question begging (i.e., the assumption of an equal unknown) corresponds to Perelman's notion, discussed above. That discussion also suggests that the criticism of a given argument in this manner is not the end of the argumentation in question, since in such cases involving a problematic proposition that is allegedly unknown, the arguer can and often will proceed to try to justify that proposition. That is precisely what happens in the present case, when a reference is made to the experiment of dropping rocks from the top of the mast on a ship moving forward, the result being supposedly that such bodies fall behind and strike the deck some distance backward from the foot of the mast. This experiment is mentioned by Simplicio at the end of the passage quoted in the Appendix, but in the passage immediately following (not included here), Salviati criticizes this subargument.[9]

Salviati objects partly that the impossibility of the dropped rock to move with mixed motion is not the only possible explanation of the alleged experimental result: this result might happen because the horizontal motion imparted by the ship to the rock is violent motion, which would be dissipated after the rock is left to itself; or it might happen because of air resistance, which would oppose the horizontal motion acquired by the rock. He also objects to the Aristotelian claim about the results of the ship experiment; he argues that this alleged experiment does not in fact happen this way, but rather that bodies fall to the foot of the mast regardless of whether the ship is moving or standing still. This criticism renders

---

[9] Galilei 2008, 226-33. Cf. Finocchiaro (1980a, 387-91; 1997, 160-70, 323-25; 2010b); and chapter 3.2 above.

the Aristotelian premise about the impossibility of mixed motion both unjustified and false.

Thus, on the one hand, the argument from apparent vertical fall and the impossibility of mixed motion is *ad hoc* and question-begging (in the sense of assuming an equal unknown). On the other hand, if one strips the argument of such rhetorical baggage, by constructing a longer argument that includes the ship-experiment justification of the impossibility of mixed motion, then the argument becomes inferentially inductively weak and substantively contains a false premise.

A third issue emerges directly out the second one just discussed. Now the question is whether there are other rhetorical considerations made by Galileo in his critique, besides the one just discussed, relating to the *ad hocness* or question begging of the revised argument from vertical fall. In fact, the rhetorical import of Galileo's critique has not escaped the attention of some scholars. In particular, Paul Feyerabend famously exploited it to support his anarchist methodology that anything goes in science, and more generally in argumentation, attributing to Galileo various pejorative techniques such as deceptive propaganda, psychological tricks, and sophistry (Feyerabend 1975, 81-92). Although Feyerabend misunderstands both the logical structure and the rhetorical import of the Galilean meta-argument, he does perceptively and insightfully see that Galileo is struggling to take into account and reach a proper balance of two aims, one involving logical rationality, the other rhetorical persuasion.[10] In the language of the pragma-dialectical school, Galileo is involved in strategic maneuvering.[11] When seen in such a light the Galilean critique turns out to be highly instructive. To unravel this issue I want to make three points.

First, it is certainly a clever move for Galileo to begin his critique of the argument from vertical fall by interpreting it to refer to *actual* vertical fall. To be sure, it must be admitted that the original Aristotelian argument simply interchanges apparent and actual fall without distinguishing them; so Galileo could have begun with either version of the argument. The fault of the argument from *apparent* vertical fall (the "revised" argument, in Sagredo's first speech in the quoted passage) is that it depends on a premise which is as much in need of proof as the conclusion at issue. This fault is less serious than the circularity of the argument from actual vertical fall. It is indeed more effective to start with the more serious criticism, and so Galileo's first rhetorical move is a judicious one.

Second, having focused on the actual vertical fall version, Galileo had in front of him a formally valid instance of denying the consequent (*modus tollens*). Of the two premises of this argument, propositions (2) and (3) above, Galileo could have questioned either one. If he had questioned the conditional proposition, the circularity could not have been exhibited as easily since, as it will be shown below, the justification of this conditional premise is circular only insofar as it depends ultimately on the nonconditional premise of the original *modus tollens,* namely the proposition (3) that bodies really fall vertically. It is certainly rhetorically effective, but certainly not improper, to exhibit the failure of an argument *in the easiest possible way.*

Third, when Galileo comes around to discussing the apparent vertical fall version of the argument, this is made to look like a *revised* version of the original argument, and hence to some extent as *ad hoc*, as we have seen. This seems to be a merely rhetorical fault, rather than a logical one. However, the argument is also charged as being dependent on a premise which is as much in need of proof as its conclusion. This feature, though not purely

---

[10] For a detailed criticism, as well as some appreciation, of Feyerabend's account, see Finocchiaro 1980a, 180-201.

[11] Cf. Eemeren (2009, 2010) and Eemeren and Houtlosser (2002b, 2003a, 2003b).

Galileo on the Motion of the Earth   227

logical, is not purely rhetorical either. For although questions of the comparative knowability of propositions are context-dependent, the context is an epistemological or logical one, rather than one dependent merely on persuasion or on who is speaking to whom. Hence, this feature of Galileo's meta-argument may be deemed logical in Toulmin's sense of applied logic. Finally, when such an epistemological flaw is remedied by the addition of the ship-experiment evidence, there emerge two different flaws, inductive weakness and substantive falsehood. Now, the sequential order of such criticisms is surely a rhetorical matter of persuasion, even though their content is logical. But there is no reason to confuse the two perspectives, or to attribute to Galileo such a confusion; rather his strategy can be seen to be proper if we keep in mind the context, namely that different criticisms apply to different arguments (or different versions or segments of the same arguments), and do so in relation to still other arguments.

13.4 Structure of Circular and Question-begging Arguments

Next, we shall explore some of these issues even more deeply, seeking an even deeper understanding and evaluation of the Aristotelian ground-level argument and the Galilean meta-argument. This will involve a greater focus on what Toulmin called the structure of arguments.

With the above discussions behind us, we shall now feel freer to engage in abstractions and reconstructions. First of all, let us use the following abbreviations:

M = the earth *moves* (with daily axial rotation);
S = bodies are *seen* to fall vertically (to a terrestrial observer);
V = bodies do actually fall *vertically* (in physical reality);
W = bodies ejected vertically upward land some distance *westward* of the place of ejection; and
X = freely falling bodies move with mixed motion that is simultaneously toward and around the center.

Moreover, I shall use the methodical numbering system introduced earlier (chapter 2) for keeping track of various claims in argumentation, and for indicating their place in the network that makes up the propositional macrostructure of arguments. Recall that he key idea is that if a given claim is labeled [n], then the premises that directly support it are labeled [n1], [n2], [n3], etc.; and if claim [nm] is part of some subargument, then the premises directly supporting it are labeled [nm1], [nm2], [nm3], etc. Such numbering is also associated with the visual representation of complex arguments by means of structure diagrams in the shape of tree roots.

The original segment of the argument from actual vertical fall is:

(A) (111)   if M, then W.
    (11)   so, if M, then not-V;
    (12)   V;
    (1)   so, not-M.

When Galileo asks the Aristotelian how he knows that V is true, the following justification emerges:

(B) (12221) if not-M and S, then V;
    (1222) so, if not-M, then if S then V;
    (1221) not-M;
    (122)   so, if S, then V;

(121)   S;
(12)   so, V;

Putting these two together, we get the full argument from actual vertical fall:

(C) (12221) if not-M and S, then V;
    (1222)   so, if not-M, then if S then V;
    (1221)   not-M;
    (122)   so, if S, then V;
    (121)   S;
    (12)   so, V;
    (111)   if M, then W;
    (11)   so, if M, then not-V;
    (1)   so, not-M.

Here, in (C), the last four lines constitute the initial segment of the argument. Propositions (121) and (122) constitute the justification of V. Since (122) is not true if M is, it needs a justification, which in the context can only be provided by (1221) and (1222). The final premises on which the final conclusion, not-M, rests are (12221), (1221), (121), and (111). Unfortunately, one of them, (1221), is identical with the final conclusion. The original segment (A) is question-begging in the sense that fuller argument (C) is circular, i.e., it uses the conclusion as one of the premises. The role of the crucial consideration "if M and S, then not-V" is to motivate the justification of (122), and thus force the Aristotelian to use (1222) and not-M, as grounds.

This full argument (C), however, is not the same as the original one (A), given in Salviati's first speech, and containing only the last four lines of this reconstruction. It is only by logical analysis that the speakers can construct the fuller deduction from the original segment. Since the rhetorical uselessness follows (obviously) from the structure of this deduction, that shows that the rhetorical analysis is being grounded on the logical one. It follows that one cannot carry out the former in abstraction of the latter, although two distinct claims are being made from the two points of view. Moreover, it is also obvious that there is no way of identifying the begging of the question by just looking at the initial segment of the argument: "(111), so (11); but (12); so (1)"; that is, by just looking at Salviati's first speech in the Galilean text. One also needs the other parts of the passage which provide evidence that this segment is really part of the longer one.

How much context need one provide in one's analysis? After all, in Galileo's book the discussion does not end where our quotation ends (see Appendix). I think the answer must be that one needs as much context as is necessary to justify the claim one is making in one's meta-argument. Although some passages may be too short to allow any such meta-argumentation to get started, the present passage is obviously sufficient since it provides a good illustration of a key concept, *petitio principii*, which is crucial for an understanding of the relationship between argumentation and demonstration, between rhetoric and logic.

Galileo does not ask the Aristotelian why proposition (111) is supposed to be true, and exactly how it implies (11). Let *us* pursue these questions, beginning with the latter. First the Aristotelian would probably say:

(112)   if W, then not-V;
(111)   if M, then W;
(11)   so, if M, then not-V.

Then Simplicio could be asked 'why (112)?', and his answer would probably be:

Galileo on the Motion of the Earth   229

(1122)   if W, then not-S;
(1121)   if not-S, then not-V;
(112)   so, if W, then not-V.

Proposition (1122) would be obvious or self-evident. But 'why (1121)?'. Probable answer:

(11211)   if V, then S;
(1121)   so, if not-S, then not-V.

But why (11211)? Because:

(112112)   if not-M, then if V then S;
(112111)   not-M;
(11211)   so, if V, then S.

Thus, the justification of (112) would also contain not-M as a premise. This would indicate a *second* circularity in the original argument. That is, combining all these deductions in the (11) strand, we get:

(D) (112112)   if not-M, then if V then S;
    (112111)   not-M;
    (11211)   so, if V, then S;
    (1121)   so, if not-S, then not-V;
    (1122)   if W, then not-S;
    (112)   so, if W, then not-V;
    (111)   if M, then W;
    (11)   so, if M, then not-V.
    (12)   V;
    (1)   so, not-M

Let us now ask whether (111) is true. The Aristotelian would say so because:

(E) (1112)   if V, then if M then W;
    (1111)   V;
    (111)   so, if M, then W.

Here the first premise if obvious, but in order to justify the second one, Simplicio would probably give the same justification as for (12). This would yield a *third* circularity for the argument from vertical fall.

It should be noted that both this third circularity and the first one are the result of the attempt to justify V empirically, by grounding it on S. But the Aristotelian might resort to an *a priori* justification. He could argue, for example, that bodies really fall vertically because it is their nature to move in a "real" straight line toward the center of the earth. This would be circular insofar as "actual vertical fall" here means "actual rectilinear motion toward the center of the earth." Moreover, insofar as there is a reference to the "nature" of bodies we would have a disguised refusal to give a justification. Hence the final conclusion would be resting on a proposition equally in need of support; and this would be begging the question in the other sense of attempting to prove the unknown by means of what is equally unknown.

The Aristotelian could also argue that bodies actually fall vertically because it is their nature to move in a "real" straight line toward the center of *the universe*, and this center happens to coincide with the center of the earth. Then Galileo would question this second premise. The Aristotelian could support it *a priori* or *a posteriori*. The empirical argument, as Galileo shows elsewhere, is itself circular. The *a priori* argument would presuppose that the earth does not rotate on its axis while located at the center of the universe; this would render circular the original objection to the motion of the earth.

Finally, let us combine these three argumentative strands, each of which instantiates begging the question, into a single argument, called (F). Rather than even writing it out in symbols, let us merely draw the diagram that represents the combined structure of the whole (Figure 13.2).

Argument F: Figure 13.2

```
                               1
                      ┌────────┴────────┐
                     11                 12
                  ┌───┴───┐          ┌───┴───┐
                 112     111        121     122
               ┌──┴──┐  ┌─┴─┐              ┌─┴─┐
             1121  1122 1111 1112        1221 1222
              │         ┌┴┐                     │
            11211     11111 11112            12221
           ┌──┴──┐         ┌┴┐
        112111 112112   111121 111122
                                │
                             1111221
```

The three circularities stem from the identity of proposition (1) with, respectively, (1221), (112111), and (111121).

Let us now examine the *revised* Aristotelian argument, namely the argument from apparent vertical fall. The original segment of this argument is the following:

(G) (22)   if M, then not-S;
    (21)   S;
    (2)   not-M.

Here the justification of the major (conditional) premise requires the following subargument:

(H) (22122)   if M and S, then not-V;
    (22121)   if not-V, then X;
    (2212)   so, if M and S, then X;
    (2211)   not-X;
    (221)   so, not-(M and S);
    (22)   so, if M, then not-S.

Combining the two together, we get the full argument from apparent vertical fall:

(I) (22122) if M and S, then not-V;
(22121) if not-V, then X;
(2212) so, if M and S, then X;
(2211) not-X;
(221) so, not-(M and S);
(22) so, if M, then not-S;
(21) S;
(2) not-M.

The basic problem here is that proposition (2211) is as much in need of support as (2). The argument begs the question, not in the sense that its final conclusion is one of the final premises, but in the sense that among its final premises there is a proposition (2211 = not-X) that is no better known than the conclusion (2 = not-M). And as we saw before, if this problematic proposition (2211 = not-X) is justified by means of the ship experiment, then it becomes criticizable as groundless and false.

Now, we can also re-examine a question we have touched on before. How is this argument (I) related to the original full argument (C). From the logical point of view (indeed from a mere visual inspection in the present discussion), they are different arguments. However, rhetorically speaking, they are really two versions of the same *ambiguous* argument. The ambiguity in this root argument derives from the confusion of apparent vertical fall and actual vertical fall, which ambiguity is present in Aristotle's original statement of the argument and which (for the sake of the meta-argument) is retained even in Galileo's statement at the beginning of the passage under consideration. So one of the things which Galileo accomplishes in this passage is to distinguish these two meanings of vertical fall.

This does not mean, however, that the Aristotelian argument from vertical fall is committing a fallacy of equivocation. For I take equivocation to be the fallacy committed when the premises contain a term with two possible meanings such that: if it is used in one sense, one of the premises is clearly false, although it together with the other premises implies the conclusion; whereas if the term is used in the other sense, the premises clearly do not imply the conclusion, although admittedly the previously problematic premise is clearly true.[12] Such exploitation of ambiguity is simply not happening in this argument. Instead what is happening is that the term "vertical fall" is being used with two possible meanings in the root argument: "vertical fall; if earth moves then no vertical fall; so, earth does not move"; and if "vertical fall" means "actual vertical fall," this argument turns out to beg the question because the full argument (C) of which it is part is circular; whereas, if "vertical fall" means "apparent vertical fall," then the root argument begs the question because it becomes part of a fuller argument (I) containing a premise which is no better known than the conclusion.

Nevertheless, what we have here is a beautiful illustration of another crucial difference between the logical and the rhetorical points of view, between Perelman's demonstration and argumentation. In Perelman's own words, this is "another difference of paramount importance between argument and formal proof. The standard logical calculi are formulated in artificial languages in which any one sign has one, and only one, meaning; in natural languages the same word often has different meanings" (Perelman 1979, 144). Although Galileo is not using any formal calculus, he is using logical analysis to strip Aristotle's original "rhetorical" argument of some of its "rhetoric," so to speak. That is, Galileo is trying to turn the problem at hand from one where only rhetorical analysis would

---

[12] For an elaboration and some examples, see Finocchiaro (1980a, 354, 370-71, 376, 379-80, 410-11; 1997, 317-18, 322, 325; 2005a, 136-45).

be relevant to one susceptible to some logical analysis. Of course, Galileo is not eliminating completely the rhetorical aspects of the problem, nor would he want to; but he is using logic so that the rhetorical problem is moved to a different point. This means that the line of separation between rhetoric and logic shifts as knowledge grows; or to be more exact, more and more arguments enter the domain of logic after crossing the borderline with rhetoric; or again, since I believe that the two are not separate domains but distinct aspects, what happens is that there is a growth of problems of the kind where the question of acceptance becomes reduced to the question of establishing truth.

Let us now construct some other arguments that explore more systematically the consequences of the most important propositions which are indisputable in this context. This can be done in terms of the following semi-formal derivation, where the claim on each line is followed by a justification (<within pointed brackets>).

(J) (J1)   if M and S, then not-V;   <empirical intuition, cf. Figure 13.1(b)>
    (J2)   if not-M and S, then V;   <empirical intuition, cf. Figure 13.1(a)>
    (J3)   if M and V, then not-S;   <empirical intuition, cf. Figure 13.1(c)>
    (J4)   if not-M and V, then S;   <empirical intuition, cf. Figure 13.1(a)>
    (J5)   S;   <simple observation>
    (J6)   if S and M, then not-V   <from (J1)>
    (J7)   if S, then if M then not-V   <from (J6)>
    (J8)   if M, then not-V   <from (J7) and (J5)>
    (J9)   if V, then not-M   <from (J8)>
    (J10) if S and not-M, then V   <from (J2)>
    (J11) if S, then if not-M then V   <from (J10)>
    (J12) if not-M, then V   <from (J11) and (J5)>
    (J13) V iff not-M   <from (J12) and (J9)>
    (J14) if M, then if S then not-V   <from (J1)>
    (J15) if M, then not-(if S then V)   <from (J14); problematic step; see discussion below>
    (J16) if (if S then V), then not-M   <from (J15)>
    (J17) if not-M, then if S then V   <from (J2)>
    (J18) (if S then V) iff not-M   <from (J17) and (J16)>
    (J19) if M, then if V then not-S   <from (J3)>
    (J20) if M, then not-(if V then S)   <from (J19); problematic step; see discussion below >
    (J21) if (if V then S), then not-M   <from (J20)>
    (J22) if not-M, then if V then S   <from (J4)>
    (J23) (if V then S) iff not-M   <from (J22) and (J21)>
    (J24) not-M iff (S iff V)   <from (J23) and (J18)>
    (J25) M iff not-(S iff V)   <from (J24)>.

Proposition (J1) is Galileo's crucial consideration, and can be seen to be true upon reflection with the help of one's empirical intuition. This was (hopefully) accomplished by means of Figure 13.1 above. Similarly, (J2), (J3), and (J4) are obvious truths. Both Galileo and the Aristotelians would accept them. Proposition (J5) is an empirical fact, ascertainable though simple observation, and it is certainly shared by both sides.

Proposition (J13) tells us that to say that falling bodies actually fall vertically is equivalent to saying that the earth stands still. Hence it seems difficult to avoid circular reasoning as long as one grounds the immobility of the earth on actual vertical fall. Besides allowing Galileo to criticize the original Aristotelian argument, (J13) is useful to him as a step toward his ultimate aim of justifying the motion of the earth. For he now knows that to do this he has to deny that bodies actually fall vertically. His formulation, elsewhere (Galilei 2008, 230-33), of the principle of inertia (or conservation of motion) is a step toward the justification of that denial.

The steps from (J14) to (J15) and from (J19) to (J20) are problematic. For, if these claims are interpreted as sentence schemata, and the "if-then" in them is interpreted as the "material conditional" of truth-functional logic, then the second schema in each pair is not a logical consequence of the first. This can be seen from the following assignment of truth values, which renders the first member of such pairs true and the second false: assign truth to M and V and falsehood to S for the step from (J14) to (J15); and assign truth to M and S and falsehood to V for the step from (J19) to (J20).

To address this problem, I would say the following. Partly I would want to say that I am reconstructing Galileo's reasoning, and if it is invalid, that is not the fault of my reconstruction; and if one objects that I ought to use the Principle of Charity, I can assure the critic that I have already done so, and I challenge him to find a more accurate reconstruction. Partly I would say that I am advocating a practice-oriented logical theory (à la Toulmin), which denies as an oversimplification the hierarchical principle that logical norms are primary, independent of practice, and to be imposed upon practice; instead it advocates the Kantian-inspired interactionist principle that although argumentative practice without logical norms is blind, logical norms without argumentative practice are empty. Partly, there is no good reason why the "material" interpretation of the "if-then" should be accepted here, and I have deliberately not symbolized it by means of the "horseshoe" symbol of formal deductive logic, but retained the more vague natural language expression. Similarly, note that such problematic inferences are assuming a principle that seems rhetorically sound, namely that: if (if S then not-V), then not-(if S then V); for this amounts to saying that if one can infer not-V from S, then one cannot infer V from S. Partly, I would say that perhaps the "if-then" of these schemata should be interpreted as entailment or formal implication; then "not-(if S then V)" would mean that S does not entail V, which does follow from "S entails not-V", under the plausible assumption that we are dealing with self-consistent sentences. Finally, one could try to solve the problem by adapting a probabilistic interpretation of conditionals, along the lines of Adams's (1975, 69-102) account: that is, conditionals are assigned probabilities rather than truth values; their probability is not the probability of being true, but the ratio of the probability of the conjunction of antecedent and consequent to the probability of the antecedent (Adams 1975, 5ff); and inferences involving conditionals are best evaluated in terms of the probabilistic criterion that it should be impossible for the premises to be probable while the conclusion is improbable (Adams 1975, 1).

Proposition (J18) is a somewhat weaker assertion than (J13). It may be interpreted as saying that to infer actual vertical fall from apparent vertical fall is equivalent to assuming the immobility of the earth. Hence, as previously noted, as long as one gives that kind of empirical justification of actual vertical fall when justifying the stability of the earth, one is arguing in a circle.

Proposition (J24) tells us that to say that the earth stands still is equivalent to regarding S and V as equivalent. Hence, justifying not-M involves a failure to distinguish S from V. And (J25) tells us that to distinguish apparent from actual vertical fall is to assert that the earth moves. Now, the first thought that comes to mind is to use these facts to prove the motion of the earth by arguing that since it is one thing for bodies to be seen to fall vertically and another for them to actually fall vertically, it would follow that the earth moves, by (J25). This would be an *a priori* justification, a kind of transcendental deduction of the earth's motion, so to speak. Hence, Galileo could be claimed to have, with his criticism of the Aristotelian argument, not merely refuted the argument, but also its conclusion, i.e. proved the motion of the earth. The fact that Galileo did not claim so much should make us suspect that there is something wrong with this suggestion.

The problem with the suggestion must be that the distinction between S and V is not subsisting in a Platonic heaven, but is something dependent on M, an empirical fact. The consequences of this are somewhat surprising. For it would seem that the nonequivalence between S and V is a conceptual matter, accessible by pure thought. But if this were so, then one would be able to prove the motion of the earth in the manner indicated, which seems an incorrect procedure. Hence the distinction between S and V is either not conceptual, or some conceptual distinctions are empirical at least in the sense of having an empirical origin. The last alternative seems preferable. In other words, the use and misuse to which (J25) could be put suggests for us an asymmetry between empirical and conceptual knowledge: empirical knowledge can generate conceptual knowledge, but not vice versa.

13.5 Conclusions: Logic vs. Rhetoric, and Theorizing vs. Meta-argumentation

The preceding analysis suggests several conclusions, involving the concept of begging the question, the relationship between logic and rhetoric, and the project of doing argumentation theory as meta-argumentation.

It has emerged that there are two main types of question-begging arguments. In a first, stronger sense, an argument begs the question when one of its premises is identical to the conclusion. Arguments in which a particular proposition serves such a double duty are commonly called circular arguments; and so we may say that question-begging arguments, in the strong sense, are circular arguments. In rare cases, the problematic premise is explicitly present in a given argument. More frequently, the question-begging premise is implicit, in the sense that it becomes explicit only when the original argument is elaborated into a longer and more complex one that is needed in the context for a fuller justification of the conclusion. For example, the Aristotelian argument from actual vertical fall (A) needs to be elaborated into the fuller argument (C) for a justification of the earth's rest, and we can see by inspection that (C) is made circular by the identity of propositions (1221) and (1). The original argument (A) begs the question implicitly, and the fuller argument (C) does so explicitly.

The second, weaker type of begging the question occurs when an argument attempts to justify a conclusion by means of a premise that is no better known that the conclusion. Here the problematic premise is not literally or substantively identical to the conclusion, but rather it is identical or similar to it in epistemological status, so to speak. And again, the problematic premise can be explicit or implicit. For example, the argument from apparent vertical fall (G) begs the question in the sense that the fuller argument (I) has a premise, proposition (2211), which is no better known that the conclusion (2).

Regarding the relationship between logic and rhetoric, recall first that here I have taken these terms in the following sense. Logic, in a restricted sense, pertains to the domain of rationality, reasoning, demonstration, and relationships among truths; rhetoric pertains to the realm of persuasion, acceptance, and relationships among beliefs.[13] When so understood, they are two distinct aspects of argumentation, but they are not necessarily separate or incompatible. Often argumentation aims both at discovering and justifying the truth, and at persuading others to accept it.

Galileo's critique of the argument from vertical fall shows this kind of interaction, namely distinct aims and viewpoints that can be nevertheless be combined judiciously and soundly. For example, his initial decision was a purely rhetorical one that could not be derived from logical considerations, although it was a good and fruitful decision; that is, his

---

[13] Again, as noted earlier, for other views and useful accounts see Eemeren (2009, 2010), Eemeren and Houtlosser (2002b, 2003a, 2003b), and Johnson 2009.

decision to interpret the argument from vertical fall first as referring to actual vertical fall, so that it could be criticized as question-begging because circular, and to examine the argument from apparent vertical fall at a second stage of analysis, and criticize it as begging the question for using a premise no better known than the conclusion. And this decision was good because the former defect is more serious than the second. Similarly, the same applies to Galileo's decision to show the question-begging circularity of the original *modus-tollens* step of the argument from actual vertical fall by questioning the justification of the minor premise V rather than that of the conditional premise "if M, then not-V." This was a good decision because the question-begging character of V can be proved more easily and quickly than that of the conditional premise. Analogously, something similar can be said of the decision to introduce the argument from apparent vertical fall as a revision of the argument from actual vertical fall, so as to criticize it also as being *ad hoc*. This decision enabled Galileo to attribute to the latter argument two defects, instead of just one: not only is it question-begging in the sense of being based on a premise no better known than the conclusion, but also it is an arbitrary revision of the earlier argument. Finally, if and to the extent that the argument from apparent vertical fall is freed of the latter two defects, then it becomes vulnerable to criticisms of weak inferential link and false or unjustified premise.

At a deeper level, this material substantiates the very important point, stemming from Perelman (cf. chapter 1 above) that question-begging arguments are defective rhetorically, but not logically; indeed, not only are they not defective logically, but also they must possess logical merit in order to exhibit the rhetorical defect. Galileo's critique shows how the argument from vertical fall is rhetorically ineffective *because* it is logically unobjectionable, or to be more precise, because it is technically valid from the point of view of formal deductive logic. Question-begging circularity can come about only if we have formal deductive validity; and question-begging assumption of an equal unknown can be a problem only if the assumption is accompanied by a valid derivation of the conclusion. Such arguments are persuasively *ineffective* insofar as they are formally deductively *valid*. Thus, the justification of the *negative* rhetorical evaluation has to be based on a *positive* logical assessment of validity, which in turn is based on a logical interpretation of the circular structure for question-begging arguments of the first kind, and on a logical assessment of the equal epistemological status of conclusion and problematic premise for question-begging arguments of the second kind.

And this brings us to the topic of meta-argumentation. Galileo's critique is a meta-argument in the simple sense that it is primarily an argument about the argument from vertical fall, which latter argument may be called the ground-level or object argument. The critique advances some claims about the structure, merits, and defects of the argument from vertical fall, and it justifies them with reasons describing various aspects of the argument and appealing to various principles. These principles are not explicitly discussed, but merely appealed to; they do not become the primary subject matter of the discussion.

Thus, the Galilean critique is *not* a meta-argument in the sense that it leaves the ground level, moves to the meta-level, and discusses a new topic stemming from, but different than, the original topic.[14] For example, such would be the case if the speakers in the dialogue went on to discuss whether there is really anything wrong with begging the question, circular reasoning, and supporting the unknown by means of what is equally unknown.

Another interesting and important meta-level digression in our case study would be to

---

[14] This seems to be the notion of meta-argument, or at least of the meta-level, which would also include meta-dialogues, advanced and elaborated by Krabbe 2003 and Krabbe and van Laar 2011.

discuss the logical authority of Aristotle, which is an issue raised by Simplicio on an earlier occasion, when Salviati criticized as begging the question (because circular) Aristotle's empirical argument for the coincidence of the centers of the earth and of the universe.[15] Simplicio had objected that such criticism must be untenable because Aristotle was the founder of the science of logic, and so he could not have committed such a logical error.[16] Salviati's reply is that Aristotle's authority as a logician is not sacrosanct, and in any case it pertains to the domain of logical theory rather than ratiocinative practice; his authority as a logical practitioner cannot be grounded on his undisputed authority as a logical theorist, but only on the correctness of his concrete reasoning; the latter can be tested only by the analysis of his actual arguments, which analysis one should thus be free to carry out. This argument about the role of the logical authority of Aristotle is a meta-argument in the sense of an argument at the meta-level, occasioned by the original ground-level argument and leaving it (temporarily) behind. Such discussions, which I call methodological reflections (Finocchiaro 1997, 335-56; 2005a, 96-100), are important in argumentation and more generally in the search for truth.

There is another fruitful methodological reflection that could have been elaborated in this Galilean passage, but was not, although it is elaborated elsewhere. That is, when in the present passage Simplicio justifies actual vertical fall (V) by means of apparent vertical fall (S), he could have insisted that the major premise of this subargument was the principle that sense experience is normally reliable, namely, that normally the human senses tell us the truth; and then the discussion could have shifted to the question of the truth and applicability of such a principle of empiricism, occasioning the distinction between naïve and critical empiricism. The empiricist principle is precisely the subject of a methodological reflection later in the *Dialogue*, where the anti-Copernican argument involved is the objection from the alleged deception of the senses, and where the observation of apparent vertical fall on a rotating earth is given by Simplicio as an example of (allegedly) a deception of the senses.[17] Although such a methodological discussion would be relevant, as well as important and interesting in its own right, it is obvious that it does not come up in the Galilean critique of the argument from vertical fall. What we have there instead is a much more concrete and directly relevant type of argumentation.

Methodological reflections typically involve arguments about methodological principles, which are general rules about the conduct of inquiry, search for truth, and acquisition of knowledge. Such reflections may even be meta-arguments if and when the methodological principles under discussion are principles of argumentation. However, not all arguments about methodological principles are meta-arguments,[18] nor are all meta-arguments about methodological principles. Some meta-arguments are simply about particular arguments or classes of arguments. Galileo's critique of the argument from vertical fall is such a non-methodological meta-argument. It argues about the concrete

---

[15] Galilei (1967, 32-38; 1997, 83-90); cf. Finocchiaro (1980a, 32-34, 107-8, 353-56; 1997, 83-90); cf. also Arnauld and Nicole 1996, 190.

[16] Simplicio's words are sufficiently quaint and eloquent to deserve quotation: "Please, Salviati, speak of Aristotle with more respect. How can you ever convince anyone that he could have committed a serious error like assuming as known what is in question, given that he was the first, only, and admirable explainer of syllogistic forms, demonstration, fallacies, the methods for exposing sophisms and paralogisms, and in short the whole of logic? Gentlemen, one must first understand him perfectly, and then try to impugn him" (Galilei 1997, 87).

[17] Galilei 1997, 212-20; cf. Finocchiaro (1980a, 124-25; 1997, 344-48).

[18] Although they would probably be in a Toulminian approach to epistemology and methodology, because then all epistemological and methodological principles would be interpreted as principles of argumentation about arguments for knowledge claims and arguments for procedural claims.

details of the object argument.

Meta-argumentation is a form of theorizing about arguments that structures itself as argumentation. In this sense, all meta-argumentation is theorizing, but not all theorizing is meta-argumentation. The latter is easy to understand when theorizing is about such things as numbers, atoms or molecules, and historical events; such theorizing may or may not take the form of argumentation, but there is no reason why it should be meta-argumentation, about arguments. The more difficult and interesting case is theorizing about arguments that is not meta-argumentation. Can there be such a thing?

An elaborate answer to this question is beyond the scope of the present chapter, but here I can say the following. Argumentation becomes relevant primarily in the context of the testing, justification, and criticism of theoretical claims. However, before such meta-argumentation becomes possible, the theoretical claims must be conceived and formulated, and it is not clear that such concept formation, verbal expression, and propositional explicitization involve argumentation, at least in any important and essential way. Moreover, even after the theoretical claims have been justified through meta-argumentation, that is not the end of inquiry, but rather such activities as clarification and systematization may come into play; and again, it does not seem that argumentation is essentially involved in such clarification and systematization.

Thus, meta-argumentation may not be the only form of theorizing, but it is one. What is it like? To learn and say more about it, following the applied-logic approach, and in accordance with the meta-argumentation project being implemented in this book, we can study the particular example of the Galilean critique of the argument from vertical fall. To this end, let us reconstruct this Galilean meta-argumentation more explicitly. Although empirical accuracy will be the paramount aim, another guiding idea is the question of whether, how, and to what extent such meta-argumentation enables us to stay close to the original ground-level arguments.

For this purpose, we need some additional abbreviations. Continuing with the previously used alphabetical sequence of letters to refer to various arguments and argument structures, let (K) be the original Aristotelian arguments as found in the previously quoted passage from Aristotle's *On the Heavens*:

(K) "It is clear, then, that the earth must be at the center and immovable, not only for the reasons already given, but also because heavy bodies forcibly thrown quite straight upward return to the point from which they started, even if they are thrown to an infinite distance."[19]

Next, let us label (L) the Galilean statement of the argument from vertical fall, as found in Salviati's first speech in the passage quoted in the Appendix. This, or course, may be regarded as a paraphrase, but also an elaboration, of the original Aristotelian argument:

(L) a most certain argument for the earth's immobility is based on the fact that we see bodies which have been cast upwards return perpendicularly by the same line to the same place from which they were thrown, and that this happens even when the motion reaches a great height; this could not happen if the earth were moving because, while the projectile moves up and down separated from the earth, the place of ejection would advance a long way toward the east due to the earth's turning, and in falling the projectile would strike the ground that much distance away from the said place. [Galilei 2008, 223]

Now it also will be useful to have a new propositional abbreviation:
Z = bodies fall vertically

---

[19] Aristotle, *On the Heavens*, book 2, chapter 14, 296b22-27 (1941, 434).

for a claim which, as we saw earlier, is vague and ambiguous insofar as it does not yet distinguish between actual and apparent vertical fall. Then, the argument in (L) can be reconstructed as follows:

(M) (311)   if M, then W.
    (31)   so, if M, then not-Z;
    (32)   Z;
    (3)   so, not-M.

These new abbreviations, together with the previous ones, enable us to succinctly state Galileo's main meta-argument as follows:

(N) (41) The argument from vertical fall is argument M. (421) Argument M can be interpreted as either argument A or argument G, because (4211) proposition Z can mean either V or S. (422) If argument M is interpreted as argument A, then it begs the question; for (4221) A presupposes argument C, since (42211) proposition V needs a justification, and (42212) the only justification of V available in the context is argument B; but (4222) argument C is circular, because (42221) its final premise 1221 is identical to its final conclusion 1. (423) If argument M is interpreted as argument G, then it also begs the question; for (4231) G presupposes argument I, since (42311) proposition 22, if M then not-S, now needs a justification, and (42312) the only justification of it available in the context is argument H; but (4232) argument I begs the question because (42321) its final premise 2211 is no better known than its final conclusion 2. Therefore, (42) argument M begs the question. Consequently, (4) the argument from vertical fall is ineffective, i.e., it is an unconvincing way of showing that the earth stands still, not-M.

The structure of this argument is already clear from the systematic labeling of its constituent propositions, which I have used. Thus, its visual representation is easily given by the diagram in Figure 13.3.

Figure 13.3

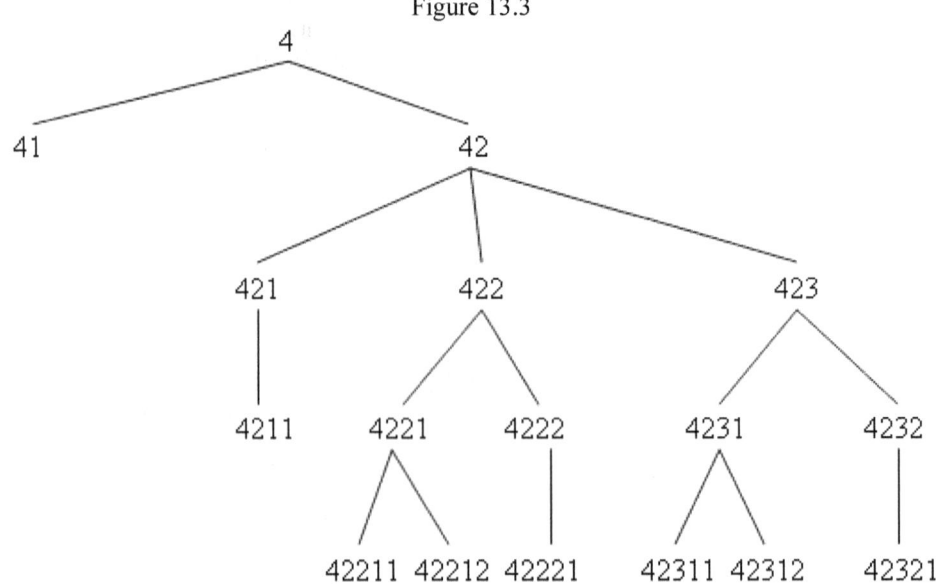

More importantly, it should be noted that although my reconstruction (N) tries to be relatively explicit, explicitness is a relative matter, and so inevitably I have left out many subarguments and propositions that could be added and would be needed for greater explicitness. It will be fruitful to examine some of this implicit, or latent, structure.

The step from 42221 to 4222 presupposes the following definition of circularity:

(42222) a circular argument is one in which the conclusion is identical to one of the premises.

The step from 4222 and 4221 to 422 presupposes the following definition of begging the question, or to be more precise, the following connection between circularity and question begging:

(4223)   a circular argument is one type of question-begging argument.

The step from 42321 to 4232 presupposes the following definition of a special case of begging the question:

(42322) one type of question begging is an argument in which a premise is no better known than the conclusion.

The step from 41 and 42 to 4 presupposes the following evaluative principle:

(43)   question begging arguments are ineffective.

These claims should be familiar by now. We have encountered them already as theoretical conclusions suggested by our case study. Now we are trying to be more explicit and systematic regarding such relation of "suggestion." What has emerged from our more detailed analysis is this: if we take Galileo's ratiocinative practice as our starting point (in accordance with the Toulminian applied-logic approach); and if we notice that Galileo's argument is essentially a meta-argument (in accordance with our meta-argumentation project); and if our observational description of this argumentative reality and our intuitive judgment convince us that this meta-argument is essentially right; and if the deeper analysis and reconstruction of this meta-argument brings out various theoretical claims inherent in this practice; then we can take such theoretical claims as justified (at least in part), insofar as they receive empirical grounding in argumentative reality.

With regard to the question of the interaction between logic and rhetoric, at least some of the conclusions could be obtained by further lengthening or complicating argument N. For example, start with the claims that (4222) argument C is circular, and that (4232) argument I begs the question. Combine them with the just elaborated theoretical claims, to get that arguments C and I are ineffective. Add to this that they are rhetorically flawed, from the point of view of persuasion. Add also that they are logically valid, in the sense of formal deductive logic. Conclude that logic and rhetoric are distinct points of view. Then elaborate this last conclusion in indefinite ways that need not be pursued here.

Finally, it should be obvious from all this that such meta-argumentation enables us to penetrate the depth and the essence of the arguments under consideration in a very effective and perhaps unique manner, *without* leaving behind the original ground-level arguments and moving to another (the meta) level to discuss a new topic.

13.6 Summary

This chapter has been an analysis of Galileo's critique of Aristotle's geostatic argument from vertical fall. This argument claimed that the earth cannot rotate because on a rotating earth bodies could not fall vertically, and it was advanced by Aristotle in his book *On the Heavens*. It was one of the many arguments against the earth's motion that had to be overcome in order for the Copernican Revolution to be possible, and so this critique is only one of the many which Galileo undertook. It is found near the beginning of Day II of his *Dialogue on the Two Chief World Systems, Ptolemaic and Copernican*. Galileo's criticism is that if the argument refers to actual vertical fall, then it begs the question by being circular; and that if it refers to apparent vertical fall, it is an *ad hoc* revision of the original argument and begs the question by assuming a premise no better know than the conclusion.

My analysis has pointed out three main things. First, Galileo's critique is instructive insofar as it uses and illustrates the concepts of begging the question, circularity, and assuming an equal unknown, which are extremely important because they embody the problem of deductive validity vs. persuasive ineffectiveness. Second, Galileo's critique is enlightening regarding the general problem of the relationship between logic and rhetoric, for it is full of strategic maneuvering with regard to both logical rationality and rhetorical persuasion, and my analysis shows that he manages to balance these two aims judiciously, without conflating them or perverting either (*pace* Paul Feyerabend). Third, Galileo's critique is obviously a famous meta-argument, and so my analysis also illustrates that such meta-arguments can have interesting theoretical import.

Appendix: Galileo's Argument about Vertical Fall
(from Galilei 2008, 222-226)

Salviati. So we can now go on to the fourth argument, which should be discussed at great length since it is based on an observation from which most of the remaining arguments then derive their strength. Aristotle says that a most certain argument for the earth's immobility is based on the fact that we see bodies which have been cast upwards return perpendicularly by the same line to the same place from which they were thrown, and that this happens even when the motion reaches a great height; this could not happen if the earth were moving because, while the projectile moves up and down separated from the earth, the place of ejection would advance a long way toward the east due to the earth's turning, and in falling the projectile would strike the ground that much distance away from the said place. Here we may also include the argument from the cannon ball shot upwards, as well as another one used by Aristotle and Ptolemy, namely that one sees bodies falling from great heights move in a straight line perpendicular to the earth's surface. Now, to begin to untie these knots, I ask Simplicio how Aristotle and Ptolemy would prove, if someone denied it, that bodies falling freely from on high move in a straight and perpendicular line, namely, in the direction of the center.

Simplicio. By means of the senses: they assure us that the tower is straight and perpendicular; they show us that the falling rock grazes it without inclining so much as a hairbreadth to one side or the other; and they show that the rock lands at the foot of the tower exactly under the place from which it was dropped.

Salviati. But if by chance the terrestrial globe were rotating and consequently were also carrying the tower along with it, and if the falling rock were still seen to graze the edge of the tower, what would its motion have to be?

Simplicio. In that case one would rather have to speak of "its motions"; for there would be one that would take it from above downwards, and it would have to have another in order to follow the course of the tower.

Salviati. Therefore, its motion would be a compound of two, namely, one with which it grazes the edge of the tower, and another one with which it follows the tower; the result of this compound would be that the rock would no longer describe a simple straight and perpendicular line, but rather an inclined, and perhaps not straight, one.

Simplicio. I am not sure about its not being straight; but I understand well that it would have to be inclined and different from the straight perpendicular one it would describe on a motionless earth.

Salviati. Therefore, from just seeing the falling rock graze the tower, you cannot affirm with certainty that it describes a straight and perpendicular line unless you first assume the earth to be standing still.

Simplicio. That is correct; for if the earth were moving, the rock's motion would be inclined and not perpendicular.

Salviati. Here, then, is the paralogism of Aristotle and Ptolemy made clear and evident, and discovered by yourself; the argument is assuming as known what it is trying to prove.

Simplicio. In what way? To me it seems to be a syllogism in proper form and not a fallacy of question begging.

Salviati. Here is how. Tell me: does not the demonstration regard the conclusion as unknown?

Simplicio. Yes, unknown, for otherwise it would be superfluous to demonstrate it.

Galileo on the Motion of the Earth

Salviati. But, should not the middle term be known?

Simplicio. That is necessary, for otherwise it would be an attempt to prove the unknown by means of what is equally unknown.

Salviati. Is not the conclusion to be proved, and which is unknown, the proposition that the earth stands still?

Simplicio. It is.

Salviati. Is not the middle term, which must be already known, the straight and perpendicular fall of the rock?

Simplicio. That is the middle term.

Salviati. But, did we not just conclude that we can have no knowledge that this fall is straight and perpendicular unless we first know that the earth is standing still? Therefore, in your syllogism the certainty of the middle term is inferred from the uncertain conclusion. So you see the type and the seriousness of the paralogism.

Sagredo. On behalf of Simplicio, I should like to defend Aristotle, if possible, or at least to understand better the strength of your inference. You say: seeing the rock graze the tower is not enough to become certain that its motion is perpendicular (which is the middle term of the syllogism) unless one assumes that the earth stands still (which is the conclusion to be proved); for, if the tower were moving together with the earth and the rock grazed it, the rock's motion would be inclined and not perpendicular. However, I will answer that, if the tower were moving, it would be impossible for the falling rock to graze it; hence, from seeing the falling rock graze it one infers that the earth is motionless.

Simplicio. That is correct. For, if the falling rock should graze the tower while the latter was carried along by the earth, the rock would have to have two natural motions (namely, straight toward the center and circular around the center); and this is impossible.

Salviati. Therefore, Aristotle's defense consists in its being impossible, or at least in his having regarded it as impossible, that the rock could move with a motion mixed of straight and circular; for, if he had not regarded it as impossible that the rock could move simultaneously toward the center and around the center, he would have understood that it could happen that the falling rock could graze the tower when it is moving as well as when it is standing still; consequently, he would have realized that from this grazing nothing could be inferred regarding the motion or the rest of the earth. However, this does not in any way excuse Aristotle because he should have said so if he had had this thought in mind, it being such a key point in his argument; moreover, one cannot say either that this effect is impossible or that Aristotle regarded it as impossible. The first cannot be said because I will soon show that it is not only possible but necessary. Nor can one say the second, for Aristotle himself grants that fire goes naturally upward in a straight line and turns by participation with the diurnal motion, which is transferred by the heavens to all of the element fire and to most of the air; if, then, he did not regard it as impossible to mix straight upward motion with the circular one communicated to fire and air by the inside of the lunar orb, much less should he regard it as impossible to mix the rock's straight downward with the circular one that would be natural for the whole terrestrial globe of which the rock is a part.

Simplicio. It does not look that way to me; for, if the element fire turns together with the air, it is very easy and indeed necessary that a particle of fire rising from the earth and going through the rotating air should receive the same motion since it is such a rarefied and light body and most ready to move; but it is completely incredible that a very heavy rock or cannon ball falling through the air should let itself be carried along by it or anything else. Furthermore, there is the very appropriate experiment of the rock dropped from the top of a ship's mast; that is, when the ship is standing still it falls at the foot of the mast, but when the ship is going forward it falls away from the same place at a distance equal to that traversed by the ship during the rock's fall (which amounts to many cubits when the ship's course is fast).

Salviati. There is a great disparity between the case of the ship and that of the earth, if the diurnal motion should belong to the terrestrial globe ...

Conclusion
Argumentation Theory as Meta-argumentation

Let us review the preceding investigation in a succinct and focused manner that may be useful for the purpose of recapitulating what has been accomplished thus far, as well as for deciding how to proceed next. This may be done by formulating the main theses respectively elaborated in each chapter, for such theses may also be regarded as conclusions which the various chapters have attempted to justify, and which now or in the future may serve as intermediate propositions from which to develop further arguments.

0. *Meta-argumentation is argumentation about argumentation*, as distinct from argumentation about such things as natural phenomena, historical events, human actions, and supernatural or metaphysical entities; the latter may be called ground-level or object-level argumentation. (This is merely a nominal definition, formulated in the Introduction.)

1. *Meta-argumentation is a promising methodological perspective* to practice in the study of argumentation; it is also a novel subject matter, especially within the approach to logic and argumentation theory labeled "applied logic" by Toulmin and within the historical-textual approach advocated in my previous works. (This is elaborated in chapter 1.)

2. *The method of meta-argumentation can be implemented* by following appropriate principles of interpretation and of evaluation: the former focus on describing the propositional structure of the reasoning under investigation and on justifying the accuracy of these descriptions; the latter focus on criticizing the various explicitly stated propositions, their connections with each other, and their connections with other implicit propositions, and on justifying the tenability of these criticisms (chapter 2).

3. *Meta-argumentation can take several forms*, depending the nature of the conclusion of one's meta-argument: interpretation of a ground-level argument, evaluation of a ground-level argument, generalization about ground-level arguments, and theoretical claim about ground-level arguments; another form is the self-reflective presentation of a ground-level argument (chapter 3).

4-5. A critical analysis of the meta-arguments advanced by various theorists in favor of various definitions of the concept of argument yields two main conclusions. (4) *Definitions of the concept of argument can be classified depending on whether or not they regard the illative tier or the dialectical tier necessary or sufficient*: the illative definition regards the illative tier both sufficient and necessary; the hyper dialectical definition regards the dialectical tier both sufficient and necessary; the strongly dialectical definition regards the illative tier and the dialectical tier individually necessary and jointly sufficient; and the moderately dialectical definition regards each tier individually sufficient but neither necessary. (5) *The most tenable and viable definition of the concept of argument is the moderately dialectical one: an argument is a series of propositions attempting to justify one of them (the conclusion) by supporting it with reasons, or by defending it from objections, or both* (chapters 4-5).

6. A critical analysis of the meta-arguments by various theorists in favor of various common methods of criticizing arguments yields the conclusion that: *meta-argumentation is the real process underlying the application of such methods of criticism as the trivial logic-indifferent method, the method of counterexample-situation, the method of formal paraphrase, refutation by logical analogy, and evaluation by parity of reasoning* (chapter 6). That is: the trivial logic-indifferent method of proving formal invalidity is the meta-argument that an argument is formally invalid because its premises are true and its

conclusion is false; the method of counterexample-situation is the meta-argument that an argument is formally invalid because there exists some situation in which the premises are true and the conclusion is false; the method of formal paraphrase is the meta-argument that an argument A is formally invalid because it instantiates some argument form F, but F is an invalid argument form, and F captures the gist of A; a refutation by logical analogy is the meta-argument that an argument A is flawed in the sense F because A is logically analogous to argument B, and B is flawed is the sense F; an argument by parity of reasoning is the meta-argument that an argument A receives an evaluation E because A has the same logical form as argument B, and B receives evaluation E; and a formal-fallacy criticism is the meta-argument that an argument A commits a formal fallacy because it is formally invalid and it violates the agreed-upon rule to use only formally valid arguments.

7. A critical analysis of the meta-arguments by various theorists for or against various limitations of rationality for resolving deep disagreements yields the following conclusion: *meta-argumentation is one of several effective instruments for rationally resolving deep disagreements and fierce standoffs,* along with the art of moderation and compromise (codified as Ramsey's Maxim), open-mindedness, fair-mindedness, complex argumentation, *ad hominem* argument (in Johnstone's sense), and persuasive argumentation; these principles and practices are neither individually necessary, nor jointly sufficient; but they are jointly necessary and individually helpful (chapter 7).

8. A critical analysis of the meta-arguments by various theorists regarding conductive arguments yields the conclusion that *meta-argumentation is crucially involved in conductive or pro-and-con arguments*; that is, conductive arguments contain or presuppose a balance-of-considerations claim, whose justification requires meta-argumentation (chapter 8).

9. A critical analysis of various self-referential meta-arguments by various theorists yields the conclusion that *meta-argumentation has a special role in the analysis of self-referential arguments*. That is, although all argument analysis is meta-argumentation, for the case of self-referential argument the special features are: both premises and conclusion of the meta-argument use concepts used in the self-referential argument, whether we are engaged in the interpretation, or the positive evaluation, or the negative evaluation of self-referential arguments (chapter 9).

10. A critical analysis of Mill's famous argument on the freedom of thought and discussion (in *On Liberty*) yields the following conclusions: it is a meta-argument; it is cogent; its conclusion is the claim that liberty of argument is highly desirable; it is also an explicit contribution to argumentation theory; and it raises *the problem of the relationship between meta-argumentation and epistemology* (chapter 10).

11. A critical analysis of Mill's famous argument at the beginning of his essay on *The Subjection of Women* yields the following conclusions: it is primarily a meta-argument; it is cogent; its conclusion is the claim that the issue of liberation vs. subjection of women should be argued about on its merits, by means of the presentation, defense, and criticism of supporting reasons and critical objections, despite the fact that established custom and general feeling regard the issue as settled in favor of subjection; it can also be interpreted as a part of the dialectical tier, replying to the pro-subjection objection based on universal custom and feeling; and it raises *the problem of the relationships among the illative, dialectical, and meta-level tiers of argumentation* (chapter 11).

12. A critical analysis of Hume's famous critique of the theological design argument (in *Dialogues Concerning Natural Religion*) yields the following conclusions: it is a meta-argument; it is cogent; its main conclusion is that the design argument is fatally flawed, that is, that the existence of a divine designer of the universe cannot be justified; Hume's meta-argument attributes a dozen flaws to the design argument; and *the perspective of meta-*

argumentation provides a simple and elegant manner of making sense of complex and multi-faceted argumentative discussions (chapter 12).

13. A critical analysis of Galileo's famous critique of Aristotle's geostatic argument from vertical fall (in *Dialogue on the Two Chief World Systems*) yields the following conclusions: it is a meta-argument; it is cogent; its main conclusion is that the vertical-fall argument begs the question, in the sense of being circular if interpreted in terms of actual vertical fall, and in the sense of assuming *ad hoc* an equal unknown if interpreted in terms of apparent vertical fall; it embodies significant illustrations and definitions of the concept of begging the question; it embodies instructive lessons about proper strategic maneuvering with respect to the goals of the rational search for truth and the rhetorical persuasion of opponents; and it embodies an instructive lesson about *the fruitfulness and viability of doing argumentation theory as meta-argumentation* (chapter 13).

The main conclusion supported by these thirteen theses is that *argumentation theory can and ought to be practiced as meta-argumentation*. The grounding of this conclusion on the last mentioned thesis (no. 13) is obvious: the two claims are simply different ways of expressing the same basic idea. However, the same overarching conclusion can also be grounded on the other particular theses in various other ways.

One reason for doing argumentation theory as meta-argumentation is that doing so is in accordance with Toulmin's methodological norms for a revolution in logical theory (ch. 1). On the other hand, chapter 2 shows that such an approach can actually be implemented, in terms of concrete rules for interpreting and evaluating arguments. And chapter 3 clarifies that, strictly speaking, argumentation theory is one type, but not the only type, of meta-argumentation; that is, not all meta-argumentation is argumentation theory. Conversely, chapter 13 also clarifies that there are parts of argumentation theory, such as clarification and systematization, that are basically not meta-argumentation; that is, not all argumentation theory is meta-argumentation, either. In other words, we need to keep in mind the proviso or qualification that we are not indiscriminately equating argumentation theory with meta-argumentation, but rather saying that there is, can be, and ought to be a large and significant overlap between the two. A similar qualification applies, and even more so, insofar as we speak of logic and logical theory, instead of argumentation theory.

Furthermore, chapters 4-9 show that argumentation theory, as actually practiced, can *de facto* be interpreted as consisting of meta-arguments, for these chapters construe in terms of meta-argumentation a considerable body of theoretical work, by leading theorists, such as Allen, Barker, Blair, Eemeren, Ennis, Fogelin, Freeman, Goldman, Govier, Hitchcock, Johnson, Johnstone, Krabbe, Wellman, and Woods. But I think that this part of the book accomplishes something more: it shows that the meta-argumentative interpretation of current theoretical work *facilitates* the *evaluation* of such work, for it is obvious that my own analysis was not purely interpretive, and did not refrain from exploiting this interpretation for evaluative purposes. And there is a third aspect of my analysis: my meta-argumentative interpretation and evaluation pave the way for the drawing of important lessons, i.e., for theoretical conclusions that can be contextually justified by argument. In fact, each of my theses 4-9 is an important theoretical claim derived from my critical interpretation of corresponding theoretical work on definitions of argument, argument criticism, deep disagreements, conductive arguments, and self-referential arguments.

Finally, there is a fourth aspect of what chapters 4-9 accomplish. That is, not only we can interpret theoretical work in terms of meta-argumentation; not only we can evaluate such theoretical work from the point of view of meta-argumentation; not only can we thus illustrate doing theory as meta-argumentation; but also it turns out on several occasions that meta-argumentation is substantively involved in the derived theoretical claims; that is, the theoretical claims are sometimes *about* the nature and role of meta-argumentation. This is most striking for theses 6, 7, and 8, which make claims about the role of meta-argumentation in criticism, deep disagreements, and conductive arguments.

Chapters 10-13 show something analogous with regard to famous meta-arguments. That is, some classics of the history of thought can be understood or interpreted as meta-arguments; they can be evaluated or assessed from such a point of view; and they can be utilized or exploited to yield theoretical claims or formulate theoretical problems. Let us elaborate this third point.

In some cases, the theoretical claims become apparent through a slight reformulation, rephrasing, or redescription of the topic under consideration. This is clearest in Mill's meta-argument about liberty of thought and discussion; for this topic is easily viewed as liberty of argument, and so it is relatively explicit that his conclusion is that liberty of argument is desirable; but this conclusion is a theoretical claim.

In other cases, such theoretical claims are claims that are assumed or presupposed, implicitly or explicitly; or more pedantically expressed, they are propositions that are part of the explicit or latent structure of the reconstructed meta-arguments. Such is the case for Mill's endorsement of the moderately dialectical conception of argument, and for his thesis that argumentation is an essential method in the search for truth and the acquisition of knowledge. This is also the case for Galileo's claims that one type of question-begging argument is a circular argument; that another type of question-begging argument is one with a final premise whose epistemological status is no firmer that the final conclusion; and that question-begging arguments are valid from the point of view of formal deductive logic, but are ineffective from the perspective of rhetorical persuasion. And the same applies to Hume's meta-arguments, especially his critical ones charging various parts of the design argument with such flaws as hasty generalization, fallacy of composition, misapplication of generalization, infinite regress, and fallacy of incomplete evidence; here the main difference is that the larger number of meta-arguments and the greater variety of charges decrease the explicitness of the theoretical claims involved.

Furthermore, the formulation or illustration of theoretical problem is perhaps an even more important yield from famous meta-arguments. For example, Mill's meta-argument on liberty of argument is so filled with epistemological principles that it suggests an incisive formulation of the problem of the relationship between epistemology and argumentation theory; and Galileo's meta-argument on the earth's motion and vertical fall is so full of implicit and explicit rhetorical moves that it provides a clear illustration of the problem of the relationship between logic and rhetoric.

Similarly, Mill's meta-argument against the subjection of women provides a concrete and crisp illustration of the problem of the relationship among the illative, the dialectical, and the meta-level tiers of argumentation. The illustration is occasioned by the fact that two of his meta-arguments (one involving causal undermining, the other predictive extrapolation) can be easily interpreted as reasons supporting the liberation of women at the illative level, as well as replies defending this same conclusion from objections. And here it may be useful to elaborate a general formulation of this problem.

Suppose we define an argument as an attempt to justify a conclusion by supporting it with reasons, or by defending it from objections, or both; this is, of course, the moderately dialectical definition advocated in this book, and emerged explicitly in chapter 4-5. And suppose we define the illative tier as that part or aspect of an argument consisting of the support of the conclusion with reasons; and suppose we define the dialectical tier as that part or aspect of an argument consisting of the defense of the conclusion from objections; these suppositions again represent my twist to or revision of the traditional conceptions.

Then we can also say that an argument has a third tier. In the definition just given, this is reflected in the phrase "attempt to justify." That is, besides having a proposition supported by reasons and/or defended from objections, there must be an arguer's intention to do this; the conclusion must be purportedly, or allegedly, or supposedly supported and/or defended. As we saw in chapter 4, this corresponds to what Johnson called the functional or teleological aspect of argumentation; and we also saw that even Copi's purely illative conception contained this element, with its talk of the conclusion being "claimed to follow" from the premises and of the premises being "regarded as providing support" (Copi and Cohen 1994, 5). What I want to stress here is that this is a metacognitive element: the arguer normally makes such self-reflective judgments, and the argument normally contains such meta-argumentative claims, although this meta-argumentative aspect can be implicit or explicit, or more or less implicit or explicit.

By analogy with the illative and dialectical tiers, I have been calling this aspect of an argument its meta-argumentative tier. However, the analogy may not extend too far, in the sense that the meta-argumentative tier appears to be a necessary condition for argumentation, whereas I have been arguing (in accordance with the moderately dialectical conception) that the illative and dialectical tiers are each sufficient, but not individually necessary. Moreover, it seems that the meta-argumentative tier is not sufficient, for the simple reason that it cannot subsist by itself; that is, the appropriate metacognitive claim can come into being only when applied to some illative claim that a reason supports a conclusion, or to some dialectical claim that an objection does not refute or undermine a conclusion. Thus, what is sufficient, but not necessary, is either one of the following two conjunctions: a conjunction of an illative tier and its corresponding meta-argumentative tier, or a conjunction of a dialectical tier and its corresponding meta-argumentative tier. And this in turn means that the illative and dialectical tiers are not really sufficient individually, and so, strictly speaking, the moderately dialectical definition as previously formulated may need revision.

However, rather than solving this problem, let it be one placed on the agenda for future investigations. Also on that agenda, as we have seen, is the problem of the relationship between rhetorical persuasion and logical argumentation, which emerged from our analysis of Galileo's meta-argument on vertical fall (chapter 13). In this regard, I believe that the context in which this problem emerged suggests a way of proceeding in further work on the problem; that would be to study other important historical examples of such strategic maneuvering. One of these is the argument advanced by William the Silent's *Apology* (1581), in which he defended himself from the proclamation issued against him a year earlier by Philip II, King of Spain. Such an example does not come out the blue, so to speak, but has already been studied from this point of view by Eemeren and Houtlosser (1999b, 2000, 2002b, 2003a, 2003b). Moreover, William's argument happens to be an explicit meta-argument, as I have argued elsewhere (Finocchiaro 2012). Thus, we have at least two reasons for placing the analysis of William's *Apology* on the agenda for future analysis: its potential for theoretical lessons on the interplay of logic and rhetoric, and its being a "famous meta-argument."

Finally, a few words about the problem of the relationship between epistemology and argumentation theory, which emerged earlier in our analysis of Mill's meta-argument on liberty of argument (chapter 10). A suggestion made then can now be motivated more strongly. The plan would be to study the arguments of those scholars who advocate an epistemic approach to argumentation theory (e.g., Biro, Freeman, Goldman, Lumer, Pinto, and Siegel). Their arguments would be bound to be meta-arguments, and now we can say that they could be examined in the same manner in which other chapters have examined scholarly meta-arguments on other topics: chapters 4-5 for the arguments about conceptions of argument, chapter 7 for the arguments about deep disagreements, and chapter 8 for the arguments about conductive arguments. I would like to believe that it is impossible to have read those chapters without detecting that they follow what I am calling the meta-argumentation approach. The promise of such a future project is a deeper understanding of such arguments (at the interpretive level); a facilitated assessment of them (at the evaluative level); and a derivation of theoretical lessons from them (at the level of constructive meta-argumentation).

Bibliography

(Besides works actually cited in the body of this book and in the notes, this bibliography includes a few other works which I found to be especially relevant, useful, insightful, or provocative.)

Aberdein, A. 2006. "The Uses of Argument in Mathematics." In Hitchcock and Verheij 2006, 327-40.
Aberdein, A. 2007. "The Informal Logic of Mathematical Proof." In Bendegam & Kerkove 2007, 135-51.
Adams, D.M. 2005. "Knowing When Disagreements Are Deep." *Informal Logic* 25: 65-78.
Adams, E.W. 1975. *The Logic of Conditionals*. Dordrecht: Reidel.
Adler, J. 2004a. "Reconciling Open-mindedness and Belief." *Theory and Research in Education* 2: 127-42.
Adler, J. 2004b. "Shedding Dialectical Tiers." *Argumentation* 18: 279-93.
Agassi, J. 1977. *Towards a Rational Philosophical Anthropology*. The Hague: Martinus Nijoff.
Aiken, H.D. 1948. Introduction. In Hume 1948, vii-xvii.
Allen, D. 1990. "Trudy Govier's *Problems in Argument Analysis and Evaluation.*" *Informal Logic* 12: 43-62.
Allen, D. 1993. "Relevance, Conduction, and Canada's Rape-shield Law." *Informal Logic* 15: 105-22.
Allen, D. 2011. "Conductive Arguments and the Toulmin Model: A Case Study." In Blair and Johnson 2011, 167-90.
Alston, W. 2005. *Beyond "Justification."* Ithaca: Cornell University Press.
Angeles, P.A. 1981. *Dictionary of Philosophy*. New York: Barnes & Noble.
Angell, R.B. 1964. *Reasoning and Logic*. New York: Appleton.
Aquinas, T. 1952. *The Summa theologica.* 2 vols. Chicago: Encyclopedia Britannica.
Aristotle. 1941. *Basic Works*. Ed. R. McKeon. New York: Random House.
Arnauld, A., and P. Nicole. 1996. *Logic or the Art of Thinking*. Trans. J.V. Buroker. Cambridge: Cambridge University Press. Original edition, 1662.
Astroh, M., D. Gerhardus, and G. Heinzmann, eds. 1997. *Dialogische Handeln: Eine Festscrift für Kuno Lorenz*. Spektrum Akademischer Verlag.
Barker, S.F. 1957. *Induction and Hypothesis*. Ithaca: Cornell University Press.
Barker, S.F. 1989. "Reasoning by Analogy in Hume's *Dialogues.*" *Informal Logic* 11: 173-84.
Barth, E.M. 1982. "A Normative-Pragmatical Foundation of the Rules of Some Systems of Formal$_3$ Dialectics." In Barth and Martens 1982, 159-70.
Barth, E.M. 1985a. "A New Field: Empirical Logic." *Synthese* 63: 375-88.
Barth, E.M. 1985b. "Toward a Praxis-Oriented Theory of Argumentation." In *Dialogue: An Interdisciplinary Approach*, ed. M. Dascal, 73-86. Amsterdam: John Bejamins.
Barth, E.M. 1987. "Logic to Some Purpose: Theses Against the Deductive-Nomological Paradigm in the Science of Logic." In Eemeren, Grootendorst, Blair, and Willard 1987, 33-45.
Barth, E.M. 2002. "A Framework for Intersubjective Accountability: Dialogical Logic." In Gabbay, Johnson, Ohlbach, and Woods 2002, 225-93.
Barth, E.M., and E.C.W. Krabbe. 1982. *From Axiom to Dialogue*. Berlin: Walter de Gruyter.

Barth, E.M., and E.C.W. Krabbe, eds. 1992. *Logic and Political Culture*. Amsterdam: Royal Netherlands Academy of Arts and Sciences.
Barth, E.M., and J.L. Martens, eds. 1982. *Argumentation: Approaches to Theory Formation*. Amsterdam: John Benjamins.
Barth, E.M., J. Vandormael, and F. Vandamme, eds. 1992. *From an Empirical Point of View: The Empirical Turn in Logic*. Ghent: Communication and Cognition.
Battersby, M.E. 1989. "Critical Thinking as Applied Epistemology." *Informal Logic* 11: 91-100.
Battersby, M.E., and S. Bailin. 2011. "Guidelines for Reaching a Reasoned Judgment." In Blair and Johnson 2011, 145-57.
Bendegam, J.P. van, and B. van Kerkove, eds. 2007. *Perspectives on Mathematical Practices*. Dordrecht: Kluwer.
Benthem, J. van. 2009. "One Logician's Perspective on Argumentation." *Cogency* 1(2): 13-25.
Benthem, J. van., F.H. van Eemeren, R. Grootendorst, and F. Veltman, eds. 1966. *Logic and Argumentation*. Amsterdam: North-Holland.
Berg, J. 1987. "Interpreting Arguments." *Informal Logic* 9: 13-21.
Bergmann, M. 2004. "Epistemic Circularity: Malignant and Benign." *Philosophy and Phenomenological Research* 69: 709-27.
Bickenbach, J.E., and J.M. Davies. 1977. "Conductive Reasoning." In: J.E. Bickenbach and J.M. Davies, *Good Reasons for Better Arguments*, 321-26. Peterborough: Broadview Press.
Biro, J., and H. Siegel. 1992. "Normativity, Argumentation and an Epistemic Theory of Fallacies." In Eemeren, Grootendorst, Blair, and Willard 1992, 85-103.
Biro, J., and H. Siegel. 2006a. "In Defense of the Objective Epistemic Approach to Argumentation." *Informal Logic* 26: 91-101.
Biro, J., and H. Siegel. 2006b. "Pragma-dialectic versus Epistemic Theories of Arguing and Arguments." In Houtlosser and Rees 2006, 1-10.
Blair, J.A. 1995. "Premise Adequacy." In Eemeren, Grootendorst, Blair, and Willard 1995a, 191-202.
Blair, J.A. 1998. "The Limits of the Dialogue Model of Argument." In Hansen, Tindale, and Colman 1998.
Blair, J.A. 2002. "'Argument' and 'Logic' in Logic Textbooks." In Hansen, Tindale, Blair, Johnson, and Pinto 2002.
Blair, J.A. 2003. "Towards a Philosophy of Argument." In Blair, Farr, Hansen, Johnson, and Tindale 2003.
Blair, J.A. 2006. "Informal Logic's Influence on Philosophy Instruction." *Informal Logic* 26: 259-86.
Blair, J.A. 2011. "Conductive Reasoning/Arguments: A Map of the Issues." In Blair and Johnson 2011, 1-9.
Blair, J.A, D. Farr, H.V. Hansen, R.H. Johnson, and C.W. Tindale, eds. 2003. *Informal Logic at 25: Proceedings of the Windsor Conference*. Windsor (ON): Ontario Society for the Study of Argumentation. CD-ROM; ISBN: 0-9683461-3-8.
Blair, J.A., and R.H. Johnson, eds. 1980. *Informal Logic: The First International Symposium*. Inverness, CA: Edgepress.
Blair, J.A., and R.H. Johnson. 1987. "Argumentation as Dialectical." *Argumentation* 1: 41-56.
Blair, J.A., and R.H. Johnson, eds. 2011. *Conductive Argument: An Overlooked Type of Defeasible Reasoning*. London: College Publications.
Boella, G., D.M. Gabbay, L. Van der Torre, and S. Villata. 2009. "Meta-Argumentation

Modelling I: Methodology and Techniques." *Studia Logica* 93: 297-355.
Brandom, R.B. 1994. *Making It Explicit: Reasoning, Representing, and Discursive Commitment*. Cambridge: Harvard University Press.
Brandom, R.B. 2000. *Articulating Reasons: An Introduction to Inferentialism*. Cambridge: Harvard University Press.
Brooks, D. 2009. "The Hardest Call." *New York Times*, December 18.
Brown, H.I. 1979. *Perception, Theory and Commitment*. Chicago: University of Chicago Press.
Brown, H.I. 1987. *Observation and Rationality*. New York: Oxford University Press.
Campolo, C. 2005. "Treacherous Ascents: On Seeking Common Ground for Conflict Resolution." *Informal Logic* 25: 37-50.
Campolo, C. 2007. "Commentary on Vesel Memedi: 'Resolving Deep Disagreement'." In Hansen, Tindale, Blair, Johnson, and Godden 2007.
Campolo, C., and D. Turner. 2002. "Reasoning Together." *Argumentation* 16: 3-19.
Carnap, R. 1950. *Logical Foundations of Probability*. Chicago: University of Chicago Press.
Carroll, L. 2000. *The Annotated Alice, The Definitive Edition: Alice's Adventures in Wonderland & Through the Looking-Glass*. Ed. M. Gardner. New York: Norton.
Cattani, A. 2001. *Botta e risposta: L'arte della replica*. Bologna: Il Mulino.
Chichi, G.M. 2002. "The Greek Roots of *Ad Hominem* Argument." *Argumentation* 16: 333-48.
Chittleborough, P., and M.E. Newman. 1993. "Defining the Term 'Argument'." *Informal Logic* 15: 189-207
Cohen, D.H. 2001. "Evaluating Arguments and Making Meta-arguments." *Informal Logic* 21: 73-84.
Cohen, D.H. 2005. *Arguments and Metaphors in Philosophy*. Lanham: University Press of America.
Cohen, D.H. 2007. "Commentary on Maurice Finocchiaro: 'Famous Meta-Arguments: Part I, Mill and the Tripartite Nature of Argumentation'." In Hansen, Tindale, Blair, Johnson, and Godden 2007.
Cohen, L.J. 1986. *The Dialogue of Reason: An Analysis of Analytical Philosophy*. Oxford: Clarendon Press.
Cohen, M., and E. Nagel. 1934. *Introduction to Logic and Scientific Method.* New York: Harcourt, Brace and Company.
Copi, I.M. 1968. *Introduction to Logic*. 3rd edn. New York: MacMillan.
Copi, I.M. 1986a. *Informal Logic*. New York: MacMillan.
Copi, I.M. 1986b. *Introduction to Logic*. 7th edn. New York: MacMillan.
Copi, I.M., and C. Cohen. 1990. *Introduction to Logic*. 8th edn. New York: MacMillan.
Copi, I.M., and C. Cohen. 1994. *Introduction to Logic.* 9th edn. New York: MacMillan.
Corner, A., and U. Hahn. 2007. "Evaluating the Meta Slope: Is There a Slippery Slope Argument Against Slippery Slope Arguments?" *Argumentation* 21: 349-59.
Costantini, S. 2002. "Meta-reasoning: A Survey." In *Computational Logic*, ed. A.C. Kakas and F. Sadri, 253-88. Berlin: Springer.
Craig, R.T. 1996. "Practical-theoretical Argumentation." *Argumentation* 10: 461-74.
Crawshay-Williams, R. 1957. *Methods and Concepts of Reasoning: An Inquiry into the Structure of Controversy*. London: Routledge and Kegan Paul.
Crenshaw, K. 1998. "A Black Feminist Critique of Anti-discrimination Law." In *The Politics of Law*, ed. David Kairy, 3rd edn. New York: Basic Books.
Croce, B. 1917. *Logic as the Science of the Pure Concept*. Trans. D. Ainslie. London: Macmillan. Original edition, 1909.

Cummings, L. 2002. "Hilary Putnam's Dialectical Thinking ." *Argumentation* 16: 197-229.
Dauer, F. 1974. "The Diagnosis of an Argument." *Metaphilosophy* 5: 113-32.
Davson-Galle, P. 1992. "Arguing, Arguments, and Deep Disagreements." *Informal Logic* 14: 147-56.
Dewey, J. 1910. *How We Think*. Lexington: Heath.
Dewey, J. 1938. *Logic: The Theory of Inquiry*. New York: Holt.
Doss, S. 1985. "Three Steps toward a Theory of Informal Logic." *Informal Logic* 7: 127-35.
Dove, I. 2007. "On Mathematical Proofs and Arguments: Johnson and Lakatos." In Eemeren, Blair, Willard, and Garssen 2007, 347-51.
Dove, I. 2009. "Towards a Theory of Mathematical Argument." *Foundations of Science* 14: 137-52.
Doury, M. 2009. "Argument Schemes Typologies in Practice: The Case of Comparative Arguments." In Eemeren and Garssen 2009, 141-55.
Dryzek, J.S., and S. Niemeyer. 2006. "Reconciling Pluralism and Consensus as Political Ideals." *American Journal of Political Science* 50: 634-49.
Eemeren, F.H. van. 1987. "For Reason's Sake." In Eemeren, Grootendorst, Blair, and Willard 1987, 201-15.
Eemeren, F.H. van, ed. 2002. *Advances in Pragma-Dialectics*. Amsterdam: Sic Sat.
Eemeren, F.H. van. 2009. "Strategic Manoeuvering between Rhetorical Effectiveness and Dialectical Reasonableness." In Ribeiro 2009, 55-74.
Eemeren, F.H. van. 2010. *Strategic Maneuvering in Argumentative Discourse*. Amsterdam: John Benjamins.
Eemeren, F.H. van, J.A. Blair, C.A. Willard, and A.F. Snoeck Henkemans, eds. 2003a. *Anyone Who Has a View*. Dordrecht: Kluwer.
Eemeren, F.H. van, J.A. Blair, C.A. Willard, and A.F. Snoeck Henkemans, eds. 2003b. *Proceedings of the Fifth Conference of the International Society for the Study of Argumentation*. Amsterdam: Sic Sat.
Eemeren, F.H. van, J.A. Blair, C.A. Willard, and B. Garssen, eds. 2007. *Proceedings of the Sixth Conference of the International Society for the Study of Argumentation*. Amsterdam: Sic Sat.
Eemeren, F.H. van, and B. Garssen, eds. 2009. *Pondering on Problems of Argumentation*. Dordrecht: Springer.
Eemeren, F.H. van, and B. Garssen, eds. 2012. *Topical Themes in Argumentation Theory*. Dordrecht: Springer.
Eemeren, F.H. van, B. Garssen, D. Godden, and G.R. Mitchell, eds. 2011. *Proceedings of the Seventh International Conference of the International Society for the Study of Argumentation*. Amsterdam: Rozenberg Publishers & Sic Sat Publishers. CD ROM: ISBN 978 90 3610 243 8.
Eemeren, F.H. van, and R. Grootendorst. 1984. *Speech Acts in Argumentative Discussions*. Dordrecht: Foris.
Eemeren, F.H. van, and R. Grootendorst. 1988. "Rationale for a Pragma-dialectical Perspective." *Argumentation* 2: 271-91.
Eemeren, F.H. van, and R. Grootendorst. 1992. *Argumentation, Communication, and Fallacies*. Hillsdale: Lawrence Erlbaum.
Eemeren, F.H. van, and R. Grootendorst. 1993. "The History of the *Argumentum ad Hominem* since the Seventeenth Century." In Krabbe, Dalitz, and Smit 1993, 49-68.
Eemeren, F.H. van, and R. Grootendorst, eds. 1994. *Studies in Pragma-Dialectics*. Amsterdam: International Centre for the Study of Argumentation.
Eemeren, F.H. van, and R. Grootendorst. 1995. "Perelman and the Fallacies." *Philosophy*

and Rhetoric 28: 122-33.
Eemeren, F.H. van, and R. Grootendorst. 2004. *A Systematic Theory of Argumentation.* New York: Cambridge University Press.
Eemeren, F.H. van, R. Grootendorst, J.A. Blair, and C.A. Willard, eds. 1987. *Argumentation Across the Lines of Discipline.* Dordrecht: Foris.
Eemeren, F.H. van, R. Grootendorst, J.A. Blair, and C.A. Willard, eds. 1992. *Argumentation Illuminated.* Amsterdam: Sic Sat.
Eemeren, F.H. van, R. Grootendorst, J.A. Blair, and C.A. Willard, eds. 1995a. *Analysis and Evaluation.* Amsterdam: Sic Sat.
Eemeren, F.H. van, R. Grootendorst, J.A. Blair, and C.A. Willard, eds. 1995b. *Perspectives and Approaches.* Amsterdam: Sic Sat.
Eemeren, F.H. van, R. Grootendorst, J.A. Blair, and C.A. Willard, eds. 1999. *Proceedings of the Fourth International Conference of the International Society for the Study of Argumentation.* Amsterdam: Sic Sat.
Eemeren, F.H. van, R. Grootendorst, S. Jackson, and S. Jacobs. 1993. *Reconstructing Argumentative Discourse.* Tuscaloosa: University of Alabama Press.
Eemeren, F.H. van, R. Grootendorst, and T. Kruiger. 1984. *The Study of Argumentation.* New York: Irvington Publishers.
Eemeren, F.H. van, R. Grootendorst, and A.F. Snoeck Henkemans, eds. 1996. *Fundamentals of Argumentation Theory.* Mahwah: Lawrence Erlbaum.
Eemeren, F.H. van, and P. Houtlosser. 1999a. "Delivering the Goods in Critical Discussion." In Eemeren, Grootendorst, Blair, and Willard 1999, 163-63.
Eemeren, F.H. van, and P. Houtlosser. 1999b. "William the Silent's Argumentative Discourse." In Eemeren, Grootendorst, Blair, and Willard 1999, 168-71.
Eemeren, F.H. van, and P. Houtlosser. 2000. "The Rhetoric of William the Silent's *Apologie.*" In Suzuki et al. 2000, 37-40.
Eemeren, F.H. van, and P. Houtlosser, eds. 2002a. *Dialectic and Rhetoric.* Dordrecht: Kluwer.
Eemeren, F.H. van, and P. Houtlosser. 2002b. "Strategic Maneuvering: Maintaining a Delicate Balance." In Eemeren and Houtlosser 2002a, 131-59.
Eemeren, F.H. van, and P. Houtlosser. 2003a. "More about Fallacies as Derailments of Strategic Maneuvering: The Case of *Tu Quoque.*" In Blair, Farr, Hansen, Johnson, and Tindale 2003.
Eemeren, F.H. van, and P. Houtlosser. 2003b. "Strategic Manoeuvering: William the Silent's *Apologie.*" In Komlósi et. al. 2003, 177-85.
Eemeren, F.H. van, and P. Houtlosser. 2007a. "Countering Fallacious Moves." *Argumentation* 21: 243-52.
Eemeren, F.H. van, and P. Houtlosser. 2007b. "The Relationship between Johnstone's Ideas about Philosophical Argument and the Pragma-dialectical Theory of Argumentation." *Philosophy and Rhetoric* 40: 51-70.
Einstein, A. 2005. *Relativity: The Special and General Theory.* New York: Pi Press.
Eisend, M. 2006. "Two-sided Advertising: A Meta-analysis." *International Journal of Research in Marketing* 23: 187-98.
Ennis, R.H. 1996. "Critical Thinking Dispositions." *Informal Logic* 18: 165-82.
Ennis, R.H. 2001. "Argument Appraisal Strategy: A Comprehensive Approach." *Informal Logic* 21: 97-140.
Ennis, R.H. 2004. "Applying Soundness Standards to Qualified Reasoning." *Informal Logic* 24: 23-39.
Epstein, R.L. 2002. *Critical Thinking.* 2nd edn. Belmont: Wadsworth.
Feigl, H., and G. Maxwell, eds. 1962. *Minnesota Studies in the Philosophy of Science,* vol.

3. Minneapolis: University of Minnesota Press.
Feigl, H., and M. Scriven, eds. 1956. *Minnesota Studies in the Philosophy of Science,* vol. 1. Minneapolis: University of Minnesota Press.
Feigl, H., M. Scriven, and G Maxwell, eds. 1958. *Minnesota Studies in the Philosophy of Science*, vol. 2. Minneapolis: University of Minnesota Press.
Feldman, R. 1994. "Good Arguments." In Schmitt 1994, 159-99.
Feldman, R. 2005. "Deep Disagreements, Rational Resolutions, and Critical Thinking." *Informal Logic* 25: 13-23.
Feyerabend, P.K. 1962. "Explanation, Reduction, and Empiricism." In Feigl and Maxwell 1962, 28-97.
Feyerabend, P.K. 1975. *Against Method*. London: NLB.
Feyerabend, P.K. 1981. *Philosophical Papers*. 2 vols. Cambridge: Cambridge University Press.
Feyerabend, P.K. 1999. *Philosophical Papers*, vol. 3. Ed. J. Preston. Cambridge: Cambridge University Press.
Findlay, J. 1964. *Hegel: A Re-examination*. New York: Humanities.
Finocchiaro, M.A. 1973. *History of Science as Explanation*. Detroit: Wayne State University Press.
Finocchiaro, M.A. 1980a. *Galileo and the Art of Reasoning: Rhetorical Foundations of Logic and Scientific Method*. Boston Studies in the Philosophy of Science, vol. 61. Dordrecht: Reidel [now Springer].
Finocchiaro, M.A. 1980b. Review of Johnstone's *Validity and Rhetoric in Philosophical Argument*. *Review of Metaphysics* 34: 143-44.
Finocchiaro, M.A. 1980c. "Sztompka's Philosophy of Social Science." *Inquiry* (Oslo), 23: 357-71.
Finocchiaro, M.A. 1981. "Remarks on Truth, Problem-solving, and Methodology." *Studies in History and Philosophy of Science* 12: 261-68.
Finocchiaro, M.A. 1986. "Judgment and Reasoning in the Evaluation of Theories." In *PSA 1986: Proceedings of the 1986 Biennial Meeting of the Philosophy of Science Association*, ed. A. Fine and P. Machamer, 1: 227-35. East Lansing: Philosophy of Science Association.
Finocchiaro, M.A. 1988. *Gramsci and the History of Dialectical Thought*. Cambridge: Cambridge University Press.
Finocchiaro, M.A., ed. and trans. 1989. *The Galileo Affair: A Documentary History*. Berkeley: University of California Press.
Finocchiaro, M.A., ed. and trans. 1997. *Galileo on the World Systems: A New Abridged Translation and Guide*. Berkeley: University of California Press.
Finocchiaro, M.A. 1999. *Beyond Right and Left: Democratic Elitism in Mosca and Gramsci*. New Haven: Yale University Press.
Finocchiaro, M.A. 2003. "Dialectics, Evaluation, and Argument." *Informal Logic* 23: 19-49.
Finocchiaro, M.A. 2005a. *Arguments about Arguments: Systematic, Critical, and Historical Essays in Logical Theory*. Cambridge: Cambridge University Press.
Finocchiaro, M.A. 2005b. *Retrying Galileo, 1633-1992*. Berkeley: University of California Press.
Finocchiaro, M.A. 2006. "Reflections on the Hyper Dialectical Definition of Argument." In Houtlosser and van Rees 2006, 51-62.
Finocchiaro, M.A. 2007a. "Arguments, Meta-arguments, and Metadialogues: A Reconstruction of Krabbe, Govier, and Woods." *Argumentation* 21: 253-68.
Finocchiaro, M.A. 2007b. "Famous Meta-arguments: Part I, Mill and the Tripartite Nature

of Argumentation." In Hansen, Tindale, Blair, Johnson, and Godden 2007.

Finocchiaro, M.A. 2007c. "Mill on Liberty of Argument." In Hansen and Pinto 2007, 121-34.

Finocchiaro, M.A. 2009a. "The Galileo Affair." *Physics World*, vol. 22, no. 3, March, pp. 54-57.

Finocchiaro, M.A. 2009b. "Meta-argumentation in Hume's Critique of the Design Argument." In Ritola 2009.

Finocchiaro, M.A. 2010a. "Defending Copernicus and Galileo: Critical Reasoning and the Ship Experiment Argument." *Review of Metaphysics* 64: 75-103.

Finocchiaro, M.A. 2010b. *Defending Copernicus and Galileo: Critical Reasoning in the Two Affairs*. Boston Studies in the Philosophy of Science, vol. 280. Dordrecht: Springer.

Finocchiaro, M.A. 2011a. "Conductive Arguments: A Meta-Argumentation Approach." In Blair and Johnson 2011, 224-61.

Finocchiaro, M.A. 2011b. "Fair-mindedness vs. Sophistry in the Galileo Affair: Two Controversies for the Price of One." In *Controversies within the Scientific Revolution*, ed. M. Dascal and V. Boantza, pp. 53-73. Amsterdam: John Benjamins.

Finocchiaro, M.A. 2012a. "Deep Disagreements: A Meta-argumentation Approach." In Zenker 2012.

Finocchiaro, M.A. 2012b. "Meta-argumentation: Prolegomena to a Dutch Project." In Eemeren and Garssen 2012, 31-48.

Finocchiaro, M.A. 2013a. "Debts, Oligarchies, and Holisms: Deconstructing the Fallacy of Composition." *Informal Logic*, vol. 33, no. 1.

Finocchiaro, M.A. 2013b. *Guidebook to Galileo's* Dialogue. London: Routledge.

Fischer, T. 2011. "Weighing Considerations in Conductive Pro and Con Arguments." In Blair and Johnson 2011, 86-103.

Fischer, T. 2012. "Current Issues in Conductive Argument Weight." In Eemeren and Garssen 2012, 127-42.

Fisher, A. 1988. *The Logic of Real Arguments*. Cambridge: Cambridge University Press.

Fisher, A. 1991. "Testing Fairmindedness." *Informal Logic* 13: 31-36.

Fisher, A. 1992. "Critical Study: *Dialectics and the Macrostructure of Arguments* by James B. Freeman." *Informal Logic* 14: 193-204.

Fisher, A. 2000. Review of Ennis's *Critical Thinking. Argumentation* 14: 48-51.

Fisher, A. 2004. *The Logic of Real Arguments*. 2nd. edn. Cambridge: Cambridge University Press.

Fisher, A., and M. Scriven. 1997. *Critical Thinking: Its Definition and Assessment*. Point Reyes, CA: Edgepress.

Fogelin, R.J. 1978. *Understanding Arguments: An Introduction to Informal Logic*. New York: Harcourt Brace Jovanovich.

Fogelin, R.J. 1985. "The Logic of Deep Disagreements." *Informal Logic* 7: 1-8.

Fogelin, R.J. 2005. "The Logic of Deep Disagreements." *Informal Logic* 25: 3-11.

Franklin, J. 1987. "Non-deductive Logic in Mathematics." *British Journal for the Philosophy of Science* 38: 1-18.

Freeman, J.B. 1983. "Logical Form, Probability Interpretations, and the Inductive / Deductive Distinction." *Informal Logic* 5(2): 2-10.

Freeman, J.B. 1988. *Thinking Logically*. Englewood Cliffs: Prentice-Hall.

Freeman, J.B. 1991. *Dialectics and the Macrostructure of Arguments*. New York: Foris.

Freeman, J.B. 2005. *Acceptable Premises*. Cambridge: Cambridge University Press.

Freeman, J.B. 2011. "Evaluating Conductive Arguments in Light of the Toulmin Model." In Blair and Johnson 2011, 127-44.

Freeman, J.B. 2012. "Can Argumentation Always Deal with Dissensus?" In Eemeren and Garssen 2012, 61-76.
Friemann, R. 2002. "The Role of the Third Party in Reducing Deep Conflicts Between Ordinary People." In Hansen, Tindale, Blair, Johnson, and Pinto 2002.
Friemann, R. 2003. "Intractable Quarrels." In Eemeren, Blair, Willard, and Snoeck Henkemans 2003b, 335-39.
Friemann, R. 2005. "Emotional Backing and the Feeling of Deep Disagreement." *Informal Logic* 25: 51-64.
Gabbay, D.V., R.H. Johnson, H.J. Holbach, and J. Woods, eds. 2002. *Handbook of the Logic of Argument and Inference.* Amsterdam: Elsevier/North-Holland.
Gabbay, D.V., and H.J. Ohlbach, eds. 1996. *Practical Reasoning.* Berlin: Springer.
Gabbay, D.V., and J. Woods, eds. 2003a. *A Practical Logic of Cognitive Systems,* vol. 1, *Agenda Relevance: A Study in Formal Pragmatics.* Amsterdam: Elsevier/North-Holland.
Gabbay, D.V., and J. Woods, eds. 2003b. *A Practical Logic of Cognitive Systems,* vol. 2, *The Reach of Abduction: Insight and Trial.* Amsterdam: Elsevier/North-Holland.
Galilei, Galileo. 1890-1909. *Opere.* 20 vols. National Edition by A. Favaro et al. Florence: Barbèra. Rpt. in 1929-1939 and 1968.
Galilei, G. 1897. *Dialogo sopra i due massimi sistemi del mondo, tolemaico e copernicano.* In Galilei 1890-1909, vol. 7.
Galilei, Galileo. 1967. *Dialogue Concerning the Two Chief World Systems.* Trans. and ed. S. Drake. 2nd revised edn. Berkeley: University of California Press.
Galilei, Galileo. 1997. *Galileo on the World Systems: A New Abridged Translation and Guide.* Selected, trans., and ed. with Introduction, Notes, and Commentary by M. A. Finocchiaro. Berkeley: University of California Press.
Galilei, Galileo. 2008. *The Essential Galileo.* Trans. and ed. M.A. Finocchiaro. Indianapolis: Hackett.
Gardiner, P., ed. 1959. *Theories of History.* Glencoe: Free Press.
Godden, D.M. 2012. "Commentary on Maurice Finocchiaro's 'Deep Disagreements: A Meta-Argumentation Approach'." In Zenker 2012.
Godden, D.M, and W.H. Brenner. 2010. "Wittgenstein and the Logic of Deep Disagreement." *Cogency* 2(2):41-80.
Goddu, G.C. 2009. "What Is a 'Real' Argument." *Informal Logic* 29: 1-14.
Goodwin, D. 1992. "The Dialectic of Second-Order Distinctions: The Structure of Arguments about Fallacies." *Informal Logic* 14: 11-22.
Goldman, A.I. 1986. *Epistemology and Cognition.* Cambridge: Harvard University Press.
Goldman, A.I. 1994. "Argumentation and Social Epistemology." *Journal of Philosophy* 94: 27-49.
Goldman, A.I. 1999. *Knowledge in a Social World.* Oxford: Clarendon.
Goldman, A.I. 2003. "An Epistemological Approach to Argumentation." *Informal Logic* 23: 51-63.
Govier, T. 1980a. "Carl Wellman's *Challenge and Response.*" *Informal Logic Newsletter* 2(2): 10-15.
Govier, T. 1980b. "More on Deductive and Inductive Arguments." *Informal Logic Newsletter* 2(3): 7-8.
Govier, T. 1980c. "Assessing Arguments: What Range of Standards?" *Informal Logic Newsletter* 3(1): 2-4.
Govier, T. 1982. "Who Says There Are No Fallacies?" *Informal Logic Newsletter*, 5(1): 2-10.
Govier, T. 1985a. "Logical Analogies." *Informal Logic* 7: 27-33.

Govier, T. 1985b. *A Practical Study of Argument*. Belmont: Wadsworth.
Govier, T. 1987. *Problems in Argument Analysis and Evaluation*. Dordrecht: Foris.
Govier, T. 1989. "Critical Thinking as Argument Analysis." *Argumentation* 3: 115-26.
Govier, T. 1992. "What Is a Good Argument?" *Metaphilosophy* 23: 393-409.
Govier, T. 1995. "Critical Study of *New Essays in Informal Logic*." *Informal Logic* 17: 407-19.
Govier, T. 1998. "Arguing Forever?." In Hansen, Tindale, and Colman 1998.
Govier, T. 1999. *The Philosophy of Argument*. Newport News: Vale Press.
Govier, T. 2000. "Critical Review: Johnson's *Manifest Rationality*." *Informal Logic* 20: 281-91.
Govier, T. 2001. *A Practical Study of Arguments*. 5th edn. Belmont: Wadsworth.
Govier, T. 2002. "Should A Priori Analogies Be Regarded as Deductive Arguments?" *Informal Logic* 22: 155-57.
Govier, T. 2010. *A Practical Study of Arguments*. 7th edn. Belmont: Wadsworth.
Govier, T. 2011. "Conductive Arguments: Overview of the Symposium." In Blair and Johnson 2011, 262-76.
Gratton, C. 2010. *Infinite Regress Arguments*. Dordrecht: Springer.
Groarke, L.. 2002. "Johnson on the Metaphysics of Argument." *Argumentation* 16: 277-286.
Groarke, L. 2007. "Informal Logic." In the *Stanford Encyclopedia of Philosophy*. At http://plato.stanford.edu/entries/logic-informal/#Rhe.
Grootendorst, R. 1987. "Some Fallacies About Fallacies." In Eemeren, Grootendorst, Blair, and Willard 1987, 331-42.
Guarini, M. 2002. "On the Limits of the Woods-Hudak Reconstruction of Analogical Argument." In Hansen, Tindale, Blair, Johnson, and Pinto 2002.
Guarini, M. 2004. "A Defence of Non-deductive Reconstructions of Analogical Arguments." *Informal Logic* 24: 153-68.
Hamblin, C.L. 1970. *Fallacies*. London: Methuen.
Hansen, H.V. 1997. "Mill on Inference and Fallacies." In Walton and Brinton 1997, 125-43.
Hansen, H.V. 2002. "An Exploration of Johnson's Sense of 'Argument'." *Argumentation* 16: 263-76.
Hansen, H.V. 2005. "Does Mill Have a Theory of Argumentation?" Paper presented at the conference "The Uses of Argument," Ontario Society for the Study of Argumentation, McMaster University, Hamilton, Ontario, Canada, 18-21 May.
Hansen, H.V. 2006. "Mill and Pragma-dialectics." In Houtlosser and van Rees 2006, 97-108.
Hansen, H.V. 2011. "Notes on Balance-of-Considerations Arguments." In Blair and Johnson 2011, 31-51.
Hansen, H.V., and R.C. Pinto, eds. 1995. *Fallacies: Classical and Contemporary Readings*. University Park: Pennsylvania State University Press.
Hansen, H.V., and R.C. Pinto, eds. 2007. *Reason Reclaimed*. Newport News: Vale Press.
Hansen, H.V., C.W. Tindale, J.A. Blair, R.H. Johnson, and D.M. Godden, eds. 2007. *Dissensus and the Search for Common Ground*. Windsor: Ontario Society for the Study of Argumentation. CD-ROM. ISBN 978-0-9683461-5-0.
Hansen, H.V., C.W. Tindale, J.A. Blair, R.H. Johnson, and R.C. Pinto, eds. 2002. *Argumentation and Its Applications*. Windsor: Ontario Society for the Study of Argumentation. CD-ROM. ISBN: 0-9683461-2-X.
Hansen, H.V., C.W. Tindale, and A.V. Colman, eds. 1998. *Argumentation and Rhetoric*. St. Catharines: Ontario Society for the Study of Argumentation. CD-ROM. ISBN: 0-

9683461-0-3.
Hare, W.. 1979. *Open-mindedness and Education*. Kingston: McGill-Queen's University Press.
Hare, W. 1985. *In Defence of Open-mindedness*. Kingston: McGill-Queen's University Press.
Harman, G. 1984. "Logic and Reasoning." *Synthese* 60: 107-27.
Harman, G. 1986. *Change in View*. Cambridge: MIT. Press.
Hempel, C. 1965. *Aspects of Scientific Explanation*. New York: Free Press.
Herman, E.S., and N. Chomsky. 1988. *Manufacturing Consent*. New York: Pantheon.
Hintikka, J. 1989. "The Role of Logic in Argument." *The Monist* 72: 3-24.
Hitchcock, D. 1980. "Deductive and Inductive Types of Validity." *Informal Logic Newsletter* 2(3): 9-10.
Hitchcock, D. 1981. "Deduction, Induction, and Conduction." *Informal Logic Newsletter* 3(2): 7-15.
Hitchcock, D. 1983. *Critical Thinking*. Toronto: Methuen.
Hitchcock, D. 1985. "Enthymematic Arguments." *Informal Logic* 7: 83-97.
Hitchcock, D. 1987. "Enthymematic Arguments." In Eemeren, Grootendorst, Blair, and Willard 1987, 289-98.
Hitchcock, D. 1992. "Reasoning by Analogy: A General Theory." In *The Generalizability of Critical Thinking*, ed. S.P. Norris, 109-224. New York: Teachers College Press.
Hitchcock, D. 1994. "Validity in Conductive Arguments." In Johnson and Blair 1994, 58-66.
Hitchcock, D. 1998. "Does the Traditional Treatment of Enthymemes Rest on a Mistake?" *Argumentation* 12: 15-37.
Hitchcock, D. 2000. "The Significance of Informal Logic for Philosophy." *Informal Logic* 20: 129-38.
Hitchcock, D. 2002a. "The Practice of Argumentative Discussion," *Argumentation* 16: 287-98.
Hitchcock, D. 2002b. "Sampling Scholarly Arguments: A Test of a Theory of Good Inference." In Hansen, Tindale, Blair, Johnson, and Pinto 2002.
Hitchcock, D. 2002c. "A Note on Implicit Premisses." *Informal Logic* 22: 159-60.
Hitchcock, D. 2003. "Toulmin's Warrants." In Eemeren, Blair, Willard, and Snoeck Henkemans 2003b, 69-82.
Hitchcock, D., ed. 2005a. *The Uses of Argument*. Hamilton: Ontario Society for the Study of Argumentation.
Hitchcock, D. 2005b. "Good Reasoning on the Toulmin Model." In Hitchcock 2005a.
Hitchcock, D. 2006. "The Pragma-Dialectical Analysis of the *Ad Hominem* Fallacy." In Houtlosser and van Rees 2006, 109-19.
Hitchcock, D. 2007a. "Informal Logic and the Concept of Argument." In Jacquette 2007, 101-29.
Hitchcock, D. 2007b. "Is There an *Ad Hominem* Fallacy?" In Hansen and Pinto 2007, 187-99.
Hitchcock, D., and B. Vereij, eds. 2006. *Arguing on the Toulmin Model*. Dordrecht: Springer.
Houtlosser, P., and J.A. van Laar. 2007. "Metadialogues: Krabbe's Immanent Dialectic." *Argumentation* 21: 205-8.
Houtlosser, P., and A. van Rees, eds. 2006. *Considering Pragma-Dialectics*. Mahwah: Lawrence Erlbaum.
Hugon, P. 2008. "Arguments by Parallels in the Epistemological Works of Phya pa Chos kyi seng ge." *Argumentation* 22: 93-114.
Hume, D. 1947. *Dialogues Concerning Natural Religion*. Ed. N.K. Smith. Indianapolis:

Bobbs-Merrill.
Hume, D. 1948. *Dialogues Concerning Natural Religion*. Ed. H.D. Aiken. New York: Hafner.
Jacquette, D. 2007a. "Introduction: Philosophy of Logic Today." In Jacquette 2007c, 1-12.
Jacquette, D. 2007b. "On the Relation of Informal to Symbolic Logic." In Jacquette 2007c, 131-54.
Jacquette, D., ed. 2007c. *Philosophy of Logic*. Amsterdam: Elsevier.
Jacquette, D. 2007d. "Two Sides of Any Issue." *Argumentation* 21: 115-27.
Jakobovits, H., and D. Vermeir. 1999. "Robust Semantics for Argumentation Frameworks." Journal of Logic and Computation 9: 215-61.
Jin, R. 2011. "The Structure of Pro and Con Arguments: A Survey of the Theories." In Blair and Johnson 2011, 10-30.
Johnson, R.H. 1981a. "Charity Begins at Home." *Informal Logic Newsletter* 3(3): 4-9.
Johnson, R.H. 1981b. "Toulmin's Bold Experiment." *Informal Logic Newsletter* 3(2): 16-27; and 3(3): 13-19. Rpt. Johnson 1996, 116-52.
Johnson, R.H. 1987. "Logic Naturalized." In Eemeren, Grootendorst, Blair, and Willard 1987, 47-56.
Johnson, R.H. 1989. "Massey on Fallacy and Informal Logic." *Synthese* 80: 407-26.
Johnson, R.H. 1996. *The Rise of Informal Logic*. Newport News: Vale Press.
Johnson, R.H. 1998. "Response to Govier's 'Arguing Forever? Or: Two Tiers of Argument Appraisal'." In Hansen, Tindale, and Colman 1998.
Johnson, R.H. 2000a. *Manifest Rationality*. Mahwah: Lawrence Erlbaum.
Johnson, R.H. 2000b. "More on Arguers and Dialectical Obligations." In Tindale, Hansen, and Sveda 2000.
Johnson, R.H. 2002a. "Manifest Rationality Reconsidered." *Argumentation* 16: 311-31.
Johnson, R.H. 2002b. "Still More on Arguers and Their Dialectical Obligations." In Hansen, Tindale, Blair, Johnson, and Pinto 2002.
Johnson, R.H. 2003. "The Dialectical Tier Revisited." In Eemeren, Blair, Willard, and Snoeck Henkemans 2003b, 561-66.
Johnson, R.H. 2006. "Making Sense of Informal Logic." *Informal Logic* 26: 231-58.
Johnson, R.H. 2007a. "Anticipating Objections as a Way of Coping with Dissensus." In Hansen, Tindale, Blair, Johnson, and Godden 2007.
Johnson, R.H. 2007b. "Reply to My Commentator." In Hansen, Tindale, Blair, Johnson, and Godden 2007.
Johnson, R.H. 2007c. "Responding to Objections." In Eemeren, Blair, Willard, and Garssen 2007, 717-22.
Johnson, R.H. 2009. "Revisiting the Logical/Dialectical/Rhetorical Triumvirate." In Ritola 2009.
Johnson, R.H. 2010. "Wittgenstein's Influence on the Development of Informal Logic." *Cogency* 2(2): 81-104.
Johnson, R.H. 2011. "The Relationship between Pro/Con and Dialectical Tier Arguments." In Blair and Johnson 2011, 52-61.
Johnson, R.H., and J.A. Blair. 1977. *Logical Self-Defense*. Toronto: McGraw-Hill Ryerson.
Johnson, R.H., and J.A. Blair. 1980. "The Recent Development of Informal Logic." In Blair and Johnson 1980, 3-28.
Johnson, R.H., and J.A. Blair. 1985. "Informal Logic: The Past Five Years 1978-1983." *American Philosophical Quarterly* 22: 181-96.
Johnson, R.H., and J.A. Blair, eds. 1991. "Argument Evaluation Contest Results." *Informal Logic* 13: 167-85.
Johnson, R.H., and J.A. Blair, eds. 1994. *New Essays in Informal Logic*. Windsor: Informal

Logic Publications.

Johnson, R.H., and J.A. Blair 2000. "Informal Logic: An Overview." *Informal Logic* 20: 93-107.

Johnson, R.H., and J.A. Blair. 2002. "Informal Logic and the Reconfiguration of Logic." In Gabbay, Johnson, Holbach, and Woods 2002, 339-96.

Johnstone, H.W., Jr. 1952. "Philosophy and *Argumentum ad Hominem*." *Journal of Philosophy* 49: 489-98.

Johnstone, H.W., Jr. 1954. "Some Aspects of Philosophical Disagreement." *Dialectica* 8: 245–57.

Johnstone, H.W., Jr. 1959. *Philosophy and Argument*. University Park: Pennsylvania State University Press.

Johnstone, H.W., Jr. 1978. *Validity and Rhetoric in Philosophical Argument*. University Park: The Dialogue Press of Man & World.

Johnstone, H.W., Jr. 1989. "Argumentation and Formal Logic in Philosophy." *Argumentation* 3: 5-15.

Johnstone Jr., H.W. 1996. "Locke and Whately on the *Argumentum ad Hominem*." *Argumentation* 10: 89-97.

Johnstone, H.W., Jr. 1997. "A Bibliography, 1948-1997." *Philosophy and Rhetoric* 31: 6-19.

Jones, T. 2010. *What People Believe when They Say That People Believe: Folk Sociology and the Nature of Group Intentions*. Lanham: Lexington Books.

Jonsen, A., and S.E. Toulmin. 1988. *The Abuse of Casuistry*. Berkeley: University of California Press.

Joui, R.P. 2005. "A Citation-Based Reflection on Toulmin and Argument." *Argumentation* 19: 259-66.

Juthe, A. 2009. "Refutation by Parallel Argument." *Argumentation* 23: 133-69.

Kalish, D., and R. Montague. 1964. *Logic*. New York: Harcourt, Brace & World.

Katzav, J., and C.A. Reed. 2004. "On Argumentation Schemes and the Natural Classification of Arguments." *Argumentation* 18: 239-59.

Kauffeld, F. 2011. "Ranking Considerations and Aligning Probative Obligations." In Blair and Johnson 2011, 158-66.

Kock, C. 2003. "Multi-dimensionality and Non-deductiveness in Deliberate Argumentation." In Eemeren, Blair, Willard, and Snoeck Henkemans 2003a, 157-71.

Kock, C. 2007a. "Dialectical Obligations in Political Debate." *Informal Logic* 27: 233-48.

Kock, C. 2007b. "The Domain of Rhetorical Argumentation." In Eemeren, Blair, Willard, and Garssen 2007, 785-89.

Kock, C. 2007c. "Norms of Legitimate Dissensus." *Informal Logic* 27: 179-96.

Kock, C. 2011. "Why Argumentation Theory Should Differentiate between Types of Claims." In Blair and Johnson 2011, 62-73.

Komlósi, L.I., P. Houtlosser, and M. Leezenberg, eds. 2003. *Communication and Culture*. Amsterdam: Sic Sat.

Krabbe, E.C.W. 1995. "Can We Ever Pin One Down to a Formal Fallacy?" In Eemeren, Grootendorst, Blair, and Willard 1995a, 333-44.

Krabbe, E.C.W. 1997. "Arguments, Proofs, and Dialogues." In Astroh et al. 1997, 63-75.

Krabbe, E.C.W. 1998. "Comment on J. Anthony Blair's Paper." In Hansen, Tindale, and Colman 1998.

Krabbe, E.C.W. 1999. "The Dialectic of Quasi-Logical Arguments." In Eemeren, Grootendorst, Blair, and Willard 1999, 464-71.

Krabbe, E.C.W. 2000. "In Response to Ralph H. Johnson's 'More on Arguers and Dialectical Obligations'." In Tindale, Hansen, and Sveda 2000.

Krabbe, E.C.W. 2002. "Profiles of Dialogue as a Dialectical Tool." In Eemeren 2002, 153-67.
Krabbe, E.C.W. 2003. "Metadialogues." In Eemeren, Blair, Willard, and Snoeck Henkemans 2003b, 641-44.
Krabbe, E.C.W. 2007. "Nothing but Objections." In Hansen and Pinto 2007, 51-63.
Krabbe, E.C.W., R.J. Dalitz, and P.A. Smit, eds. 1993. *Empirical Logic and Public Debate.* Amsterdam: Rodopi.
Krabbe, E.C.W., and J.A. van Laar. 2011. "The Ways of Criticism." In Eemeren, Garssen, Godden, and Mitchell 2011, 1023-35.
Kraus, M. 2009. "Culture Sensitive Arguments." In Ritola 2009.
Kraus, M. 2012. "Cultural Diversity, Cognitive Breaks, and Deep Disagreement: Polemic Argument." In Eemeren and Garssen 2012, 91-107.
Krugman, P. 2009. "Pass the Bill." *New York Times*, December 18.
Kuhn, T.S. 1962. *The Structure of Scientific Revolutions.* Chicago: University of Chicago Press.
Kuhn, T.S. 1970. *The Structure of Scientific Revolutions.* 2nd edn. Chicago: University of Chicago Press.
Laar, J.A. van. 2002. "Equivocation in Dialectical Perspective." In Hansen, Tindale, Blair, Johnson, and Pinto 2002.
Laar, J.A. van. 2003a. "The Dialectic of Ambiguity." Doctoral Dissertation, Faculty of Philosophy, Groningen University.
Laar, J.A. van. 2003b. "The Use of Dialogue Profiles for the Study of Ambiguity." In Eemeren, Blair, Willard, and Snoeck Henkemans 2003b, 659-63.
Laar, J.A. van, and E.C.W. Krabbe. 2012. "The Burden of Criticism." In Zenker 2012.
Lakatos, I. 1976. *Proofs and Refutations.* Cambridge: Cambridge University Press.
Laudan, L. 1983. *Science and Values.* Berkeley: University of California Press.
Leff, M. 2000. "Rhetoric and Dialectic in the Twenty-first Century." *Argumentation* 14: 241-54.
Lipman, M. 1988a. "The Critical Thinker." *Teaching Thinking and Problem Solving* 10(3): 5-7.
Lipman, M. 1988b. "Critical Thinking—What Can It Be?" *Educational Leadership* 46(1): 38-43.
Lugg, A. 1986. "Deep Disagreement and Informal Logic." *Informal Logic* 8: 47-51.
Lumer, C. 2005a. "The Epistemological Theory of Argument" *Informal Logic* 25: 213-43.
Lumer, C. 2005b. "Introduction: The Epistemological Approach to Argumentation." *Informal Logic* 25: 189-212.
Machery, E., C.Y. Olivola, and M. De Blanc. 2009. "Linguistic and Metalinguistic Intuitions in the Philosophy of Language." *Analysis* 69: 689-94.
Mackie, J.L. 1993. "Causes and Conditions." In *Causation*, ed. E. Sosa and M. Tooley, 33-55. Oxford: Oxford University Press.
Malone, M., and D. Sherry. 1998. *Inference and Implication.* Dubuque: Kendall/Hunt.
Mancosu, P. 2005. "Visualization in Logic and Mathematics." In Mancosu, Jørgensen, and Pedersen 2005, 13-30.
Mancosu, P., ed. 2008. *The Philosophy of Mathematical Practice.* Oxford: Oxford University Press.
Mancosu, P, K. Jørgensen, and S. Pedersen, eds. 2005. *Visualization, Explanation and Reasoning Styles in Mathematics.* Dordrecht: Springer.
Martí, G. 2009. "Against Semantic Multi-culturalism." *Analysis* 69: 42-48.
Massey, G.J. 1975a. "Are There Good Arguments That Bad Arguments Are Bad?" *Philosophy in Context* 4: 61-77.

Massey, G.J. 1975b. "In Defense of the Asymmetry." *Philosophy in Context* 4(Supplement): 44-56.
Massey, G.J. 1981. "The Fallacy Behind Fallacies." *Midwest Studies in Philosophy* 6: 489-500.
McPeck, J.E. 1981. *Critical Thinking and Education*. New York: St. Martin's Press.
Memedi, V. 2007. "Resolving Deep Disagreement." In Hansen, Tindale, Blair, Johnson, and Godden 2007.
Memedi, V. 2011. "Intractable Disputes." In Eemeren, Garssen, Godden, and Mitchell 2011, 1259-65.
Mill, J.S. 1951. *Utilitarianism, Liberty, and Representative Government*. New York: Dutton.
Mill, J.S. 1952. *On Liberty, Representative Government, Utilitarianism*. In Great Books of the Western World, vol. 43, pp. 263-476. Chicago: Encyclopedia Britannica.
Mill, J.S. 1965. *Essential Works*. Ed. M. Lerner. New York: Bantam Books.
Mill, J.S. 1988. *The Subjection of Women*. Ed. S.M. Okin. Indianapolis: Hackett.
Mill, J.S. 1997. *The Spirit of the Age, On Liberty, The Subjection of Women*. Ed. A. Ryan. New York: Norton.
Modgil, S., and T.J.M. Bench-Capon. 2011. "Metalevel Argumentation." *Journal of Logic and Computation* 21: 959-1003.
Motta, U. 1993. "Querenghi e Galileo." *Aevum: Rassegna di scienze storiche linguistiche e filologiche* 67: 595-616.
Naess, A. 1966. *Communication and Argument*. Totowa: Bedminster Press.
Naess, A. 1982a. "An Application of Empirical Argumentation Analysis to Spinoza's 'Ethics'." In Barth and Martens 1982, 245-56.
Naess, A. 1982b. "A Necessary Component of Logic: Empirical Argumentation Analysis." In Barth and Martens 1982, 9-22.
Naess, A. 1992. "Arguing under Deep Disagreement." In Barth and Krabbe 1992, 123-31.
Natanson, M., and H.W. Johnstone, Jr., eds. *Philosophy, Rhetoric, and Argumentation*. University Park: Pennsylvania State University Press.
Nosich, G.M. 2002. "In Response to: Richard Friemann's 'Reducing Conflict Between Ordinary People'." In Hansen, Tindale, Blair, Johnson, and Pinto 2002.
Nuchelmans, G. 1993. "On the Fourfold Root of the *Argumentum ad Hominem*." In Krabbe, Dalitz, and Smit 1993, 37-48.
O'Connor, D. 2001. *Hume on Religion*. London: Routledge.
O'Keefe, D.J. 1977. "Two Concepts of Argument." *Journal of the American Forensic Association* 13: 121-28.
O'Keefe, D.J. 1982. "The Concept of Argument and Arguing." In *Advances in Argumentation Theory and Research*, ed. J.R. Cox and C.A. Willard, 3-23. Carbondale: Southern Illinois University Press.
O'Keefe, D.J. 1999. "How to Handle Opposing Arguments in Persuasive Messages: A Meta-analytic Review of the Effects of One-sided and Two-sided Messages." *Communication Yearbook* 22: 209-49.
O'Keefe, D.J. 2002. *Persuasion: Theory and Research*. 2nd edn. Thousand Oaks: Sage.
O'Keefe, D.J. 2012. "The Argumentative Structure of Some Persuasive Appeal Variations." In Eemeren and Garssen 2012, 291-306.
Okin, S.M. 1988. Editor's introduction. In Mill 1988, iv-xiv.
Oliver, J.W. 1967. "Formal Fallacies and Other Invalid Arguments." *Mind* 76: 463-78.
Pascal, Blaise. (1657-1658/1974). *Réflexions sur la géométrie en général: de l'esprit géométrique et de l'art de persuader*. Text and German translation by Jean-Pierre Schobinger. Basel and Stuttgart: Schwabe. Consulted on 17 February 2011 at:

<http://fr.wikisource.org/w/index.php?title=Sp%C3%A9cial:Livre&bookcmd=download&collection_id=8f6bfca2d0f8a5e2&writer=rl&return_to=De+l%E2%80%99esprit+g%C3%A9om%C3%A9trique+et+de+l%E2%80%99art+de+persuader>.

Passmore, J. 1961. *Philosophical Reasoning*. London: Duckworth.

Paul, R.W. 1990. *Critical Thinking*. Ed. A.J.A. Binker. Rohnert Park: Center for Critical Thinking and Moral Critique.

Pera, M. 1994. *The Discourses of Science*. Chicago: University of Chicago Press.

Perelman, Ch. 1979. *The New Rhetoric and the Humanities*. Dordrecht: Reidel.

Perelman, Ch. 1989. "Formal Logic and Informal Logic." In *From Metaphysics to Rhetoric*, ed. M. Meyer, 9-14. Dordrecht: Kluwer.

Perelman, Ch., and L. Olbrechts-Tyteca. 1958. *La Nouvelle Rhetorique: Traité de l'Argumentation*. Paris: P.U.F.

Perelman, Ch., and L. Olbrechts-Tyteca. 1969. *The New Rhetoric: A Treatise on Argumentation*. Trans. J. Wilkinson and P. Weaver. Notre Dame: University of Notre Dame.

Perkins, D.N. 1985a. "Postprimary Education Has Little Impact on Informal Reasoning." *Journal of Educational Psychology* 77: 562-71.

Perkins, D.N. 1985b. "Reasoning as Imagination." *Interchange* 16: 14-26.

Perkins, D.N. 1986. *Knowledge as Design*. Hillsdale: Lawrence Erlbaum.

Perkins, D.N. 1989. "Reasoning as It Is and as It Could Be." In *Thinking Across Cultures*, ed. D.N. Topping, D.C. Crowell, and V.N. Kobayashi, 175-94. Hillsdale: Lawrence Erlbaum.

Perkins, D.N. 2002. "Standard Logic as a Model of Reasoning." In Gabbay, Johnson, Ohlbach, and Woods 2002, 187-224.

Perkins, D.N., R. Allen, and J. Hafner. 1983. "Difficulties in Everyday Reasoning." In *Thinking, the Expanding Frontier*, ed. W. Maxwell, 177-89. Philadelphia: The Franklin Institute Press.

Perkins, D.N., M. Farady, and B. Bushey. 1991. "Everyday Reasoning and the Roots of Intelligence." In Voss, Perkins, and Segal 1991, 83-105.

Perlis, D. 1988. "Meta in Logic." In *Meta-level Architectures and Reflection*, ed. P. Maes and D. Nardi, 37-49. Amsterdam: Elsevier.

Phillips, D. 2008. "Investigating the Shared Background Required for Argument: A Critique of Fogelin's Thesis on Deep Disagreement." *Informal Logic* 28: 86-101.

Pinto, R.C. 1994. "Logic, Epistemology, and Argument Appraisal." In Johnson and Blair 1994, 116-24.

Pinto, R.C. 2001. *Argument, Inference and Dialectic*. Dordrecht: Kluwer.

Pinto, R.C. 2002. "Truth and Premiss Adequacy." In Hansen, Tindale, Blair, Johnson, and Pinto 2002.

Pinto, R.C. 2003. "Reasons." In Eemeren, Blair, Willard, and Snoeck Henkemans 2003b.

Pinto, R.C. 2006. "Evaluating Inferences." *Informal Logic* 26: 287-318.

Pinto, R.C. 2011. "Weighing Evidence in the Context of Conductive Reasoning." In Blair and Johnson 2011, 104-26.

Pollock, J. 1974. *Knowledge and Justification*. Princeton: Princeton University Press.

Pollock, J. 1986. *Contemporary Theories of Knowledge*. Totowa: Rowman & Littlefield.

Pollock, J. 1987. "Defeasible Reasoning." *Cognitive Science* 11: 481-518.

Pollock, J. 1990. "A Theory of Defeasible Reasoning." *International Journal of Intelligent Systems* 9: 33-54.

Pollock, J. 1995. *Cognitive Carpentry*. Cambridge: MIT Press.

Pollock, J. 2001. "Defeasible Reasoning with Variable Degrees of Justification." *Artificial Intelligence* 133: 233-82.

Prakken, H. 2006. "Artificial Intelligence & Law, Logic and Argument Schemes." In Hitchcock and Verheij 2006, 231-46.
Quine, W.V.O. 1961. *From a Logical Point of View*. 2nd edn. New York: Harper & Row.
Quine, W.V.O. 1970. *Philosophy of Logic*. Englewood Cliffs: Prentice-Hall.
Quine, W.V.O. 1986. *Philosophy of Logic*. 2nd edn. Cambridge: Harvard University Press.
Ramsey, F.P. 1931. *The Foundations of Mathematics and Other Logical Essays*. London: Routledge and Kegan Paul.
Reed, C. 2000. "Building Monologue." In Tindale Hansen, and Sveda 2000.
Reed, C., and D. Long. 1998. "Persuasive Monologue." In Hansen, Tindale, and Colman 1998.
Reed, C., and T.J. Norman, eds. 2004. *Argumentation Machines*. Dordrecht: Kluwer.
Reed, C., and G. Rowe. 2006. "Translating Toulmin Diagrams." In Hitchcock & Verheij 2006, 341-58.
Rees, M.A. van. 2001. "Review of Johnson's *Manifest Rationality*." *Argumentation* 15: 231-37.
Riley, J. 1998. *Mill on Liberty*. London: Routledge.
Ritola, J., ed. 2009. *Argument Cultures: Proceedings of OSSA 2009*. Windsor: Ontario Society for the Study of Argumentation. CD-ROM, ISBN 978-0-920233-51-1.
Rosen, F. 2006. "The Philosophy of Error and Liberty of Thought." *Informal Logic* 26: 121-48.
Rowland, R.C. 1987. "On Defining Argument," *Philosophy and Rhetoric* 20: 140-59.
Ryan, A. 1997. Introduction. In Mill 1997, ix-xlv.
Ryle, G. 1954. "Formal and Informal Logic." In idem, *Dilemmas*, 111-29. Cambridge: Cambridge University Press.
Salmon, M. 2002. *Introduction to Logic and Critical Thinking*. 4th edn. Wadsworth Thomson Learning.
Salmon, W.C. 1984. *Logic*. 3rd edn. Englewood Cliffs: Prentice-Hall.
Santillana, G. de. 1955. *The Crime of Galileo*. Chicago: University of Chicago Press.
Schauer, F. 2003. *Profiles, Probabilities, and Stereotypes*. Cambridge: Harvard University Press.
Schmitt, F., ed. 1994. *Socializing Epistemology*. Lanham: Rowman & Littlefield.
Scriven, M. 1956. "A Possible Distinction between Traditional Scientific Disciplines and the Study of Human Behavior." In Feigl and Scriven 1956, 330-39.
Scriven, M. 1958. "Definitions, Explanations, and Theories." In Feigl, Scriven, and Maxwell 1958, 99-195.
Scriven, M. 1959. "Truisms as the Grounds for Historical Explanations." In Gardiner 1959, 443-75.
Scriven, M. 1962. "Explanations, Predictions, and Laws." In Feigl and Maxwell 1962, 170-230.
Scriven, M. 1966. *Primary Philosophy*. New York: McGraw-Hill.
Scriven, M. 1976. *Reasoning*. New York: McGraw-Hill.
Scriven, M. 1981. "The 'Weight and Sum' Methodology." *American Journal of Evaluation* 2: 85-90.
Scriven, M. 1987. "Probative Logic." In Eemeren, Grootendorst, Blair, and Willard 1987, 7-32.
Scriven, M. 2002. "The Limits of Explication." *Argumentation* 16: 47-57.
Shelley, C. 2004. "Analogy Counterarguments." *Argumentation* 18: 223-38.
Siegel, H. 1988. *Educating Reason*. New York: Routledge.
Siegel, H. 1990. "Must Thinking Be Critical to Be Critical Thinking?" *Philosophy of the Social Sciences* 20: 453-61.

Siegel, H. 1994. "Justification by Balance and the Epistemology of Informal Logic." In Johnson and Blair 1994, 125-39.
Skyrms, B. 1966. *Choice and Chance*. Belmont: Dickenson.
Skyrms, B. 1975. *Choice and Chance*. 2nd edn. Belmont: Dickenson.
Slade, C. 1995. "Reflective Reasoning in Groups." *Informal Logic* 17: 223-34.
Slob, W.H. 2006. "The Voice of the Other." In Hitchcock and Verheij 2006, 165-80.
Smit, P.A. 1987. "An Argumentation-Theoretical Analysis of Lenin's Political Strategies." In Eemeren, Grootendorst, Blair, and Willard 1987, 317-26.
Smit, P.A. 1989. "An Argumentation-Analysis of a Central Part of Lenin's Political Logic." *Communication and Cognition* 22: 357-74.
Smit, P.A. 1992. "The Logic of Lenin's Polemics." In Barth and Krabbe 1992, 11-23.
Smith, N.K. 1947. Introduction. In Hume 1947, 1-123.
Snoeck Henkemans, A.F. 1992. *Analysing Complex Argumentation*. Amsterdam: Sic Sat.
Snoeck Henkemans, A.F. 2000. "State-of-the-Art: The Structure of Argumentation." *Argumentation* 14: 447-73.
Snoeck Henkemans, A.F. 2003. "Complex Argumentation in a Critical Discussion." *Argumentation* 17: 405-19.
Soccorsi, F. 1947. *Il processo di Galileo*. Rome: Edizioni La Civiltà Cattolica.
Sosa, E. 1997. "Reflective Knowledge in the Best Circles." *Journal of Philosophy* 94: 410-30.
Sosa, E. 2011. *Knowing Full Well*. Princeton: Princeton University Press.
Suzuki, T., Y. Yano, and T. Kato, eds. 2000. *Proceedings of the First Tokyo Conference on Argumentation*. Tokyo: Japan Debate Association.
Sztompka, P. 1979. *Sociological Dilemmas*. New York: Academic Press.
Tarski, A. 1965. *Introduction to Logic and to the Methodology of the Deductive Sciences*. 3rd edn. New York: Oxford University Press.
Thomas, S.N. 1986. *Practical Reasoning in Natural Language*. 3rd edn. Englewood Cliffs: Prentice-Hall.
Thomson, J.J. 1971. "A Defense of Abortion." *Philosophy and Public Affairs* 1: 47-66.
Tindale, C.W. 1999. *Acts of Arguing*. Albany: State University of New York Press.
Tindale, C.W. 2002. "A Concept Divided." *Argumentation* 16: 299-309.
Tindale, C.W. 2004. *Rhetorical Argumentation*. Thousand Oaks: Sage.
Tindale, C.W. 2006. "Perelman, Informal Logic, and the Historicity of Reason." *Informal Logic* 26: 341-57.
Tindale, C.W., H.V. Hansen, and E. Sveda, eds. 2000. *Argumentation at the Century's Turn*. St. Catharines: Ontario Society for the Study of Argumentation. CD-ROM. ISBN: 0-9683461-1-1.
Toulmin, S.E. 1953. *The Philosophy of Science*. London: Hutchinson.
Toulmin, S.E. 1958. *The Uses of Argument*. Cambridge: Cambridge University Press.
Toulmin, S.E. 1972. *Human Understanding*. New York: Harper.
Toulmin, S.E. 1992. "Logic, Rhetoric, and Reason." In Eemeren, Grootendorst, Blair, and Willard 1992, 3-11.
Toulmin, S.E. 2001. *Return to Reason*. Cambridge: Harvard University Press.
Toulmin, S.E. 2003. *The Uses of Argument*. Updated edn. Cambridge: Cambridge University Press.
Toulmin, S.E., and J. Goodfield. 1961. *The Fabric of the Heavens*. New York: Harper.
Toulmin, S.E., R. Rieke, and A. Janik. 1979. *Introduction to Reasoning*. New York: Macmillan.
Tully, R. 1995. "Informal Logic." In *The Oxford Companion to Philosophy*, ed. T. Honerich. Oxford: Oxford University Press.

Turing, A.M. 1950. "Computing Machinery and Intelligence." *Mind* 59: 433-60.
Turner, D. 2005. "Defending Deep Disagreement." In Hitchcock 2005, 462-64.
Turner, D., and L. Wright. 2005. "Revisiting Deep Disagreement." *Informal Logic* 25: 25-35.
Van Cleve, J. 1979. "Foundationalism, Epistemic Principles, and the Cartesian Circle." *Philosophical Review* 88: 55-91.
Verheij, B. 2006. "Evaluating Arguments Based on Toulmin's Scheme." In Hitchcock and Verheij 2006, 181-202.
Vorobej, M. 1995. "Linked Arguments and the Validity Requirement." *Argumentation* 9: 291-304.
Vorobej, M. 2006. *A Theory of Argument*. Cambridge: Cambridge University Press.
Voss, J.F., D.N. Perkins, and J.W. Segal. 1991. *Informal Reasoning and Education*. Hillsdale: Lawrence Erlbaum.
Walton, D.N. 1985. *Arguer's Positions*. Westport: Greenwood Press.
Walton, D.N. 1989. *Informal Logic*. Cambridge: Cambridge University Press.
Walton, D.N. 1990. "What is Reasoning? What is an Argument?" *Journal of Philosophy* 87: 399-419.
Walton, D.N. 1996. *Argument Structure*. Toronto: University of Toronto Press.
Walton, D.N. 1999. "Informal Logic." In *The Cambridge Dictionary of Philosophy*, ed. R. Audi. Cambridge: Cambridge University Press.
Walton, D.N. 2007. "Metadialogues for Resolving Burden of Proof Disputes." *Argumentation* 21: 291-316.
Walton, D.N. 2011. "Conductive Arguments in Ethical Deliberation." In Blair and Johnson 2011, 191-209.
Walton, D.N., and A. Brinton, eds. 1997. *Historical Foundations of Informal Logic*. Aldershot: Ashgate.
Walton, D., and E.C.W. Krabbe. 1995. *Commitment in Dialogue*. Albany: State University of New York Press.
Weinstein, M. 2002. "Critical Review of Pinto's *Argument, Inference and Dialectic*." *Informal Logic* 22: 161-80.
Wellman, C. 1971. *Challenge and Response: Justification in Ethics*. Carbondale: Southern Illinois University Press.
Wellman, C. 1975. *Morals & Ethics*. Glenview: Scott, Foresman.
Whaley, B.B. 1998. "Evaluations of Rebuttal Analogy Users." *Argumentation* 12: 351-65.
Whaley, B.B., and R.L. Holloway. 1996. "Rebuttal Analogy." *Metaphor and Symbolic Activity* 11: 161-67.
Whately, R. 1838. *Elements of Logic*. New York: William Jackson.
Willard, C.A. 1983. *Argumentation and the Social Grounds of Knowledge*. University, Alabama: University of Alabama Press.
William, Prince of Orange. 1581. *The Apologie or Defence of the Most Noble Prince William, by the Grace of God, Prince of Orange*. Delft.
Wittgenstein, L. 1969. *On Certainty*. Trans. D. Paul and G.E.M. Anscombe. Ed. G.E.M. Anscombe and G.H. von Wright. New York: Harper.
Wohlrapp, H. 1995. "Resolving the Riddle of the Non-deductive Argumentation Schemes." In Eemeren, Grootendorst, Blair, and Willard 1995a, 55-62.
Wohlrapp, H. 1998. "A New Light on Non-deductive Argumentation." *Argumentation* 12: 341-50.
Wohlrapp, H. 2008. "The Pro- and Contra-discussion (A Critique of Trudy Govier's 'Conductive Argument')." Trans. by F. Zenker from H. Wohlrapp's *Der Begriff des Arguments*, chapter 6.4, pp. 316-34, Wuerzburg: Koenigshausen und Neumann,

available at: http://www.frankzenker.de/academia.html.
Wohlrapp, H. 2011. "A Misleading Model for the Analysis of Pro- and Contra-Argumentation." In Blair and Johnson 2011, 210-23.
Woods, J. 1980. "What is Informal Logic?" In Blair and Johnson 1980, 57-68.
Woods, J. 1992. "Public Policy and Standoffs of Force Five." In Barth and Krabbe 1992, 97-108. (Rpt. Woods 2004a, 185-200.)
Woods, J. 1995. "*Ad Hominem*: From Johnstone to Locke to Aristotle." In Eemeren, Grootendorst, Blair, and Willard 1995a, 395-408. (Rpt. Woods 2004a, 111-24.)
Woods, J. 1996. "Deep Disagreements and Public Demoralization." In Gabbay and Ohlbach 1996, 650-62. (Rpt. Woods 2004a, 201-18.)
Woods, J. 2002. "In Response to Marcello Guarini's 'On the Limitations of the Woods-Hudak Reconstruction of Analogical Argument'." In Hansen, Tindale, Blair, Johnson, and Pinto 2002.
Woods, J. 2003. *Paradox and Paraconsistency*. Cambridge: Cambridge University Press.
Woods, J. 2004a. *The Death of Argument*. Dordrecht: Kluwer.
Woods, J. 2004b. "The Informal Core of Formal Logic." In Woods 2004a, 43-64.
Woods, J. 2006a. "Eight Theses Reflecting on Stephen Toulmin." In Hitchcock and Verheij 2006, 379-97.
Woods, J. 2006b. "Pragma-Dialectics: A Retrospective." In Houtlosser and van Rees 2006, 301-11.
Woods, J. 2013. *Errors of Reasoning: Naturalizing the Logic of Inference*. London: College Publications.
Woods, J., and B. Hudak. 1989. "By Parity of Reasoning." *Informal Logic* 11: 125-39. (Rpt. Woods 2004a, 253-72.)
Woods, J., A. Irvine, and D. Walton. 2000. *Argument, Critical Thinking, Logic and the Fallacies*. Toronto: Prentice-Hall.
Woods, J., R.H. Johnson, D.M. Gabbay, and H.J. Ohlbach. 2002. "Logic and the Practical Turn." In Gabbay, Johnson, Ohlbach, and Woods 2002, 1-39.
Woods, J., and D.N. Walton. 1989. *Fallacies: Selected Papers 1972-1982*. Dordrecht: Foris.
Wooldridge, M., P. McBurney, and S. Parsons. 2005. "On the Metalogic of Arguments." In *Proceedings of the Fourth International Joint Conference on Autonomous Agents and Multi-Agent Systems* (AAMAS-05), Utrecht, July 2005.
Wright, L. 1995. "Argument and Deliberation." *Journal of Philosophy* 92: 565-86.
Wright, L. 1999. "Reasons and the Deductive Ideal." *Midwest Studies in Philosophy* 23: 197-206.
Wright, L. 2001a. *Critical Thinking*. New York: Oxford University Press.
Wright, L. 2001b. "Justification, Discovery, Reason, and Argument." *Argumentation* 15: 97-104.
Wright, L. 2002. "Reasoning and Explaining." *Argumentation* 16: 33-46.
Wyatt, N. 2001. "Review of Johnson's *Manifest Rationality*." *Philosophy in Review* 21: 185-87.
Zarefsky, D. 2012. "The Appeal for Transcendence: A Possible Response to Cases of Deep Disagreement." In Eemeren and Garssen 2012, 77-90.
Zenker, F. 2009a. *Ceteris paribus in Conservative Belief Revision*. Frankfurt: Peter Lang.
Zenker, F. 2009b. "Complexity Without Insight: *Ceteris Paribus* Clauses in Conductive Argumentation." In *Concerning Argument: Proceedings of the 2007 NCA/AFA Conference on Argumentation, Alta, Utah*, ed. S. Jacobs, 810-18. Washington: National Communication Association. Independently paginated version available at: http://www.frankzenker.de/academia.html.

Zenker, F. 2011. "Deduction, Induction, Conduction: An Attempt at Unifying Natural Language Argument Structures." In Blair and Johnson 2011, 74-85.
Zenker, F., ed. 2012. *Argumentation: Cognition & Community* (Proceedings of the 9th International Conference of the Ontario Society for the Study of Argumentation, OSSA, May 18-21, 2011). Windsor: Ontario Society for the Study of Argumentation. CD-ROM, ISBN: 978-0-920233-66-5.

Index

abductive argument, 15
abortion, 85, 96, 97, 101, 103, 105, 106, 107, 109, 118
accumulation of evidence, 123, 126
action and reaction, law of, 67-68
activism, 107-9
actual vertical fall, 31, 222-23, 226, 227-39
Adams, David, 91-92, 93, 117, 118
ad hoc, 167-71, 176, 225, 225-26, 235, 239, 244
ad hominem argument, 67, 87-88, 91-92, 95, 97, 105, 110, 111-16, 116-19, 119-22, 162, 165-66, 167, 170, 176-77, 211, 243
ad rem argumentation, 111, 114-16
affirmative action, 85, 88, 96, 97, 117
affirming the consequent, 78
Allen, Derek, 136-38, 147, 157, 158, 244
ambiguity of "reason," 54
ambiguous argument, 231
ampliative argument, 15
analysis vs. theory, 9-10, 34-41
analytical approach, 9-10
analytical vs. normative, 8
analytic argument, 15
analytic invalidity, 24
anarchism, 226
Angell, Richard, 133n10
anticipating objections, 162, 171-76; vs. pre-empting objections, 173-74
Antiphon, 14
a posteriori, 202, 204, 229-30
apparent vertical fall, 31, 223-24, 225-26, 227-39; vs. actual vertical fall, 244
appeal to ignorance, 26-27
appeal to methodological principles, 92, 101-2, 103-4, 109, 117-19
applied epistemology, 3
applied logic, 3, 7, 17, 119, 124, 157, 191, 219, 227, 237, 239, 242
applied philosophy, 114
appreciating fallibility, 179-81
appreciating reasons, 181-84
approximations vs. refinements, 12-13

a priori, 9, 204, 209, 216, 217, 229; apriorist(ic) approach, 11n16, 12
Aquinas, Thomas, 4, 17, 201
argument analysis, 3, 16, 34-35, 243
argumentation theory, 1, 2, 3, 39-41, 187-89, 243, 244; vs. meta-argumentation, 15, 220, 234-40, 244
argumentative reality, 1, 4-5, 17, 219, 239
argument by analogy, 15, 80-81, 81-83, 90, 146, 202, 205, 208, 211-13, 218
argument by parity of reasoning, 81-83
argument form, 76-78
argument from evil, 214-15, 218
argument from natural motion, 221
argument-sketch, 68
argument vs. good argument, 57
Aristotle, 4, 17, 114-16, 153, 154, 219-41, 244; authority, 235-36; dialectic, 42, 44
Arnauld, Antoine, 163
artificial intelligence, 1, 3
assumption of equal unknown, 226, 229, 230, 234, 244, 245
asymmetry of favorable and unfavorable evaluation, 75
asymmetry of pros and cons, 136, 157-58
atheism, 209
audience, 166
axiologism, 107-9
axiomatic approach, 42-43

Baier, Kurt, 134
balanced approach, 193-94
balance-of-considerations argument, 123-61. *See also* conductive argument
balance-of-considerations claim, 124, 138, 144-45, 146, 147, 148-49, 157, 158, 159, 162, 187, 243
balancing conflicting reasons, 189
balancing pros and cons, 123
Barker, Stephen, 201-4, 205, 218
Barth, Else, 9, 42, 69, 84
begging the question, 13-14, 25-26, 31, 32, 51, 56, 201, 220, 221, 222, 225,

Index 269

227-40, 244, 245. *See also petitio principii*
Bentham, Jeremy, 9
Berkeley, George, 114
Beth, E. W., 77
bilaterality, 111-13, 118
Biro, John, 246
Blair, Tony, 3, 43-44, 45, 66, 162-67, 176, 179, 244
Boella, Guido, 1, 199n5
Boston, 78
Brandom, Robert, 8, 95
Brenner, William, 93-95, 117
Brooks, David, 124, 144-47, 148, 149, 150, 158, 159, 159-60, 187
burden of proof, 193-94

Campolo, Christian, 88-89, 117
capitalism vs. communism, 120-21
Carnap, Rudolf, 49
Carroll, Lewis, 6, 164
Carson City, 767
case-building, 166-67
causal undermining, 194-96, 199-200
cause of women's subjection, 194-96
causes vs. reasons, 194
center of universe, 221, 236
certainty, 180
ceteris paribus, 129, 131, 136-37, 140, 141-42, 156, 158
chaotic universe, 213-14
charity, principle of, 27, 62, 90-91, 147, 182, 189, 233
Chomsky, Noam, 103
Christianity: Christ, 185; history, 183; morality, 185
circular argument, 234-40, 245
circularity, 25-26, 31, 56, 57-58, 226, 227-39, 244; vicious vs. virtuous, 190-91
citizenship, and women's liberation, 193
clarification, 244
Cleanthes, 201-18
closed-mindedness, 98, 98-101, 110, 118. *See also* open-mindedness
code of conduct, 75-76
cognitive psychology, 3
cognitive science, 3
Cohen, Carl, 66, 246
Cohen, L. Jonathan, 132

Cohen, Morris, 164
Cold War, 120
collective bargaining 101, 105
collectivism, 107-9
communication studies, 3
communication, and argumentation, 113-14
communism vs. capitalism, 120-21
comparative approach, 8-9, 10, 219
complementary meta-arguments, 197, 199-200
complete evidence: fallacy of incomplete evidence, 211-13, 245; requirement of, 49, 139, 211-13, 217
complex argumentation, 43, 67, 68, 69, 88-89, 97, 117-19, 119-22, 167, 179, 186, 187, 204, 217-18, 218, 243, 244
complexity, 44
compromise, 122
conception of argument. *See* definition of argument
conception of concepts, 95
conceptual disagreements, 93-95, 117
conceptual explicitness, 39-40
conclusion, definition of, 18-19
conclusion-refuting criticism, 83
conduction: caveats, 149-52; conductive adequacy, 128-31, 157; conductive argument, 2, 15, 70n1, 123-61, 167, 178, 187, 219, 243; conductive definitions, 130-31, 154, 157; conductive meta-argument, 124, 129, 131, 134, 157, 219; conductivism, 151; vs. convergence, 134-35, 158; vs. deduction, 125-26, 129, 132, 133-34, 136-38, 140, 142, 158; and dialectical tier, 151; and explanation, 133; indicators, 155; vs. induction, 126, 129, 132, 134, 158; visual representation, 148-49
conductive argument, 2, 15, 70n1, 123-61, 167, 178, 187, 219, 243. *See also* conduction
confrontation stage, 171
con-reasons acknowledgment claim, 149, 187
conservation of motion, 232
conservatism vs. reformism, 184-85
considering objections, 181-84
constructive arguments, 71

constructive vs. critical meta-argument, 206
contradiction vs. contrariety, 99-101
contradictory arguments, 151
contrariety vs. contradiction, 99-101
contrary arguments, 99-101, 110
controversy, 52-53; and philosophy, 111-14
convergence, 123, 126, 132, 139-40, 148n21, 157
Copernican Revolution, 26, 34, 119-20, 153, 154, 158, 219, 239
Copernican system, 208
Copernicus, Nicolaus, 34, 120, 152-53
Copi, Irving, 44, 45, 66, 73, 80-81, 84, 164, 246
corrigibility, 179, 189
counterargument, 28, 44, 83; defined, 19-20
counterconsiderations, 70n1, 132, 133
counterexample, 24, 25, 29, 76-78
counterproductive argumentation, 193-94
critical argumentation, 48, 71
critical discussion, 167-71
critical thinking, 3
criticism, and persuasiveness, 126-27, 141
criticism of alternatives, 53-57, 60, 64, 111-13, 168, 174
criticism of arguments, 22-26, 75-84
Croce, Benedetto, 121, 163
culture-sensitive arguments, 97
cumulative support, 132
custom vs. argumentation, 193-94, 199

Dante Alighieri, 64n31
Darwin, Charles, 9
Davson-Galle, Peter, 87-88, 116, 118
deception of the senses, 236
deduction vs. induction, 130
deductive argument, 15, 201
deductive axiomatization, 44
deductive invalidity, 24
deductive logic, 6
deductive meta-argument, 210
deductive validity, 162-67; vs. persuasive ineffectiveness, 239
deductivism, 136-37, 140, 142, 158
deductivist approach, 158

deep disagreements, 2, 85-97, 97-98, 100, 111, 116-19, 122, 151n26, 162, 178, 243
de-facto vs. rational resolution, 110
defeasible reasoning or argument, 7, 15
defeater, 48
defects of arguments, 8, 223
defending alternatives, 175, 176
definitions of argument, 1, 2, 18-19, 44-48, 162, 178, 188, 242
definition of meta-argument, 1, 34
deliberative argumentation, 151, 159
Demea, 204, 209, 210, 214
democracy vs. elitism, 121
demonstration vs. argumentation, 13, 16, 220, 228, 231-32
denying the consequent, 226
description vs. prescription, 8
design argument, 3, 4-5, 17, 113-14, 201-18, 243-44, 245
determinism, 107-9
devastating criticism, 115
Dewey, John, 163, 164
dialectic: approach, 188, 191; concepts of, 42-44; definitions of argument, 42, 182; vs. dialogical approach, 44, 71; excellence, 171-76; garb, 69; vs. logic, 42; and logical theory, 97; obligations, 46, 73, 150-51, 159, 171-76; vs. rhetoric, 220n1; tier, 46, 48, 50, 150-51, 159, 179, 180, 193, 194, 243, 245; virtues vs. obligations, 175-76. *See also* dialogue
dialogue: approach, 42, 43, 65; 83-84; vs. dialectical approach, 44, 71; dialogical argumentation, 47, 66; dialogical dialectic, 44; dialogue model, 60. *See also* dialectic
Dialogue on the Two Chief World Systems (Galileo), 3, 16, 17, 26, 42, 51, 60, 71, 74, 124, 152-57, 208, 219-41, 244
Dialogues concerning Natural Religion (Hume), 3, 17, 201-18, 243
different languages, 112
disconnection 24-26
discovering truth, 179-81
disjunctive vs. conjunctive conceptions, 61
dissensus and truth, 183, 189

Index 271

divine attributes: benevolence, 214-15, 218; finiteness, 218; imperfection, 218; infinity, 210-11; intelligence, 215; mercy, 214-15; omnipotence, 214-15; perfection, 210-11; simplicity, 209, 218;
dogma vs. living truth, 181
downgrading disagreements, 102
Dryzek, John, 122
duet arguments, 43-44
dynamical approach, 142-44
dynamics of rationality, 86-87, 92, 96, 116

earth's motion, 3, 4, 17, 30-32, 34, 34-35, 36-39, 113, 124, 148n21, 152-57, 158, 159, 174-75, 208-9, 219-41, 244
earth's shape, 156
earth-heaven dichotomy, 27
Eemeren, Frans van, 2, 3, 8, 162, 167, 176, 220n1, 244, 246; hyper dialectical conception, 65-74; strategic maneuvering, 167-71
Einstein, Albert, 133n10
Eisend, Martin, 152, 159
elite vs. mass, 182-83
elitism vs. democracy, 121
empathy, 89-91, 117
empirical accuracy, 237
empirical approach, 1, 9, 10, 50-51, 119-22, 124, 143, 219
empirical content, 202-3
empirical exclusivity, 206
empirical explicitness, 206
empirical testing, 119-22
empirical vs. conceptual knowledge, 233-34
empiricism, 111, 206, 236
employment, and women's liberation, 193
enduring dissensus, 97, 100n16
Ennis, Robert, 130, 139-40, 157, 158, 244
entailment, 233
enthymeme, 133-34, 136-37
Epicurean hypothesis, 213-14
epigrammatic reasoning, 88-89, 96
epistemology, 6, 8, 9, 227-27; approach, 8, 10, 189-91; vs. argumentation, 189-91, 245, 246-47; vs. meta-argumentation, 243; modesty, 179-80, 189
equality of women, 192-200
equivocation, 25, 32, 75, 116, 231
eternity of universe, 213-14
Euclid, 9, 113
Euclidean geometry, 62
Eudoxus, 114-16
euphemisms, 103
evaluation of arguments, 22-26, 34-35, 244
evaluative arguments, 150
evaluative dialectic, 44
evil, 214-15, 215-17
existence of deep disagreements, 96
existence of God, 3, 4-5, 17, 201-18, 243-44
existential import, 78
existentialism, 111
experts vs. common people, 182-83
explanation: vs. argument, 54-55, 60; explanatory disconnection, 24-25, 26, 29, 32, 214; explanatory power, 210; nature of, 209-10; of women's subjection, 194-96. See also inference to the best explanation
explicit structure, 22
explicit vs. implicit argumentation, 1
extremism, 101, 103, 105, 106, 109, 180, 193-94

factual vs. evaluative, 8
fair-mindedness, 89-91, 92, 96-97, 100-101, 109, 110, 113, 117-19, 119-22, 153-54, 159, 175, 243
fairness, 185-86
fallacy: of composition, 207, 217, 245; conception of, 167, 176; fallacy behind fallacies, 78; fallacy criticism, 75, 84; of ignorance, 70, 198; of stereotyping, 209; theory, 167-71
fallibility, 189
famous meta-arguments, 3, 17, 178-241, 245, 246
feeling vs. argumentation, 193-94, 199
Feldman, Richard, 88, 116
female acceptance of male rule, 195-96
feminism, 192-200
Feyerabend, Paul, 226, 240
final cause, 204

final conclusion, 20, 71
final premise, 71
final reason, 20, 37
finiteness of God, 215-17
First Amendment of the US Constitution, 80
Fisher, Alec, 3
Fogelin, Robert, 2, 3, 97-98, 104, 111, 116-19, 122, 244; deep disagreements, 85-97
force, law of, 67
formal: analogy, 79; approach, 1, 11n16; dialectic(s), 42-43; disconnection, 24, 26, 29; fallacy, 75, 75-78; formal deductive logic, 6-7, 12-13, 16, 40, 151, 163, 233, 235, 245; formal-fallacy criticism, 83, 84; invalidity, 24, 242-43; semantics, 42; validity, 75-78, 222, 224, 226, 235;
foundations of mathematics, 101, 106-7
freedom of argument, 3
Freeman, James, 43, 97, 111, 130-31, 134, 154, 157, 244, 246
Freud, Sigmund, 9
Friemann, Richard, 85, 89-91, 92, 98, 100, 117
full-bodied arguments, 162-67, 176

Gabbay, Dov, 1, 199n5
Galilei, Galileo, 3, 16, 17, 34, 51, 60, 68, 87, 113, 115, 119, 120, 124, 159, 170, 174-75, 208, 211, 244, 245, 246; *Dialogue* as conductive argument, 152-57; heavenly unchangeability, 26-30; simplicity argument, 36-39; vertical fall, 30-32, 219-41
Galileo affair, 120, 158
games, 112
Gauss, Carl F., 9
generality, 39-40
generalization argument, 216
genetic fallacy, 114, 195
gestalt switch, 121
God, 39, 78
Godden, David, 93-95, 117
Gödel's theorem, 113
Goethe, Johann W. von, 181
Goldman, Alvin, 2, 47, 48-50, 50, 60, 63, 64, 66, 73, 170-71, 244, 246
Govier, Trudy, 2, 3, 46, 59, 84, 123, 130, 131, 147, 148n21, 149, 150, 151, 157, 158, 167, 244; conductive arguments, 132-44; refutation by logical analogy, 78-81
grammar, 112
Gramsci, Antonio, 120-21, 151n26
Grootendorst, Rob, 3, 8, 65-74, 167, 171
Grotius, Hugo, 9
grounding the meta-level, 199
ground-level argumentation, 1, 34, 92, 97, 109, 121-22, 144, 158, 159, 196, 197, 203, 204-6, 211, 211-12, 214-15, 215, 216, 218, 219, 223, 227, 235, 237, 242
ground-level vs. meta-level, 198-99
group rights, 96
guilt by association, 103, 106

Hamblin, Charles, 3, 71, 74, 75
handling objections, 176-77
Hansen, Hans, 46, 48, 58, 59, 62, 72, 148n21, 149n22, 178, 192
hasty generalization, 206-7, 207-8, 217, 245
health care, 124, 144-47, 150, 158, 159, 159-60, 160-61, 187
heavenly unchangeability, 26-30, 219
hefting model, 127, 135, 157
Hegel, Georg W. F., 121
Hegelian dialectic, 42, 44, 64, 107, 109, 121, 151n26
Hempel, Carl, 49
hermeneutics, 156-57
Herodes, 14
Hilbert, David, 12
historical approach, 9, 10
historical prediction, 196-97
historical-textual approach, 16, 17, 124, 157, 178, 191, 219, 242
history of argumentation, 9
Hitchcock, David, 2, 46, 51, 59, 62, 123, 128-31, 133, 134, 157, 244
holism, 94
horizontally latent structure, 22
Houtlosser, Peter, 68, 69, 162, 167-71, 176, 220n1, 246
Hudak, Brent, 81-83, 84
human nature, 5, 195
Hume, David, 3, 17, 201-18, 243, 245

Index

hyper dialectical definition of argument, 47-48, 60, 63n30, 64, 65-74, 188, 198242
hypothetico-deductive confirmation, 12, 15

idealism, 111
identification, 89-91
illative account: core vs. tier, 46, 60, 70; definition of argument, 42, 45, 48, 50, 62-64, 66, 69, 70, 72, 73, 188, 242; illative tier, 56, 179, 180, 182, 186, 193, 194, 245; illative vs. dialectical definition, 63-64; illative vs. dialectical excellence, 174; illative vs. dialectical tier, 198-99, 200; illative vs. dialectical vs. meta tiers, 243, 245, 246
illative core. *See* illative account
illative definition of argument, 42, 45, 48, 50, 62-64, 66, 69, 70, 72, 73, 188, 242. *See also* illative account
immanent dialectical approach, 16
immediate vs. later resolution, 86-87, 92
impartiality, 184-86, 189
implication vs. argument, 55-56
implicit proposition, 21-22
implicit vs. explicit argumentation, 1
incommensurability, 111, 151
incomplete arguments, 69
incomplete evidence, fallacy of, 211-13, 245
independent reasons, 21, 37-38
independent vs. linked premises, 124-25
individual freedom, 196-97, 199
inductive argument, 15, 29, 50, 80-81, 132, 201, 206, 208, 213, 215, 216, 217, 218
inductive correctness vs. deductive validity, 164-65
inductive generalization, 59, 71, 163-64, 218
inductive logic, 6, 16
inertia, law of, 67, 232
infallibility, 179-80
inference to the best explanation, 15, 24-25, 29, 32, 133, 140, 153, 156, 208, 213-14, 217, 221
inference vs. argument, 55-56
inferentialism, 95

infinite power, 214-15
infinite regress, 59, 209-10, 218, 245
infinity, 213-14
informal approach, 1
informal logic, 3, 151, 154, 155, 173
in-practice resolution, 92-93, 117
in-principle resolution, 92-93, 117
intellectual peace, 181
intelligent design, 3, 4-5, 201-18
interdependent reasons, 21
intermediate proposition, 20, 71
internal criticism, 25, 26, 28-29, 87-88, 115, 116, 119
internal disconnection, 25, 26
interpretation of arguments, 34-35, 191, 244
interpretation vs. evaluation, 172-73, 242
interpretive vs. evaluative meta-arguments, 218
intractable quarrels, 2, 85, 89-91, 97, 100, 111, 116-19, 122
INUS condition, 139-40
Iraq War, 170
irrational number, 7
irrationality, 110

Jacobs, Scott, 69
Jacquette, Dale, 152, 159
Jesus, 180
Jin, Rongdong, 148n21, 149n22
Johnson, Ralph, 2, 3, 8, 44, 45, 46, 65, 66, 69, 73, 159, 162, 168, 176-77, 179, 188, 244, 246; anticipating objections, 171-76; conductive arguments, 150-51; strongly dialectical definition of argument, 44-64
Johnson, W. E., 106
Johnstone, Henry, 2, 3, 71, 74, 85, 87-88, 91, 95, 105, 116-19, 122, 166, 167, 176, 177, 211, 243, 244; philosophical controversies, 110-16
Joshua, 153, 156
judicious approach, 135
judiciousness, 61, 185, 193-94, 226, 234-35, 240
justification of weighing, 136-38
justification vs. persuasion, 63

Kant, Immanuel, 201, 233
Kepler, Johannes, 9, 68

Kock, Christian, 97, 151, 159
Krabbe, Erik, 1, 42, 69, 75-78, 79, 81, 94, 235n14, 244
Kraus, Manfred, 97
Krugman, Paul, 144, 158, 160-61
Kuhn, Thomas, 121

Laar, Jan A. van, 75, 235n14
Las Vegas, 76-78
last gasp dialectical response, 102, 105-6
latent proposition, 21-22, 25
latent structure, 22, 26, 31, 143, 245
Lavoisier, Antoine, 9
law of revolution, 38
law of the strongest, 194-96, 199
laws of falling bodies, 68
Left vs. Right, 121
liberty of argument, 178-91, 243, 245
limitation theorem, 97, 102, 117-18
Lincoln, Abraham, 80
linked reasons, 21, 37, 38, 148n21
linked vs. independent premises, 124-25, 179, 186-87
Locke, John 87, 113, 115, 170, 211
logic: conceptions of, 162-67; vs. epistemology, 189-91; vs. formal deductive logic, 163; logical analogy, 78-81, 217; logical argumentation, 103; logical force, 126, 203; logical form, 77-78, 79-80, 81-83, 202; logical parallelism, 79-80; logical rationality, 226, 240; logical structure, 79-80. 81-83; logical theory, 3, 5-7, 8, 9, 39-40, 244; vs. philosophy of argument, 163; vs. rhetoric, 220, 228, 231-32, 234-40, 245, 246; textbooks, 162-67
Long, Derek, 44, 72
Lugg, Andrew, 86-87, 92, 116
Lumer, Christoph, 246
Luther, Martin 180

Mackie, John, 139
macrostructure, 179
Maginot Line, 74
manager argument, 139-40
manifest rationality, 61-62, 172
Manifest Rationality (Johnson), 3, 46, 47, 57, 58, 172, 173
manufacture of consent, 103
Marcus Aurelius, 180

marriage, and women's liberation, 193
Marx, Karl, 121
Marxism, 121
Marxist dialectic, 42, 107, 121
Massey, Gerald, 75, 76, 78
material conditional, 233
mathematical approach, 1, 11n16
mathematical proof, 12, 182
mathematical statements, 113
McBurney, Peter, 1
meaning, and truth, 113-14, 118
meaningful disagreements, 94
meaning of 'if-then', 233
mechanicist model, 127, 135
Medieval disputation, 183
Memedi, Vesel, 97
merits of arguments, 8, 223
meta reflection of ground level, 199
meta vs. ground levels, 200
meta-analysis, 152, 159
meta-argument by analogy, 202-4, 208, 210, 218
meta-argumentation approach, 219, 237, 239, 242, 246-47
meta-argumentative tier, 197-99, 200, 246
metacognition, 75, 91-92, 100n16, 153-54, 193, 199, 246
metadialogues, 1, 75, 82, 83-84
meta-mathematics, 16
metaphysics, 114
metareasoning, 34
meta-theory, 176
method of counterexample situation, 83, 84, 242-43
method of counterexample, 77-78, 79
method of formal paraphrase, 77-78, 79, 83, 84, 242-43
method vs. theory, 7-11, 11n16
methodical numbering system, 21, 227-39
methodological meta-arguments, 236
methodological reflection, 236
methods of argument criticism, 2, 162, 178, 242-43
military metaphor, 173-74
Mill, John Stuart, 3, 5, 17, 48, 50, 52, 66, 123, 150, 159, 174, 175, 243, 245; hyper dialectical argumentative practice, 71-74; liberty of thought and

discussion, 178-91; women's liberation, 192-200
misapplication of generalization, 208-9, 217, 245
mixed motion, 222-26
moderately dialectical definition of argument, 46-47, 53, 57, 59-64, 64, 65, 66, 69, 70, 71, 73, 188, 198, 242, 245, 246
moderation, 101, 105-6, 107, 109, 119, 122, 186, 243
modernity, 184
modus tollens, 226, 235
monolectical approach, 84
monological argumentation, 44, 47, 66
monological exercise, 69
monologue, 44, 68
moral attributes of God, 214-15, 216, 218
moral vs. natural attributes of God, 215, 217
moral vs. political controversy, 98
moral-indifference argument, 216-17, 218
morality, and theism, 204
Mosca, Gaetano, 121
multiple argumentation, 156, 158, 167

Nagel, Ernest, 164
natural deduction, 42
natural history, 9-10
natural religion, 201-18
naturalism, 107-9, 111, 114
naturalist approach, 9-10, 219
naturalized logic, 3, 16
nature of men and women, 193
nature of process of arguing, 51-52, 59
negative logic, 183
negative-evaluation approach, 168
Nevada, 76-78
New vs. Old testament, 185
New York, 78
Newton, Isaac, 9, 34, 67-68, 133n10, 208
Nicole, Pierre, 163
Niemeyer, Simon, 122
nominal definition, 242
non-cognitivism, 103, 105
nonconclusive inference, 123
nonconclusive support, 132
non-inductive meta-argument, 218

normal argumentation, 85, 87, 90, 91, 102, 104, 194, 197-98
normal reasoning, 94, 95
normative theory, 39
normative vs. descriptive, 10
normative-cum-descriptive approach, 219
Nosich, Gerald, 90
novel argument, 58

O'Keefe, Daniel, 152, 159
object level argumentation. *See* ground-level argument
objections: vs. alternative positions, 53n24; anticipating, 162, 171-76; anticipating vs. pre-empting, 173-74; considering, 181-84; defined, 19-20; handling, 176-77; pre-empting, 173-74, 176; proximity, 171-72; replying to, 57-59, 60-61, 64, 141, 167; replying vs. anticipating, 172, 173-74; salience, 171-72; strength, 171-72; strengthening, 174-75, 176
object-level argumentation. *See* gound-level argumentation
observational approach, 11-14
observational arguments, 15; heavenly unchangeability, 26-30
observational description, 219, 221-22
occult quality, 210, 218
Oliver, James, 78-79, 84
On Liberty (Mill), 3, 66, 71, 123, 150, 159, 174, 178-91, 243
On the Heavens (Aristotle), 4, 17, 237, 239
one-sidedness, 151-52, 159, 182, 184-86, 189, 193-94
open-mindedness, 61, 89-91, 92, 96-97, 98-101, 109, 110, 113, 117-19, 119-22, 153-54, 159, 175, 179, 189, 243
ordinary vs. extended arguments, 57
organism analogy, 211-13, 217
organism objection, 211-13
original argument, 58
ostensive definition, 4, 5, 6, 7, 17
"other things being equal." *See ceteris paribus*

paradigm changes, 121
paradigm example, 144

paralogism, 22
parity of reasoning, 81-83, 84, 217, 242-43
Parsons, Simon, 1
partial order, 213-14
partial truths, 184-86
partisanship, 185, 189
Pascalian minimax strategy, 101-2
Peano, Giuseppe, 12, 14
Perelman, Chaïm, 3, 13, 16, 220, 225, 231, 235
Perkins, David, 51
persecutions, 180
persuasion, 13, 91, 104, 168, 239; persuasive argumentation, 97, 117-19, 119-22, 243; persuasive disconnection, 25-26, 31; persuasive function of argument, 45; persuasive ineffectiveness, 235; persuasiveness, 126-27, 152, 155, 166, 170; vs. proof, 121; vs. rationality, 220; techniques, 85, 95, 97. *See also* rhetoric
petitio principii, 13-14, 220, 228. *See also* begging the question
Phillips, Dana, 92-93, 117
Philo, 202-18
philosophical approach, 1
philosophical commitment, 112-13
philosophical controversies, 85, 110-16, 116-19, 122
philosophical disagreements, 110-16
philosophy of argument, 16, 162-67
philosophy of logic, 111
philosophy vs. formal science, 114
physics, 154
Pinto, Robert, 246
Plantinga, Alvin, 139
Plato, 46, 102, 115
Platonic heaven, 233
plausible arguments, 15
pleasure, as chief good, 114-16
political debates, 100n16, 121-22, 151
political disagreement, 102
Pollock, John, 7, 9, 48
polytheism, 210-11
Port-Royal Logic (Arnauld and Nicole), 3. 163
practical argumentation, 151
practical logic, 3, 16
practice vs. theory, 5-7, 7-8, 17, 158, 167, 236
practicing what one preaches, 67
pragma-dialectical approach, 64-74, 167-71, 188, 226
pragmatic approach, 7-8, 10, 55-56, 219
pragmatic inconsistency, 169
predictive extrapolation, 196-97, 199-200
pre-emptive war, 174
pre-emptying objections, 173-74, 176
preliminaries to argumentation, 172
premise, defined, 18-19
premise-refuting criticism, 26, 32
presumption, 193-97, 199
presuppositional criticism, 25, 31, 32
presuppositional disconnection, 25, 26, 31
prime or minimal objection, 63n30, 69
pro-and-con argument, 2, 123-61, 183, 243. *See also* conductive argument
pro-and-con reasoning. *See* conductive argument
probable argument, 15, 29
probative vs. logical force, 126
proof vs. persuasion, 121
proponent argumentation, 48
proposals vs. propositions, 151
proposition, in an argument, 18
propositional structure, 20, 30, 124-25, 227-39, 242
pro-reasons acknowledgment claim, 148-49
proto-arguments, 57
proximity of objections, 171-72
pseudo-explanation, 218
psychotherapy, 89-91
Ptolemy, Claudius, 154, 220
Pythagorean theorem, 7, 113

qualified reasoning, 139-40
quantitative model, 127, 135
quasi-syllogism, 205, 213
Quine, Willard, 5-7, 14, 164
raccoon argument, 139
Ramsey, Frank, 101, 118
Ramsey's Maxim, 101, 104, 105, 106-9, 119, 119-22, 243
rape-shield law, 138
rationality, 189-90, 243, 244; vs. argumentation, 116-17; rational-

Index   277

mindedness, 92, 153-54; rational persuasion, 52-53, 59, 62, 87, 93-95, 97, 117-19, 122, 172; rational resolution, 85, 103, 104, 110, 116-19, 122; vs. rhetoric, 244; vs. rhetorical resolution, 170. *See also* reasoning
real possibilities, 98, 99, 103, 110, 117-18
realistic approach, 135n13
reasoning: vs. argument, 55, 162; definition of, 18; vs. instructions, 54; reason defined, 18-19; reasoning indicator, 18; reasons as causes, 54; reasons vs. criticism, 56; reasons vs. excuses, 54-55
reason-refuting criticism, 31
reason-relevance criticism, 26
reason-undermining criticism, 26, 31
reconstructed argument, 22, 30
reconstruction of an argument, 22
reductio ad absurdum, 115
reductionism, 107-9
Reed, Chris, 10, 44, 66, 72
Rees, Agnès van, 47, 65
refutation by logical analogy, 79-81, 83, 84, 242-43
refutational two-sidedness, 152
regular convex solids, 7
Reid, Thomas, 115
relativism, 103, 105
relativity of motion, 36-37
relevance, 132, 140
Reno, 76-78
replying to objections, 57-59, 60-61, 64, 141, 167
replying to vs. anticipating objections, 172, 173-74
revising theories, 170-71
rhetoric, 3, 13, 42, 45, 51-52, 62, 94, 168, 169, 233; vs. logic, 220, 228, 231-32, 234-40, 246; rhetorical definition of argument, 45-46; rhetorical persuasion, 102-3, 104-6, 116-19, 117-19, 244, 245, 246; rhetorical techniques, 103, 104
Right vs. Left, 121
role of argument, 166-67
role of meta-argumentation, 244-45
Rousseau, Jean-Jacques, 184
Rowe, Glenn, 10

Russell, Bertrand, 106

Saddam Hussein, 170
Sagredo, 220-41
salience of objections, 171-72
salience vs. appropriateness, 172
Salmon, Merrilee, 164
Salmon, Wesley, 79, 164
Salviati, 220-41
Schiavo, Terri, 96
science of logic, 236
scientific arguments, 133, 157
scientific revolutions, 121
scope of logic, 163
Scripture, 153, 156, 175, 220
Scriven, Michael, 3, 45, 134, 150, 159
self-defeating position, 115
self-referential arguments, 3, 60, 162-77, 178, 243
self-reflective argumentation, 35-36, 124
semantical criticism, 32
semantical disconnection, 25, 26
serial argument, 20
serial structure, 20
servile morality, 185
ship experiment, 219, 222-23, 225-26, 231
Siegel, Harvey, 246
simple argumentation, 88-89, 116-17
Simplicio, 220-41
simplicity: argument for terrestrial rotation, 36-39, 219; principle of, 38-39; theological objection, 39, 148n21
Simplicius, 220
skepticism, 204
Skyrms, Brian, 45, 130
Snoeck Henkemans, Francisca, 43, 65, 68, 69, 167
Soccorsi, Filippo, 154-55
social technology of persuasion, 102-3
social utility, 180
sociological dilemmas, 107-9, 121
Socrates, 102, 180; Socratic dialectics, 183, 188-89
solo arguments, 43-44
sophistry, 152, 155, 159
soul of world, 212
speculative philosophy, 114
stages of critical discussions, 168-69
standard labeling, 21, 227-39

standoffs, 82, 85, 92, 97-110, 111, 116-19, 122, 243
statistical syllogism, 15, 139, 205-6, 207, 213, 218
stereotyping, 209
sterility of argument, 61
strategic derailments, 167-71, 176
strategic maneuvering, 162, 167-71, 176, 220n1, 226, 240, 244, 246
straw-man: criticism, 27; fallacy 52
strength of objections, 171-72
strength of reasons, 140-41, 142, 158
strengthening objections, 174-75, 176
strongly dialectical definition of argument, 47, 60, 64, 65, 66, 69, 71, 72, 73, 188, 242
structural definition of argument, 44-45, 50
structure diagram, 20-22, 204n5; drawn, 28, 30, 36, 86, 104, 128, 148, 163, 230, 238
structure of argument, 8
structure vs. function of argument, 53-54, 59-60
structure vs. procedure, 142
subargument, 20, 27, 186
Subjection of Women (Mill), 3, 5, 17, 72, 74, 192-200, 243
subjection of women, 3, 5, 17
subordination of women, 192-200
superfluous argumentation, 193-99
suppression of criticism, 180
Supreme Court of the US, 80
suspending judgment, 88, 96, 116, 214
symmetry of dialectical and illative tiers, 63n30, 69-70, 74, 198
symmetry of ground and meta levels, 198-99
synthesis vs. uncritical reconciliation, 143-44
systematic labeling, 21, 124-25, 179, 204n5, 227-39
systematicity, 39-40
systematization, 244
Sztompka, Piotr, 107-9, 118, 121

Tarski, Alfred, 12, 14, 15-16
teleological argument for God's existence, 201-18

teleological definition of argument, 45, 51-52
teleological principle, 39
teleology, 246
textbook definition of argument, 44
theism, 204, 210-11, 217
theoretical meta-arguments, 2, 17, 178
theory: vs. analysis 9-10, 34-41; argumentation theory, 1, 2, 3, 39-41, 187-89, 243, 244; argumentation theory vs. meta-argumentation, 15, 220, 234-40, 244; choice of, 133, 157; logical theory, 3, 5-7, 8, 9, 39-40, 244; vs. method, 7-11, 11n16; vs. method vs. illustration, 219; vs. observation, 11-14, 15-17; vs. practice, 5-7, 7-8, 17, 158, 167, 236
theory-laden observation, 11
theory of knowledge, 6
thinking through an argument, 127
third party, 89-91, 98, 103, 117
third-level meta-argumentation, 75
Thomas, Stephen, 134
Thomson, Judith, 82-83
tides, 155-56
total evidence, requirement of. *See* complete evidence, requirement of
Toulmin, Stephen, 2, 3, 6-7, 7-17, 39, 58, 119, 124, 139, 157, 163, 178, 190, 205, 220, 223, 227, 233, 239, 242, 244; substantive model of argument criticized, 10-11, 16
trial of Galileo, 154-55, 220
trivial logic-indifferent method, 76-78, 83, 84, 157, 242-43
trivialization of logic, 6
truth: conceptions of, 189; and meaning, 113-14; as a synthesis of opposites, 184-85; vs. utility, 180
truth-functional logic, 233
truth-in-evidence: argument, 49, 60, 63; principle, 49, 64
tu quoque, 162, 167-71, 176
Turing, Alan, 48, 72
Turner, Dale, 88-89, 116
Tweedledum and Tweedledee, 6-7, 14, 164
two-sided advertising, 152
two-sidedness, 151-52, 159, 182

Index 279

types or forms of meta-argumentation, 242
ubiquity of meta-arguments, 1
universal gravitation, law of, 68
universal specification, 78, 205

validity vs. formal validity, 76-78
Van der Torre, Leendert, 1, 199n5
verbal disagreement, 112, 204
vertical fall, 3, 30-32, 219-41, 244, 246
vertically latent structure, 22
Villata, Serena, 1, 199n5
violinist-abortion argument, 82-83
visual representation, 148-49

Walton, Douglas, 8
Weddle, Perry, 139
weighing pros and cons, 125-26, 127, 129, 138, 155
weighing reasons and evidence, 135, 143-44, 150, 156-57
weight-and-sum methodology, 150, 159
Wellman, Carl, 2, 124-36, 147, 148, 149, 150, 151, 152, 154, 157-59, 167, 244
Whately, Richard, 113, 115, 118, 166, 167, 176, 177, 211
William of Orange, 68, 69, 169
William the Silent, 169, 246
Wittgenstein, Ludwig, 93, 94-95
Wohlrapp, Harald, 142-44, 148n21, 157, 158
women's liberation, 3, 5, 72-73, 192-200, 243
women's suffrage, 195
Woods, John, 2, 3, 9, 84, 85, 80, 92, 111, 117-19, 121, 122, 244; parity of reasoning, 81-83; standoffs of force five, 97-110
Wooldridge, Michael, 1
world-view dissensus, 97
Wright, Larry, 3, 88-89, 116

Zenker, Frank, 141-42, 149n22, 157, 158

www.ingramcontent.com/pod-product-compliance
Lightning Source LLC
Chambersburg PA
CBHW050131170426
43197CB00011B/1794